Structure and Mechanical Properties of Transition Group Metals, Alloys, and Intermetallic Compounds

Structure and Mechanical Properties of Transition Group Metals, Alloys, and Intermetallic Compounds

Special Issue Editor

Tomasz Czujko

MDPI • Basel • Beijing • Wuhan • Barcelona • Belgrade

MDPI

Special Issue Editor
Tomasz Czujko
Military University of Technology
Poland

Editorial Office
MDPI
St. Alban-Anlage 66
4052 Basel, Switzerland

This is a reprint of articles from the Special Issue published online in the open access journal *Materials* (ISSN 1996-1944) from 2018 to 2019 (available at: https://www.mdpi.com/journal/materials/special_issues/transition_group_metals)

For citation purposes, cite each article independently as indicated on the article page online and as indicated below:

LastName, A.A.; LastName, B.B.; LastName, C.C. Article Title. *Journal Name* **Year**, *Article Number*, Page Range.

ISBN 978-3-03921-146-3 (Pbk)
ISBN 978-3-03921-147-0 (PDF)

Contents

About the Special Issue Editor vii

Preface to "Structure and Mechanical Properties of Transition Group Metals, Alloys, and
Intermetallic Compounds" ix

Jan Huebner, Dariusz Kata, Paweł Rutkowski, Paweł Petrzak and Jan Kusiński
Grain-Boundary Interaction between Inconel 625 and WC during Laser Metal Deposition
Reprinted from: *Materials* **2018**, *11*, 1797, doi:10.3390/ma11101797 . . . . **1**

Sian Wang, Yunhe Zhang and Gaohui Wu
Interlaminar Shear Properties of Z-Pinned Carbon Fiber Reinforced Aluminum Matrix
Composites by Short-Beam Shear Test
Reprinted from: *Materials* **2018**, *11*, 1874, doi:10.3390/ma11101874 . . . . **13**

Radosław Łyszkowski, Wojciech Polkowski and Tomasz Czujko
Severe Plastic Deformation of Fe-22Al-5Cr Alloy by Cross-Channel Extrusion with
Back Pressure
Reprinted from: *Materials* **2018**, *11*, 2214, doi:10.3390/ma11112214 . . . . **27**

Sian Wang, Yunhe Zhang, Pibo Sun, Yanhong Cui and Gaohui Wu
Microstructure and Flexural Properties of Z-Pinned Carbon Fiber-Reinforced Aluminum
Matrix Composites
Reprinted from: *Materials* **2019**, *12*, 174, doi:10.3390/ma12010174 . . . . **44**

Boris B. Straumal, Anna Korneva, Askar R. Kilmametov, Lidia Lityńska-Dobrzyńska,
Alena S. Gornakova, Robert Chulist, Mikhail I. Karpov and Paweł Zięba
Structural and Mechanical Properties of Ti–Co Alloys Treated by High Pressure Torsion
Reprinted from: *Materials* **2019**, *12*, 426, doi:10.3390/ma12030426 . . . . **55**

Anna Korneva, Boris Straumal, Askar Kilmametov, Robert Chulist, Grzegorz Cios,
Brigitte Baretzky and Paweł Zięba
Dissolution of Ag Precipitates in the Cu–8wt.%Ag Alloy Deformed by High Pressure Torsion
Reprinted from: *Materials* **2019**, *12*, 447, doi:10.3390/ma12030447 . . . . **66**

Robert Kosturek, Lucjan Śnieżek, Marcin Wachowski and Janusz Torzewski
The Influence of Post-Weld Heat Treatment on the Microstructure and Fatigue Properties of
Sc-Modified AA2519 Friction Stir-Welded Joint
Reprinted from: *Materials* **2019**, *12*, 583, doi:10.3390/ma12040583 . . . . **78**

Mónica Preciado, Pedro M. Bravo, José Calaf and Daniel Ballorca
Strain Rate during Creep in High-Pressure Die-Cast AZ91 Magnesium Alloys at Intermediate
Temperatures
Reprinted from: *Materials* **2019**, *12*, 872, doi:10.3390/ma12060872 . . . . **95**

Aleksandra Szafrańska, Anna Antolak-Dudka, Paweł Baranowski, Paweł Bogusz,
Dariusz Zasada, Jerzy Małachowski and Tomasz Czujko
Identification of Mechanical Properties for Titanium Alloy Ti-6Al-4V Produced Using
LENS Technology
Reprinted from: *Materials* **2019**, *12*, 886, doi:10.3390/ma12060886 . . . . **106**

Kristina Kittner, Madlen Ullmann, Thorsten Henseler, Rudolf Kawalla and Ulrich Prahl
Microstructure and Hot Deformation Behavior of Twin Roll Cast Mg-2Zn-1Al-0.3Ca Alloy
Reprinted from: *Materials* **2019**, *12*, 1020, doi:10.3390/ma12071020 . . . . **125**

Anna Antolak-Dudka, Paweł Płatek, Tomasz Durejko, Paweł Baranowski,
Jerzy Małachowski, Marcin Sarzyński and Tomasz Czujko
Static and Dynamic Loading Behavior of Ti6Al4V Honeycomb Structures Manufactured by
Laser Engineered Net Shaping (LENS^{TM}) Technology
Reprinted from: *Materials* **2019**, *12*, 1225, doi:10.3390/ma12081225 . . . . **140**

Maurício Silva Nascimento, Givanildo Alves dos Santos, Rogério Teram,
Vinícius Torres dos Santos, Márcio Rodrigues da Silva and Antonio Augusto Couto
Effects of Thermal Variables of Solidification on the Microstructure, Hardness, and
Microhardness of Cu-Al-Ni-Fe Alloys
Reprinted from: *Materials* **2019**, *12*, 1267, doi:10.3390/ma12081267 . . . . **160**

Dariusz Garbiec, Volf Leshchynsky, Alberto Colella, Paolo Matteazzi and Piotr Siwak
Structure and Deformation Behavior of Ti-SiC Composites Made by Mechanical Alloying and
Spark Plasma Sintering
Reprinted from: *Materials* **2019**, *12*, 1276, doi:10.3390/ma12081276 . . . . **170**

Tomasz Durejko, Magdalena Łazińska, Julita Dworecka-Wójcik, Stanisław Lipiński,
Robert A. Varin and Tomasz Czujko
The Tribaloy T-800 Coatings Deposited by Laser Engineered Net Shaping (LENS^{TM})
Reprinted from: *Materials* **2019**, *12*, 1366, doi:10.3390/ma12091366 . . . . **186**

Yunhe Zhang, Sian Wang, Xiwang Zhao, Fanming Wang and Gaohui Wu
In Situ Study on Fracture Behavior of Z-Pinned Carbon Fiber-Reinforced Aluminum Matrix
Composite via Scanning Electron Microscope (SEM)
Reprinted from: *Materials* **2019**, *12*, 1941, doi:10.3390/ma12121941 . . . . **199**

About the Special Issue Editor

Tomasz Czujko graduated in 1991 with an MSc in Technical Physics from the Faculty of Chemistry and Technical Physics of the Military University of Technology, Warsaw, Poland. In 1998 he received his PhD in Materials Science and Engineering from the Military University of Technology. From 2000 to 2001 he carried out research as a Post-Doctoral Fellow/Research Associate at the University of Waterloo, Ontario, Canada. From 2004 to 2009 he was affiliated with the Department of Mechanical and Mechatronics Engineering, University of Waterloo, where he was a Research Associate Professor. Between 2009 and 2010 he carried out research as a Research Officer at CanmetENERGY, Hydrogen Fuel Cells and Transportation, Natural Resources Canada. Since 2010 he has been affiliated with the Faculty of Advanced Technology and Chemistry of the Military University of Technology. His research has covered fields including the structural characterization of materials, the quantitative characterization of microstructures in polycrystalline and nanocrystalline materials, materials for hydrogen storage, the relationship between structure and properties of materials, and additive manufacturing.

Preface to "Structure and Mechanical Properties of Transition Group Metals, Alloys, and Intermetallic Compounds"

The mechanical properties of a material are defined as the reaction of the material to an applied load. The mechanical properties of metal determine the range of usability of the material—the material's ability to mold to the right shape and limit the useful life that can be expected. Considering the diversity of commercially available materials, mechanical properties are also used to classify and identify materials. The properties considered most often are strength, toughness, hardness, hardenability, brittleness, ductility, creep and slippage, and elasticity and fatigue.

The mechanical properties of polycrystalline structural materials, such as transition group metals, alloys and intermetallic compounds, are significantly affected by their microstructure, including phase composition, grain shape and size, grain boundary distribution, dislocation density, dispersed particles and solutes, and internal stresses. Therefore, studies of the relationships between microstructure and mechanical properties are of great practical importance. The development of metallic/intermetallic constructive material with the desired structure results in beneficial combinations of mechanical properties. Various thermo-mechanical treatments are widely used to produce metallic materials achieving the preferred microstructure owing to the diverse mechanisms of its evolution. Knowledge of the effects of applied techniques and processing windows on the structural changes in the metals, alloys, and intermetallic compounds allows for the development of structural material manufacturing methods with enhanced mechanical properties.

Almost all structural materials are anisotropic and, for this reason, their mechanical properties vary with orientation. The changes in properties can be due to crystallographic or morphological texture from casting, forming or cold working processes, thermo-mechanical treatment, the controlled alignment of fiber reinforcement, etc. Moreover, temperature, rate of loading, and environment affect the mechanical properties of materials.

Tomasz Czujko
Special Issue Editor

materials

MDPI

Article

Grain-Boundary Interaction between Inconel 625 and WC during Laser Metal Deposition

Jan Huebner [1,*]**, Dariusz Kata** [1]**, Paweł Rutkowski** [1]**, Paweł Petrzak** [2] **and Jan Kusiński** [2]

[1] Faculty of Materials Science and Ceramics, Department of Ceramics and Refractories, AGH University of Science and Technology, al. Mickiewicza 30, 30-059 Krakow, Poland; kata@agh.edu.pl (D.K.); pawelr@agh.edu.pl (P.R.)
[2] Faculty of Metals Engineering and Industrial Computer Science, Department of Surface Engineering and Materials Characterisation, AGH University of Science and Technology, al. Mickiewicza 30, 30-059 Krakow, Poland; ppetrzak@agh.edu.pl (P.P.); kusinski@agh.edu.pl (J.K.)
* Correspondence: huebnerj@agh.edu.pl; Tel.: +48-663-132-761

Received: 26 July 2018; Accepted: 17 September 2018; Published: 21 September 2018

Abstract: In this study, the laser metal deposition (LMD) of the Inconel 625–tungsten carbide (WC) metal matrix composite was investigated. The composite coating was deposited on Inconel 625 substrate by powder method. A powder mixture containing 10 wt% of WC (5 μm) was prepared by wet mixing with dextrin binder. Coating samples obtained by low-power LMD were pore- and crack-free. Ceramic reinforcement was distributed homogenously in the whole volume of the material. Topologically close-packed (TCP) phases were formed at grain boundaries between WC and Inconel 625 matrix as a result of partial dissolution of WC in a nickel-based alloy. Line analysis of the elements revealed very small interference of the coating in the substrate material when compared to conventional coating methods. The average Vickers hardness of the coating was about 25% higher than the hardness of pure Inconel 625 reference samples.

Keywords: metal matrix composites; laser metal deposition; Inconel 625; additive manufacturing; laser processing

---

1. Introduction

A constantly increasing need for improvement in the field of energy harvesting has resulted in much research focused on developing innovation. Materials that can be used in high temperatures are promising because of possible applications in powerplants, and the aerospace and chemical industry [1–3]. The easiest way to improve the effectiveness of gas turbines used in engines is to elevate their work temperature. Today, widely used metallic materials allow for the production of turbine blades that are able to operate in temperatures in the range of 650–1200 °C. Additionally, these parts are constantly exposed to chemical and mechanical factors [4,5]. During their work, an aggressive environment causes microcavities in the material that may cause the complete destruction of working parts. Turbine blades are made of heat-resistant steel or nickel/cobalt-based superalloys. Because of the harsh environment and relatively short lifespan of the blades, any opportunity to regenerate damaged element is very attractive [6–9].

Nickel and cobalt superalloys are efficiently used in the production of parts for high-temperature applications. Their properties, excellent weldability, high plasticity, and corrosion/wear-resistance in high temperatures [10–16], are suitable for use in high-temperature environments. In order to improve the quality and lifespan of the used materials, metal matrix composites (MMC) with carbide reinforcement were proposed. In MMCs, the desirable properties of metals and ceramics are fused to obtain improved material. The combination of coating and substrate material can be designed to enhance specific properties: corrosion, oxidation, erosion, and high wear resistance [4,5,17,18].

Production of whole parts from materials that are characterized by good wear resistance is expensive due to the high cost of alloying elements such as Ni, Co, Mo, V, and W. A possible solution is surface modification by the deposition of protective coating. Deposition technology has had a huge impact on the fretting resistance of coatings. Surface layers produced from the same material but by different techniques vary in physical and performance properties. Industrial production of metal–ceramic composites can be done by laser metal deposition [18,19], thermal spraying [19–21] and plasma arc welding.

Some research investigated the laser processing of Inconel–carbide systems. It is reported that the mechanical properties of LMD-deposited materials are improved when compared to pure alloys [22–26]. Thanks to rapid solidification, the microstructure of the material is much finer when compared to conventional methods, such as Tungsten Inert Gas. Pure Inconel alloys are strengthened by the γ' and γ'' phases, which precipitate from austenitic γ-Ni during heat treatment. The δ phase (Ni_3Nb) can also precipitate from alloys with a high Nb content. Phase δ helps refine the grain size and impeding dislocation in the structure. Introduction of tungsten carbide (WC) grains with a diameter over 10 μm into the system [27,28] shows higher wear resistance and hardness of the material. It was reported that big carbide particles could not be distributed homogenously throughout the whole volume of the sample. This results in non uniform distribution of hardness. Metal–carbide systems are commonly deposited using two separate powder feeders with regulated feed ratios that could not provide the same carbide content in the whole volume [27,28]. Introduction of a small amount of TiC nanoparticles into Inconel [29–31] resulted in a modified microstructure of the material. Hardness was improved due to grain-size refinement. Such a microstructure improves tensile properties of the composite.

In this research, we propose the use of nickel-based MMC protective coatings with ceramic reinforcement with a diameter about 5 μm. Rapid prototyping, laser metal deposition (also called laser cladding), was used as a deposition method [32,33]. Due to precise heating of a small part of material surface, it is possible to avoid interference in the substrate microstructure and chemical composition. High-energy density of a laser beam enables fast heating followed by rapid cooling and solidification coating. As a result, the obtained microstructure is characterised by fine grains and is resistant to erosion [27,28,34]. In this study, the grain-boundary Inconel 625–WC interaction was investigated during laser metal deposition. The phenomenon that occurs on the interface between these two different types of materials is crucial for understanding the nature of the LMD process. Moreover, obtained results can lead to easier implementation of this method to other composite systems. In this work, interaction between WC grains (5 μm) and a nickel-based superalloy during high-temperature LMD was investigated. The particle size of 5 μm was chosen in order to achieve uniform carbide distribution in the whole volume of the material. Additionally, this grain size allowed to avoid dissolution of a significant amount of WC in the metal matrix [35–38]. The powder mixture of Inconel 625 and WC was used to enhance homogenous distribution of carbide in the material.

2. Materials and Methods

The experiments were performed by use of a powder mixture instead of two separate powders. It was obtained by homogenization of commercially available WC and Inconel 625. Inconel 625 powder, with an average particle diameter of 104 μm, and WC powder, with an average particle diameter of 5 μm, were mixed in a 9:1 mass proportion that resulted in 10 wt% of WC in mixture. Powders were initially homogenized for 90 min in a ball mill using WC balls in a weight ratio of 1:1 (grinding media:powder). In order to improve adhesion between Inconel 625 and WC particles, 0.25 wt% of resin was added and homogenization was repeated. Further addition of 0.25 wt% of dextrin was needed to achieve desired adhesion between powders after process. The morphology of the powder mixture at each step is shown in Figure 1.

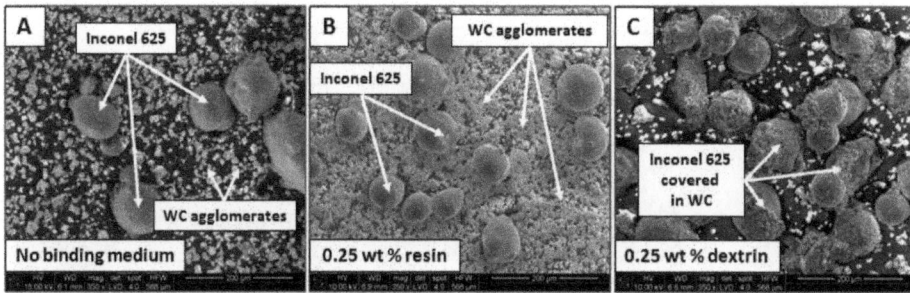

Figure 1. Scanning electron microscopy (SEM) images of powder mixture after each stage of homogenization: (**A**) no binder, (**B**) 0.25 wt% of resin binder, (**C**) 0.25 wt% of dextrin binder.

JK Laser Company model JK2000FL equipped with ytterbium-doped fiber was used to perform the laser metal deposition of the composite. Metal matrix composite coating was obtained by powder-mixture deposition on the Inconel 625 substrate. It was chosen to prevent additional impurities in the samples. To obtain a coating of 10 × 10 mm and of about 1 mm thickness, 10 subsequent tracks with a width of about 1 mm were deposited in 6 sublayers. The powder mixture was transported in protective atmosphere of argon from the powder feeder to the laser head and then sprayed onto the substrate. The powder particles melted due to exposure to the high-energy laser beam. The formation of a melt pool containing both the powder mixture and substrate material was observed. To avoid the decomposition of the ceramic reinforcement (WC melting point at 2870 °C), low power of the laser (320 W) was used. This enabled melting the Inconel 625 matrix (melting point at1340 °C). A radiation pyrometer monitored temperature changes during the LMD process at one point on the surface. Solidification of the material proceeded as the laser head moved with a set scanning speed. Samples were prepared by LMD according to the parameters presented in Table 1. Additionally, pure Inconel 625 reference samples were prepared. The schematic representation of process is shown in Figure 2.

Table 1. Laser metal deposition process parameters.

Parameter	Value
Diameter of laser beam spot (μm)	500 ± 5
Wavelength of laser beam (nm)	1063 ± 10
Nominal laser power (W)	320 ± 5
Scanning speed (mm/s)	10
Powder feed rate (g/min)	7.78 ± 0.10
Number of sublayers	6
Track length (mm)	10.00 ± 0.05
Temperature of the melt pool (°C)	1662 ± 10

Samples were cut in parallel and perpendicularly to the deposited tracks and then ground and polished. In order to observe the microstructure, samples were electrochemically etched in a 10% CrO_3 water solution.

Scanning electron microscopy (SEM) observations and energy-dispersive X-ray analysis were performed on a HITACHI S-3500N microscope equipped with an EDS NORAN 986B-1SPS analyzer, and an FEI Inspect S50 microscope with an EDAX EDS analyzer. To check the phase composition of the samples, X-ray diffraction analysis was performed using a PANalitycal X-ray Diffractor (XRD) equipped with Cu tube and X-pert HighScore software. The angle of the XRD ranged from 5° to 90° with a 0.008° measurement step. Transmission electron microscopy (TEM) observations were performed using a 200 kV JEAOL JEM-2010ARP microscope. Additionally, TEM-EDS analysis was done to check the elemental composition of the precipitates. Hardness was measured with a Future-Tech FM-700 hardness tester with a Vickers indenter under a load of 200 g for 15 s.

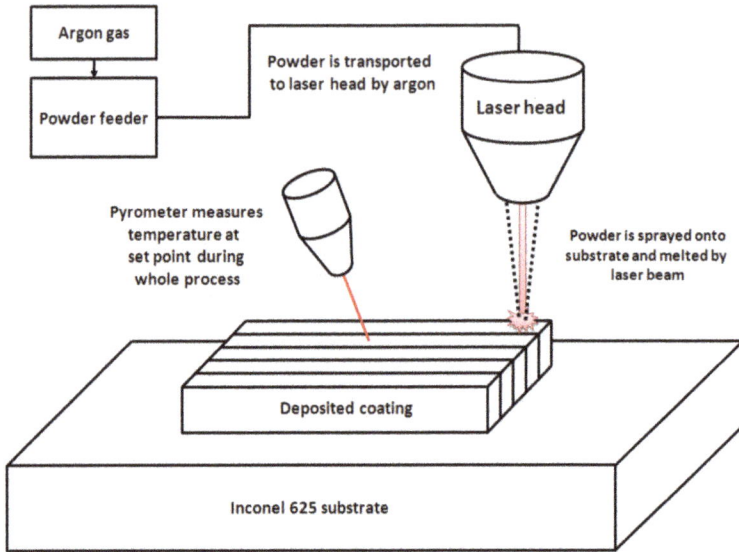

Figure 2. Schematic representation of the laser metal deposition process.

3. Results

X-ray diffraction analysis is presented in Figure 3. The performed XRD analysis revealed the presence of two major phases: $Ni_{0.85}W_{0.15}$ and $Ni_6Mo_6C_{1.06}$. This indicates that WC grains could be dissolved in a Ni matrix. The same carbide behavior was reported in different types of similar austenitic structures [30,31]. During that process, W diffused inside the γ-Ni matrix, while carbon remained in the intergranular region. It enabled the formation of a small amount of secondary carbides. The level of detection in the XRD technique varied from about 5%; thus, there is a possibility that other undetected phases are present in the material.

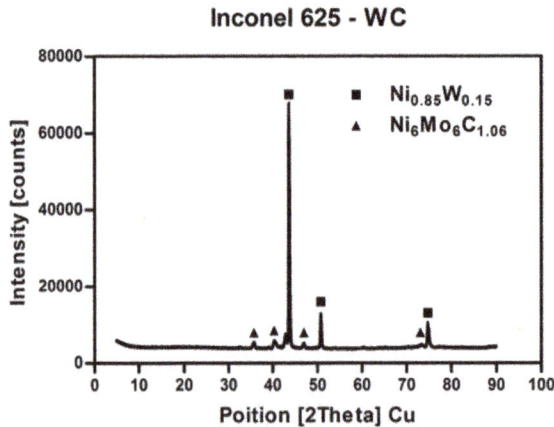

Figure 3. X-ray diffraction analysis of obtained Inconel 625–WC composite.

Figure 4 depicts the morphology of the prepared samples at various magnifications. As shown in Figure 4A,B, three areas can be recognized. The composite coating obtained by LMD has a fine uniform microstructure. The transitional area constitutes the boundary between composite coating and base

4

material. The bottom area represents the substrate Inconel 625. Its microstructure is characterized by a larger grain size than the coating. Additionally, Figure 4B shows that the structural orientation of the coating grains tends to reflect the grain orientation in the substrate. Figure 4C presents the area where the direction of grain growth was changed because of a slightly different temperature gradient during the process.

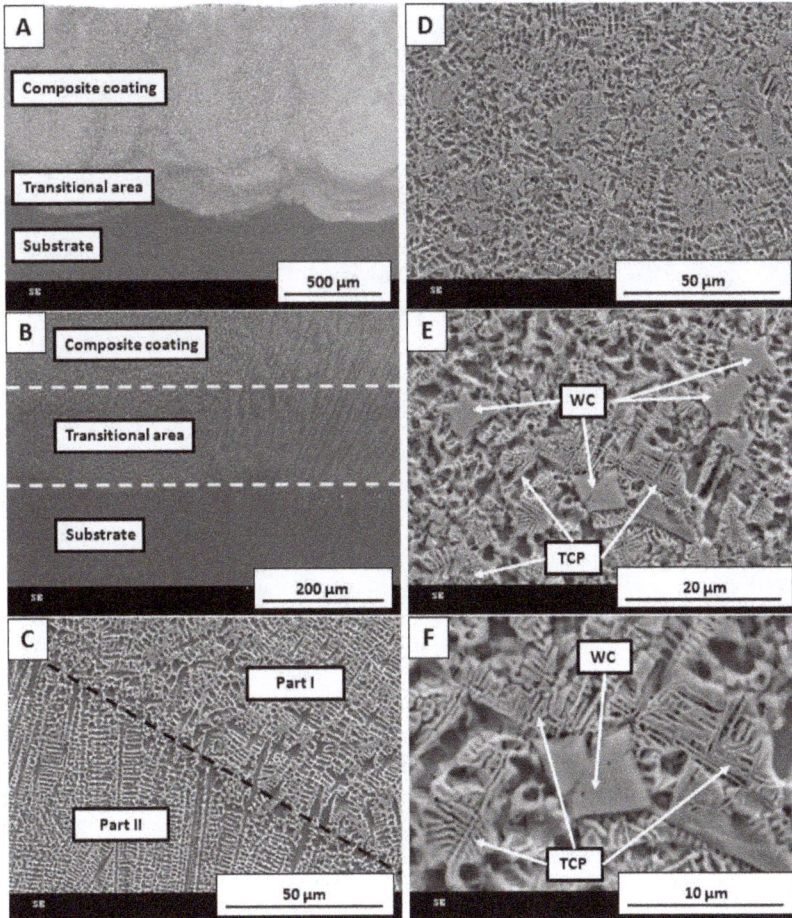

Figure 4. SEM images of different parts of the sample: (**A,B**) cross-section of the coating–substrate boundary, (**C**) area where direction of grain growth was altered, (**D–F**) partially dissolved WC grains with characteristic fishbone-like structure of topologically close-packed (TCP) phases.

Figure 4D–F shows the part of the sample where partially dissolved WC was visible. When compared to Figure 4C, the microstructure of the material was altered. Metallic grains were equiaxial, and WC was located at their boundaries. According to Figure 4E, it is evident that coarse WC was partially dissolved in metallic matrix, forming a typical eutectic microstructure. In Figure 4F, it can be seen that the dissolution process started at WC grain tips and propagated into the metal matrix. It formed a fishbone-like structure typical for topologically close-packed (TCP) phases.

SEM-EDS maps are shown in Figure 5. An increased amount of Mo, Nb, and C is visible at grain boundaries. This is the result of element segregation that normally occurs after long exposure to elevated temperature. The solidification process of Inconel alloys exhibits the tendency of individual elements to segregate into a dendrite axis and interdendritic spaces depending on their partition coefficient k.

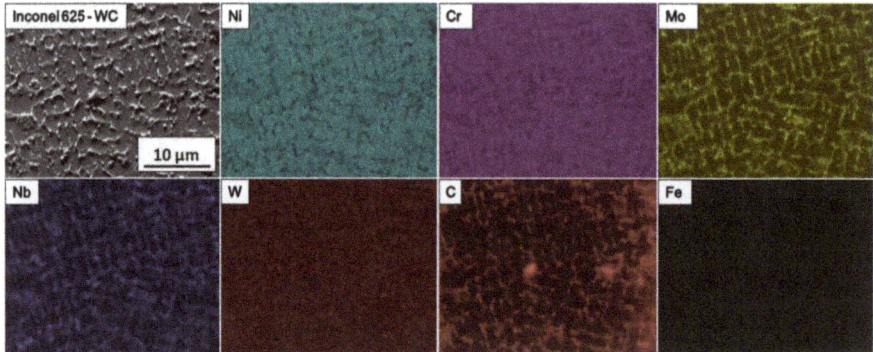

Figure 5. SEM-EDS elemental maps—segregation of the elements in the metal matrix composite (MMC) coating.

The K coefficient is experimentally determined by X-ray microanalysis of the element concentration in the dendrite core or cell (C_{core}). It represents the average concentration (C_0) of a given element in the analyzed coating area according to the following equation [9,22,26]:

$$k = C_{core}/C_0, \tag{1}$$

The elements for which parameter k < 1 tends to segregate at interdendritic spaces, and those for which the value of parameter k > 1 diffuses into the dendrite axis. The k coefficient for Fe, Cr, W, and Co is close to 1, in comparison to $0.80 \div 0.85$ for Mo and about 0.5 for Nb. Phases formed from elements that tend to segregate at the interdendritic spaces crystallize the last. As a result of rapid cooling, these elements often form M_xC_y carbides or intermetallic phases during the eutectic reaction. Rapid cooling during the LMD process leads to grain refinement, which results in a fine microstructure.

Because of the high temperature of LMD and rapid cooling, deposited material remained in a non-equilibrium state, which allowed for the formation of TCP phases at grain boundaries. They mostly contained Mo, Nb and C (Figure 5). Typically, TCP phases are formed in nickel alloys after long heat treatment; however, the nature of laser deposition technology and the addition of ceramics in WC form induced their appearance in the coating. Main alloying elements Ni and Cr were present in the metal matrix, while Fe and W were spread equally throughout the whole volume of sample. This shows that W diffused inside the metal matrix due to high temperature occurring.

As seen in Figure 6, the highest LMD temperature reached 1662 °C. This is much higher than the melting point of Inconel 625, 1340 °C. Thanks to the excellent wettability of WC by Ni, the appearance of a nickel-based liquid was the reason for ceramic particle dissolution. Sudden changes in temperature caused the fast melting of the Inconel 625–WC powder mixture, followed by recrystallization due to rapid solidification.

Figure 7 presents the element line distribution in the material. The measured thickness of the coating was about 1200 μm. The line analysis of the cross-section shows differences in the content of the elements depending on the distance from the sample surface. The amount of Ni in the coating was 50 wt%, while in the substrate material it was 60 wt%. Chromium level remained at about 18 wt% regardless of distance from the sample surface. Mo content slightly deviated from the average, with a level in the coating of 8 wt%–9 wt%. The amount of Nb did not exhibit any significant differences depending on the transition from coating to substrate. Deviations for both Mo and Nb were caused by element segregation, which occurred mostly in the part of the sample affected by laser processing.

W was present only in the coating and transitional area. It stayed at about 10 wt%, decreasing slightly with a larger distance from the sample surface, and completely disappeared at about 1500 μm. Carbon content was the same in the whole volume of the sample. The amount of Fe was very small in the coating and slightly rose in the transitional area and substrate.

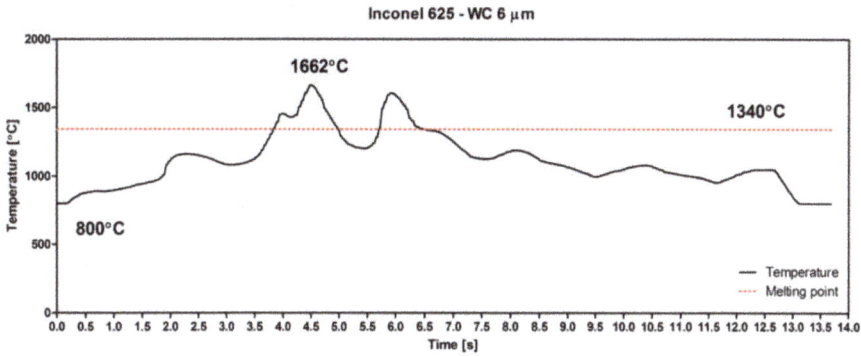

Figure 6. Temperature curve of deposition of a single layer of Inconel 625–WC coating during laser metal deposition.

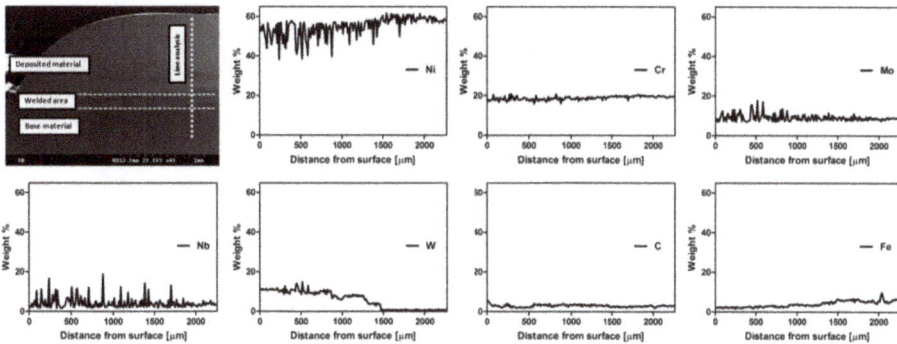

Figure 7. SEM images showing edge of prepared MMC sample.

The EDS point analysis performed by TEM is shown in Figures 8 and 9. Four different areas were investigated. Element concentration is presented in Table 2. High amount of Ni in area #EDS1was about two times higher in comparison to three other investigated regions. This states that light-gray areas in the images represents metal matrix of the composite. Points #EDS2, #EDS3 and #EDS4 contains high amount of W: 44, 50, and 53 wt% respectively. Together with a slightly increased concentration of Mo and Nb this indicates that dark areas represent undissolved WC particles. The presence of Nb and Mo resulted from element segregation at the grain boundaries during solidification of the material.

Figure 9 presents partially dissolved WC particles located at the grain boundaries of the metal matrix. Figure 9A depicts the bright field image of material. The light-gray area is identified as γ-Ni, while dark-gray areas represent the WC and TCP phases. It can be observed that carbides have smooth edges that were caused by partial dissolution in the Ni. The dark field of the same area is shown in the Figure 9B, with a complex diffraction pattern in Figure 9C–F. The main reflexes originated from γ-Ni matrix grains with FCC crystallographic orientation of [-112], [-113], [-114] as presented in Figure 9D–F. This indicates that matrix-grain growth proceeded similarly. The differences were caused by the introduction of additional particles into the material. Secondary reflexes visible in Figure 9C originated from WC with a crystallographic orientation of [-4311].

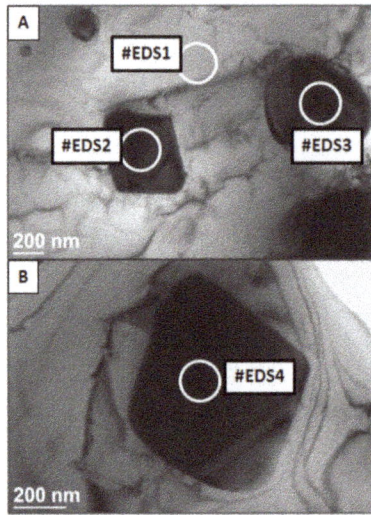

Figure 8. Transmission electron microscopy (TEM) image with four areas analyzed by EDS: (**A**) small precipitates (**B**) large precipitate.

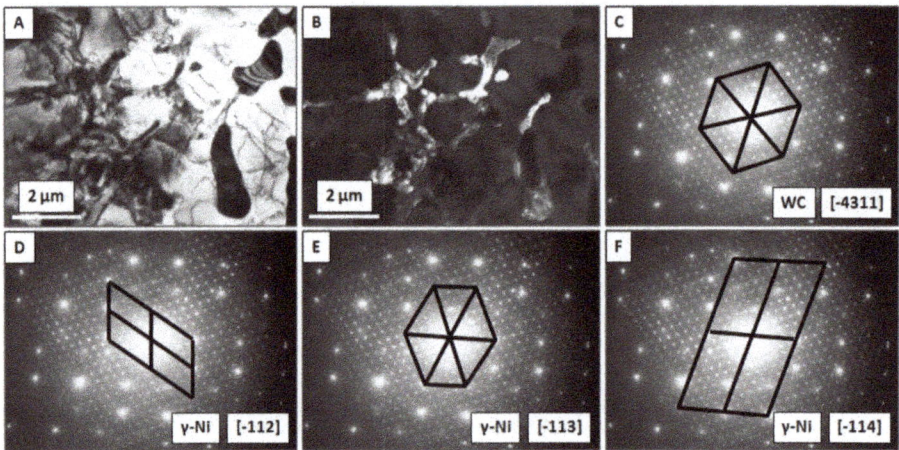

Figure 9. TEM images showing WC dissolution in the γ-Ni matrix: (**A**) bright-field image, (**B**) dark-field image, (**C**) diffraction pattern of WC with [-4311] crystallographic orientation, (**D–F**) diffraction pattern of γ-Ni grains with [-112], [-113], [-114] crystallographic orientation.

Table 2. TEM-EDS point analysis in areas shown in Figure 8.

Point	Ni	Cr	Mo	Nb	W	Fe	Total
#EDS1	73	13	3	0	10	0	100
#EDS2	35	11	8	2	44	0	100
#EDS3	29	10	8	2	50	1	100
#EDS4	30	6	7	3	53	1	100

As shown in Figure 10, the hardness of the obtained composite coating was higher than in pure Inconel 625. The presence of carbide particles significantly increased the hardness of laser-clad material from 399 ± 14 HV$_{0.2}$ for Inconel 625, to 502 ± 20 HV$_{0.2}$ for the composite. Introduction of hard WC particles and the formation of a small amount of TCP phases at the grain boundaries increased the overall hardness of the composite. For Inconel 625, sample hardness measured in a distance of 0 to 400 μm from the surface was below average. In deeper parts of the coating, it rose to a maximum of 425 $HV_{0.2}$ and gradually declined, with the distance to a minimum of 334 $HV_{0.2}$ at about 1400 μm. For the Inconel 625–WC composite, hardness was constant, in the range of 0–800 μm distance, where it increased to a maximum of 542 $HV_{0.2}$. It was followed by a smooth decrease to 288 $HV_{0.2}$. This emphasizes that properties of the sample in the transitional area were affected by diffusion between two materials characterized by different hardness values.

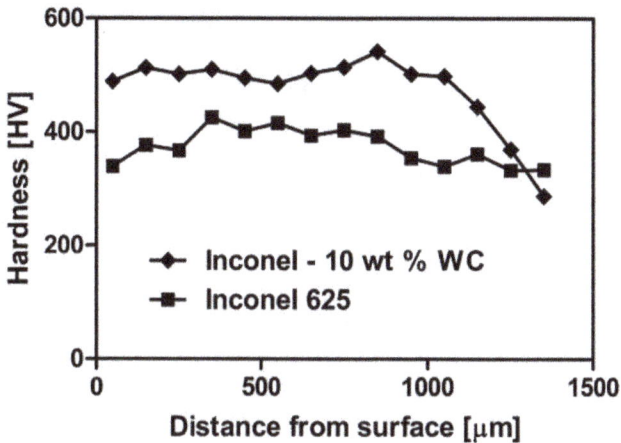

Figure 10. Vickers hardness of Inconel 625–WC coating and pure Inconel 625.

4. Discussion

The analysis of the obtained material proved that the LMD-produced Inconel 625–WC composite was crack- and pore-free. Ceramic particles were well-connected to the nickel-based matrix. TCP phases provided additional "anchoring" of the reinforcement in metal. Deposition of the Inconel 625–WC powder mixture resulted in a homogeneous distribution of ceramic particles in the obtained samples when compared to non mixed powders [27–29].

The results show that the introduction of WC modifies the microstructure and hardness of the obtained coating. The grain size of WC allowed for observation of interesting processes on the WC–Inconel 625 boundary. Good wettability of WC by a nickel-based alloy and rapid heating and cooling during LMD resulted in surface dissolution of the ceramics. It began in sharp tips of the grains and formed a fishbone-like eutectic structure at the metal–ceramic interface. Thanks to that, it was possible to observe how WC grains were reacting with the metal matrix. Samples were kept in a temperature above the Inconel 625 melting point for a very short time (up to 2 s). Due to rapid cooling, the sample microstructure remained partially dissolved, which is difficult to achieve when using conventional coating-deposition techniques. Because of the rapid nature of the process, the material remained in a physicochemical non-equilibriumbrium state.

The WC presence in the obtained material was revealed by TEM-EDS analysis, which confirmed the assumption that grains of selected sizes survive laser processing. TEM diffraction patterns showed that the coating microstructure is complex. Crystallographic orientation of the Inconel 625 grains is similar and differences are caused by the inclusion of other phases in the material.

The presence of the WC and TCP phases was beneficial for material's hardness properties. The overall hardness of the coating is about 25%higher than that of pure Inconel 625 obtained by the same technique. The linear decline of hardness was observed in deeper parts of the samples because of mixing between substrate and powder mixture during laser processing. This was confirmed by

SEM-EDS linear element-distribution analysis. The amount of Ni, Nb, Mo and W decreased in the transitional area in comparison to coating.

Introduction of WC caused grain-size refinement. It also strengthened the composite microstructure. However, it can negatively affect corrosion resistance [32,33]. Uniform distribution of carbide particles in the whole volume of the coating is expected to improve coating wear resistance. The appearance of TCP phases can further improve wear resistance and hardness, but weaken elastic properties when compared to nano-reinforcement [35–37].

5. Conclusions

- LMD allowed us to obtain crack- and pore-free homogeneous material.
- Initially prepared Inconel 625–WC powder mixture resulted in uniform distribution of reinforcement in the whole volume of the material.
- WC grain size of 5 μm is suitable to survive the LMD process.
- Partial dissolution of WC in nickel-based matrix resulted in the appearance of TCP phases at the ceramic–metal interface.
- Composite hardness was improved by about 25% in comparison to pure Inconel 625 obtained by the same technique and parameters.

Author Contributions: Conceptualization, D.K.; Formal analysis, J.H. and J.K.; Funding acquisition, J.H.; Investigation, J.H., P.R. and P.P.; Methodology, J.H., P.R. and P.P.; Project administration, J.H.; Supervision, D.K. and J.K.; Writing—original draft, J.H.; Writing—review & editing, D.K. and J.K.

Funding: This research was funded by National Science Center Poland, PRELUDIUM 13grant "Protective composite coatings for high temperature applications obtained by laser-cladding method" number [UMO-2017/25/N/ST5/02319]. The funders had no role in the design of the study; in the collection, analyses, or interpretation of data; in the writing of the manuscript; or in the decision to publish the results.

Conflicts of Interest: The authors declare no conflict of interest.

References

1. Teppernegg, T.; Klünsner, T.; Kremsner, C.; Tritremmel, C.; Czettl, C.; Puchegger, S.; Marsoner, S.; Pippan, R.; Ebner, R. High temperature mechanical properties of WC-Co hard metals. *Int. J. Refract. Met. Hard Mater.* **2016**, *56*, 139–144. [CrossRef]
2. Dinda, G.P.; Dasgupta, A.K.; Mazumder, J. Laser aided direct metal deposition of Inconel 625 superalloy: Microstructural evolution and thermal stability. *Mater. Sci. Eng. A* **2009**, *509*, 98–104. [CrossRef]
3. Zhou, S.; Huang, Y.; Zeng, X.; Hu, Q. Microstructure characteristics of Ni-based WC composite coatings by laser induction hybrid rapid cladding. *Mater. Sci. Eng. A* **2008**, *480*, 564–572. [CrossRef]
4. Riddihough, M. Stellite as a Wear Resistant Material. *Tribology* **1970**, *3*, 211–215. [CrossRef]
5. De Damborena, J.; Lopez, V.; Vazquez, A.J. Improving High-temperature Oxidation of Incoloy 800H by Laser Cladding. *Surf. Coat. Technol.* **1994**, *70*, 107–113. [CrossRef]
6. Rottwinkel, B.; Nölke, C.; Kaierle, S.; Wesling, V. Crack repair of single crystal turbine blades using laser cladding technology. *Procedia CIRP* **2014**, *22*, 263–267. [CrossRef]
7. Tosto, S.; Pierdominici, F.; Blanco, M. Laser Cladding and Alloying of a Ni-Base Superalloy on Plain Carbon Steel. *J. Mater. Sci.* **1994**, *29*, 504–509. [CrossRef]
8. Ayers, J.D.; Schaefer, R.J.; Robey, W.P. A Laser Processing Technique for Improving the Wear Resistance of Metals. *JOM* **1981**, *33*, 19–23. [CrossRef]
9. DuPont, J.N. Solidification of an Alloy 625 Weld Overlay. *Metall. Mater. Trans. A* **1996**, *27*, 3612–3620. [CrossRef]
10. Davis, J.R. Directionally Solidified and Single-Crystal Superalloys. In *ASM Specialty Handbook: Heat-Resistant Materials*; ASM International: Materials Park, OH, USA, 1997; pp. 255–271. ISBN 0-87170-596-6.
11. Seth, B.B. *Superalloys—The Utility Gas Turbine Perspective*; TMS: Warrendale, PA, USA, 2000; pp. 3–16.
12. Sims, C.T.; Stoloff, N.S.; Hagel, W.C. *Superalloys II: High-Temperature Materials for Aerospace and Industrial Power*; Wiley: New York, NY, USA, 1987; ISBN 978-0-471-01147-7.

13. DuPont, J.N.; Lippold, J.C.; Kiser, S.D. *Welding Metallurgy and Weldability of Nickel Based Alloys*; Wiley: New York, NY, USA, 2009; ISBN 978-0-470-08714-5.

14. Petrzak, P.; Blicharski, M.; Dymek, S.; Solecka, M. Electron microscopy investigation of Inconel 625 weld overlay on boiler steel. *Solid State Phenom.* **2015**, *231*, 113–118. [CrossRef]

15. Costa, L.; Vilar, R. Laser powder deposition. *Rapid Prototyp. J.* **2009**, *15*, 264–279. [CrossRef]

16. Jasim, K.M.; Rawlings, R.D.; West, D.R.F. Thermal Barrier Coatings Produced by Laser Cladding. *J. Mater. Sci.* **1990**, *25*, 4943–4948. [CrossRef]

17. Steen, W.M.; Powell, J. Laser Surface Treatment. *Mater. Des.* **1981**, *2*, 157–162. [CrossRef]

18. Ghosal, P.; Majumder, M.C.; Chattopadhyay, A. Study on direct laser metal deposition. *Mater. Today* **2017**, *5*, 12509–12518. [CrossRef]

19. Przybyłowicz, J.; Kusinski, J. Laser Cladding and Erosive Wear of Co-Mo-Cr-Si Coatings. *Surf. Coat. Technol.* **1999**, *125*, 13–18. [CrossRef]

20. Grunling, H.W.; Schneider, K.; Singheiser, L. Mechanical Properties of Coated Systems. *Mater. Sci. Eng.* **1987**, *88*, 177–189. [CrossRef]

21. Hurricks, P.L. Review Paper: Some Aspects of the Metallurgy and Wear Resistance of Surface Coatings. *Wear* **1972**, *22*, 291–320. [CrossRef]

22. Zhou, S.; Lei, J.; Dai, X.; Guo, J.; Gu, Z.; Pan, H. A comparative study of the structure and wear resistance of NiCrBSi/50 wt.% WC composite coatings by laser cladding and laser induction hybrid cladding. *Int. J. Refract. Met. Hard Mater.* **2016**, *60*, 17–27. [CrossRef]

23. Huang, S.W.; Samandi, M.; Brandt, M. Abrasive wear performance and microstructure of laser clad WC/Ni layers. *Wear* **2004**, *256*, 1095–1105. [CrossRef]

24. Moussaoui, K.; Rubio, W.; Mousseigne, M.; Sultan, T.; Rezai, F. Effects of Selective Laser Melting additive manufacturing parameters of Inconel 718 on porosity, microstructure and mechanical properties. *Mater. Sci. Eng. A* **2018**, *735*, 182–190. [CrossRef]

25. Yuan, K.; Guo, W.; Li, P.; Wang, J.; Su, Y.; Lin, X.; Li, Y. Influence of process parameters and heat treatments on the microstructures and dynamic mechanical behaviors of Inconel 718 superalloy manufactured by laser metal deposition. *Mater. Sci. Eng. A* **2018**, *721*, 215–225. [CrossRef]

26. Zhang, D.; Feng, Z.; Wang, C.; Wang, W.; Liu, Z.; Niu, W. Comparison of microstructures and mechanical properties of Inconel 718 alloy processed by selective laser melting and casting. *Mater. Sci. Eng. A* **2018**, *724*, 357–367. [CrossRef]

27. Nagarathnam, K.; Komvopoulos, K. Microstructural Characterization and In Situ Transmission Electron Microscopy Analysis of Laser-Processed and Thermally Treated Fe-Cr-W-C Clad Coatings. *Metall. Trans. A* **1993**, *24*, 1621–1629. [CrossRef]

28. Cooper, K.P.; Slebodnick, P. Recent Developments in Laser Melt/Particle Injection Processing. *J. Laser Appl.* **1989**, *1*, 21–29. [CrossRef]

29. Bi, G.; Sun, C.N.; Nai, M.L.; Wei, J. Micro-structure and Mechanical Properties of Nano-TiC Reinforced Inconel 625 Deposited using LAAM. *Phys. Procedia* **2013**, *41*, 828–834. [CrossRef]

30. Zhang, B.; Bi, G.; Wang, P.; Bai, J.; Chew, Y.; Nai, M.S. Microstructure and mechanical properties of Inconel 625/nano-TiB2 composite fabricated by LAAM. *Mater. Des.* **2016**, *111*, 70–79. [CrossRef]

31. Jiang, D.; Hong, C.; Zhong, M.; Alkhayat, M.; Weisheit, A.; Gasser, A.; Zhang, H.; Kelbassa, I.; Poprawe, R. Fabrication of nano-TiCp reinforced Inconel 625 composite coatings by partial dissolution of micro-TiCp through laser cladding energy input control. *Surf. Coat. Technol.* **2014**, *249*, 125–131. [CrossRef]

32. DuPont, B.C.; Lugscheider, E. Comparison of Properties of Coatings Produced by Laser Cladding and Conventional Methods. *Mater. Sci. Technol.* **1992**, *8*, 657–665. [CrossRef]

33. Heigel, J.C.; Gouge, M.F.; Michaleris, P.; Palmer, T.A. Selection of powder or wire feedstock material for the laser cladding of Inconel 625. *J. Mater. Process. Technol.* **2016**, *231*, 357–365. [CrossRef]

34. Frenk, A.; Kurz, W. High Speed Laser Cladding, Solidification Conditions and Microstructure of Cobalt-Based Alloy. *Mater. Sci. Eng. A* **1993**, *173*, 339–342. [CrossRef]

35. Abioye, T.E.; Farayibi, P.K.; McCartney, D.G.; Clare, A.T. Effect of carbide dissolution on the corrosion performance of tungsten carbide reinforced Inconel 625 wire laser coating. *J. Mater. Process. Technol.* **2016**, *231*, 89–99. [CrossRef]

36. Papaefthymiou, S.; Bouzouni, M.; Petrov, R.H. Study of Carbide Dissolution and Austenite Formation during Ultra-Fast Heating in Medium Carbon Chromium Molybdenum Steel. *Metals* **2018**, *8*, 646. [CrossRef]
37. Silva, F.J.G.; Santos, J.; Gouveia, R. Dissolution of Grain Boundary Carbides by the Effect of Solution Annealing Heat Treatment and Aging Treatment on Heat-Resistant Cast Steel HK30. *Metals* **2017**, *7*, 251. [CrossRef]
38. Huebner, J.; Kata, D.; Kusiński, J.; Rutkowski, P.; Lis, J. Microstructure of laser cladded carbide reinforced Inconel 625 alloy for turbine blade application. *Ceram. Int.* **2017**, *43*, 8677–8684. [CrossRef]

![materials logo] *materials*

MDPI

Article

Interlaminar Shear Properties of Z-Pinned Carbon Fiber Reinforced Aluminum Matrix Composites by Short-Beam Shear Test

Sian Wang [1], Yunhe Zhang [1,*] and Gaohui Wu [2]

[1]   College of Mechanical and Electrical Engineering, Northeast Forestry University, Harbin 150040, China;
      daowsa@nefu.edu.cn
[2]   School of Materials Science and Engineering, Harbin Institute of Technology, Harbin 150080, China;
      wugh@hit.edu.cn
*   Correspondence: yunhe.zhang@nefu.edu.cn; Tel.: +86-451-8219-2843 or +86-156-6352-6798;
      Fax: +86-451-8219-2843

Received: 27 August 2018; Accepted: 29 September 2018; Published: 1 October 2018

Abstract: This paper presents the effect of through-thickness reinforcement by steel z-pins on the interlaminar shear properties and strengthening mechanisms of carbon fiber reinforced aluminum matrix composites (Cf/Al) with a short beam shear test method. Microstructural analysis reveals that z-pins cause minor microstructural damage including to fiber waviness and aluminum-rich regions, and interface reaction causes a strong interface between the stainless steel pin and the aluminum matrix. Z-pinned Cf/Al composites show reduced apparent interlaminar shear strength due to a change in the failure mode compared to unpinned specimens. The changed failure mode could result from decreased flexural strength due to microstructural damage as well as increased actual interlaminar shear strength. Fracture work is improved significantly with a z-pin diameter. The strong interface allows the deformation resistance of the steel pin to contribute to the crack bridging forces, which greatly enhances the interlaminar shear properties.

Keywords: metal matrix composites; Z-pin reinforcement; delamination; carbon fiber; strengthening mechanisms

---

1. Introduction

Carbon fiber reinforced aluminum matrix composites (Cf/Al) have increasingly been used for a variety of automotive and aerospace applications because of their high specific modulus, high specific strength, high thermal conductivity, and low coefficient of thermal expansion [1–5]. However, Cf/Al composites have been reported to have high residual stress, leading to the separation of the interface and delamination cracks during the forming process [4,5]. Delamination cracks cause relatively low interlaminar properties of Cf/AI composites, despite their excellent in-plane mechanical properties. Composites containing these cracks between laminates would suffer from delamination failures when there are out-of-plane loads, causing the severe decline of mechanical properties and earlier structural failure. Thus, it is significantly crucial to improve the interlaminar shear properties of Cf/AI composites for their future applications.

The z-pinning method has been advocated as a simple and effective method to enhance the delamination resistance of composites [6–8]. Many studies [6–10] have demonstrated that z-pins can improve the delamination fracture toughness, interlaminar shear strength, impact damage tolerance, and delamination fatigue resistance of composites. z-pins are also effective at increasing the ultimate strength, fatigue performance, and damage tolerance of bonded composite joints [11–13].

There are two types of z-pins: Fibrous z-pin and metal z-pin. The effect of fibrous z-pins, which are typically unidirectional carbon fiber rods, on the interlaminar properties has been studied in detail [6,13–15]. The literature shows that the fiber z-pin generates a traction force in the bridging zone that reduces the strain energy at the crack tip, thus improving the delamination toughness. In addition, the carbon fiber z-pin is very effective at transforming the crack growth from fast propagation to stable propagation in polymer matrix composites. Less is known about metallic z-pins and their effects on the interlaminar properties of z-pinned composites. Pingkarawat and Mouritz [16] found that the mode I delamination toughness and fatigue strength of carbon-epoxy composites are related to stiffness, strength and fatigue resistance of the material of the z-pin. Zhang et al. [5] reported that the interlaminar shear strength of Cf/Al composites was enhanced by 70–230% using stainless steel z-pins as reinforcements and found that the interfacial reaction layer between the metal pin and the Al alloy was controlled by the z-pin diameter. However, a detailed investigation on the strengthening mechanism of the metal z-pin was not conducted. Ko et al. [13] found that jagged stainless steel pins can increase the static strength and fatigue strength of composite single-lap joints by 11.8–65.8% in different environments, such as various temperatures and relative humidity. The improvement of the mechanical properties was attributed to the transfer of the fastening force between the reinforcing pin and the matrix material through friction. Although the effectiveness of stainless steel z-pins to enhance the interlaminar properties of composites has been proven, their interlaminar strengthening mechanisms have not been systematically addressed and understood.

Recent studies [14–17] on delamination fracture mechanisms have focused on observation and analysis of the fractographic results of z-pinned composites using optical microscopy and scanning electron microscopy (SEM). These methods could not work well in comprehensively analyzing the interlaminar strengthening mechanisms of z-pinned composites. As an excellent nondestructive inspection method based on absorption contrast, X-ray radiography is a better choice to observe z-pin's deforming and composites' delamination, with z-pin located at their original place in the composites [18,19].

The purpose of this paper is to study the Mode II interlaminar shear properties of Cf/Al composites reinforced with stainless steel z-pins by the short-beam test method. The influence of stainless steel z-pins parameters, including the volume content, diameter, and interval, on the interlaminar shear properties of composites is discussed. X-ray radiography is used to observe deformed specimens and to evaluate the role of the steel pins in the Mode II interlaminar fracture process. This study would enhance the understanding of Mode II interlaminar strengthening mechanisms of z-pinned Cf/Al composites.

2. Experimental

2.1. Materials

The matrix in this study was a 5056Al alloy purchased from Northern Light Alloy Company Ltd., Harbin, China; it had the following chemical composition (wt.%): Al: 94.89%, Mg: 4.80%, Fe: 0.12%, Mn: 0.07%, Si: 0.06% and Cr: 0.06%. M40 carbon fiber (purchased from Toray Industries Inc., Tokyo, Japan) and AISI 321 (Shanghai Baosteel Group Corporation, Shanghai, China) stainless steel were used as the reinforcing material and the z-pin, respectively. The properties of the M40 fibers, matrix alloy, and AISI 321 are presented in Table 1.

Table 1. Basic properties of 5056Al alloy, fibers and z-pins.

Materials	Density (g/cm^3)	Tensile Strength (MPa)	Elastic Modulus (GPa)	Elongation to Fracture (%)
5056Al	2.64	314	66.7	16.0
M40	1.76	4410	377.0	1.2
AISl321	7.85	1905	198.0	2.0

2.2. Sample Preparation

The z-pinned and unpinned Cf/Al composites were fabricated by the pressure infiltration technology. The Cf/Al composite without z-pin reinforcement was produced as the control material to study the changes to the interlaminar properties of the z-pinned composites and the interlaminar strengthening mechanisms of stainless steel pins. The fabrication procedure of investigated composites is shown in Figure 1. Preforms were made using the stainless steel and the carbon fiber, which was unidirectionally winded to the specific shape by a CNC Winding Machine. The preheating temperature of the steel mold with the preform was 500 ± 10 °C. Then the molten 5056Al alloy was infiltrated into the preform under pressure at 780 ± 20 °C, and the infiltration pressure was kept at 0.5 MPa. The composites fabricated were then annealed at 330 °C for 0.5 h.

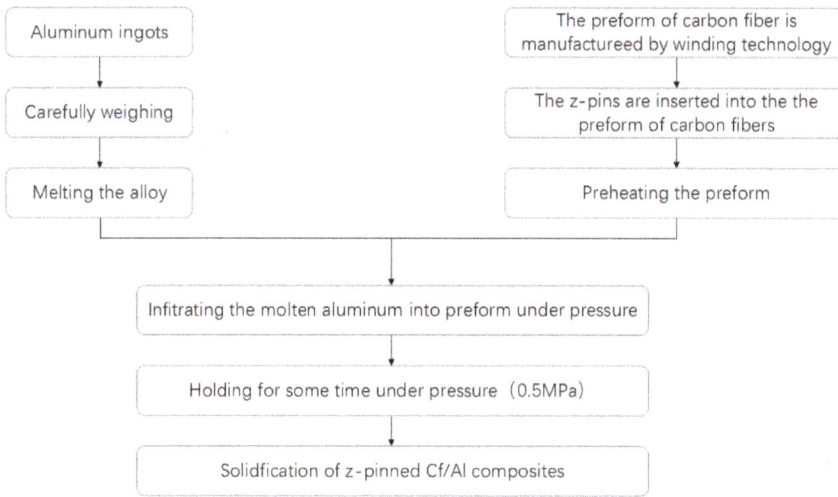

Figure 1. Fabrication procedure of z-pinned Cf/Al composites.

Several types of z-pinned Cf/Al composite specimens and the unpinned Cf/Al composite specimens were made for a short-beam interlaminar shear test (Table 2). The influence of the z-pin diameter and volume content on the interlaminar properties was studied. Composites were reinforced with stainless steel pins with diameters of 0.3, 0.6, and 0.9 mm and had volume contents of 0.25%, 0.5%, and 1.0%. For the fixed 0.3 mm diameter of z-pins, the volume contents were varied between 0.25%, 0.5%, and 1.0% to study the influence of z-pin volume content on the interlaminar properties. For these 0.3 mm diameter z-pins, the intervals between the pins were 5.3, 3.8, and 2.7 mm. With a fixed 2.0% volume content of z-pins, the z-pins' diameters were varied between 0.3, 0.6, and 0.9 mm to study the influence of z-pin diameter on the interlaminar properties. The intervals between the pins were 2.7, 5.3, and 8.0 mm for the diameter of 0.3, 0.6, and 0.9 mm, respectively. The volume content of carbon fiber in the unpinned and z-pinned composites was approximately 55%.

Table 2. Short-beam shear interlaminar test matrix of laminate specimens.

z-Pin Volume Content (%)	z-Pin Diameter (mm)	z-Pin Space (mm)	Interlaminar Shear Properties
-	-	-	Yes
0.25	0.3	5.3	Yes
0.5	0.3	3.8	Yes
1	0.3	2.7	Yes
1	0.6	5.3	Yes
1	0.9	8.0	Yes

2.3. Characterization Techniques

The interlaminar shear strength of the unpinned and z-pinned composites was determined using the short beam shear test based on the classical beam theory. The specimens were machined into their dimensions of 30 mm long, 5.3 mm wide and 5 mm thick. The span of the support points was 20 mm (the span length was equivalent to 4 times the thickness). The tests were performed in accordance with ASTM D 2344 [20] at room temperature using an Instron-5569 universal electronic tensile testing machine with a cross-head speed of 0.5 mm/min. The loading setup of the specimens is shown in Figure 2. The specimens were placed on the central position of two cylindrical supports, and a cylindrical head was used to apply a force at the center of the specimens until failure. Cylindrical supports radius and cylindrical head radius are 5 mm and 10 mm, respectively. The apparent interlaminar shear strengths of z-pinned and unpinned composites are calculated using

$$\tau = 3P/4bh \tag{1}$$

where τ is the apparent interlaminar shear strength, P is the maximum load, b and h are the width and thickness of the specimen, respectively. Five specimens of each type of composite were tested, and the interlaminar shear strength was averaged from these five replicates.

The microstructures of the z-pinned Cf/Al composites were characterized by the S-4700 SEM (Royal Dutch Philips Electronics Ltd., Amsterdam, Netherlands). The transformation characteristics of z-pins and failure modes were observed by a SEFT225 X-ray camera (GE Sensing & Inspection Technologies GmbH, Ahrensburg, Germany) to investigate the failure mechanism of the z-pinned Cf/Al composites.

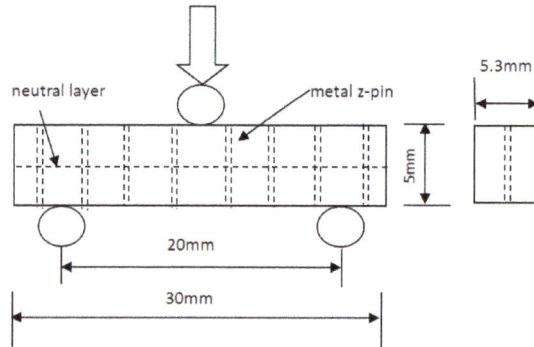

Figure 2. Schematic of short beam shear testing.

3. Results and Discussion

3.1. Microstructure

Figure 3 shows the microstructure of z-pinned Cf/Al composites. The interface layer between AISI 321 steel and Cf/Al is clearly visible, indicating a good combination of the two materials. The Cf/Al composites reinforced by different diameters of steel pin show a high denseness without defects such as porosity and cracks. Compared to a Cf/Al composite without a z-pin, a fusiform aluminum-enriched region is formed around the z-pin in z-pinned composites. During the sample preform preparation phase, the z-pin squeezed the surrounding carbon fibers and a void around the z-pin was left; this gap was then filled by Al at a later stage of melt Al being poured into the preform. The matrix enrichment is equivalent to the increase of z-pin diameter to further increase the interlaminar strength of the material. However, the yielding of the fiber could result in a certain angle between the fiber orientation and the direction of the force of the composite material, and thus the decrease of in-plane properties of the laminate including tensile, compressive and bending properties [6].

In addition, Figure 3 also shows that, as the z-pin's diameter increases, the thickness of the interfacial reaction layer between the steel and the matrix gradually decreases. This is because as the z-pin's diameter increases, the volume becomes larger, requiring more heat to achieve the same increase in the temperature. But, the contact time between the z-pin and the Al liquid in the sample preparation process is too limited to complete the heat exchange. Within the same period, the z-pin with smaller diameters has a larger temperature increase, stronger atomic diffusion ability, and a greater degree of interface reaction.

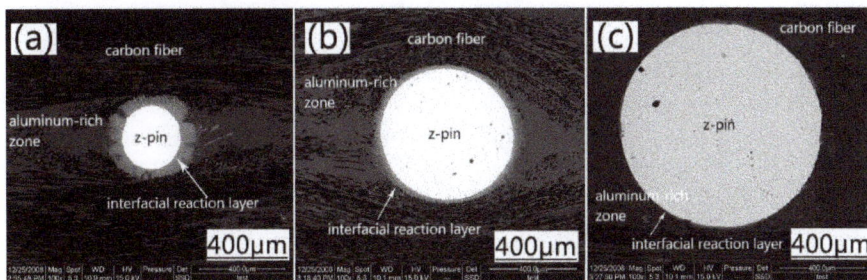

Figure 3. Microstructures of z-pinned Cf/Al composites with different diameters of metal pins (**a**) φ0.3 mm, (**b**) φ0.6 mm and (**c**) φ0.9 mm.

3.2. Apparent Interlaminar Shear Strength

The effect of the z-pins volume content on the apparent interlaminar shear strength of the Cf/Al composites is shown in Figure 4. The strength decreases with increasing volume content of the stainless steel pins. Table 3 shows the apparent interlaminar shear strength experimental values for each group of z-pinned Cf/Al composites and unpinned Cf/Al composites as well as their standard deviations. The very small standard deviations confirm that adding metal z-pins affects the apparent interlaminar shear strength of Cf/Al composites in a statistically significant way. This is consistent with the results from other researchers. For example, Mouritz et al. [9] reported the similar experimental results in carbon-epoxy composites.

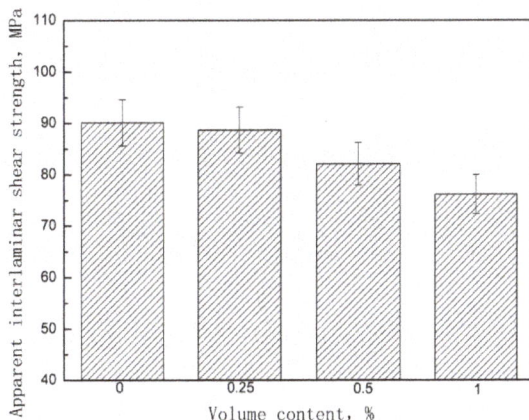

Figure 4. Effect of metal pin content on apparent interlaminar shear strength of z-pinned Cf/Al composites using stainless steel pins with diameters of 0.3 mm.

Table 3. Apparent interlaminar shear strength of unpinned Cf/Al composites and z-pinned Cf/Al composites using stainless steel pins with diameters of 0.3 mm measured by short-beam shear test.

Specimen Number	Apparent Interlaminar Shear Strength (MPa)			
	0 vol%	0.25 vol%	0.5 vol%	1 vol%
1	89.3	88.7	82.1	77.7
2	92.3	90.0	86.8	76.2
3	89.9	91.5	90.3	79.8
4	89.1	87.4	78.4	75.6
5	90.4	86.2	83.2	76.9
Average	90.2	88.8	84.2	77.2
Standard deviation	1.15	1.87	4.07	1.46

This apparent strength degradation is believed to be related to a change in the fracture mode of the investigated composite with different volumes of stainless steel pins. The failure mode mainly depends on the mechanical properties of the composite and the span-to-thickness ratio of the specimen. We can see that for the composite without z-pins, shear delamination and tensile fracture are caused by the combination of shear and bending stress. The shear delamination is induced by the low interlaminar shear strength of the composites without z-pins. However, for the z-pinned composite, the failure occurred only by fracture along the specimen cross section. The stress state in short beam shear test specimens was complex, involving a combination of compressive, tensile, flexural and interlaminar shear stress. The failure of specimens often occurred by flexural and interlaminar shear stress and stainless steel z-pins were effective in enhancing the actual interlaminar shear strength. Hence, the flexural strength was also the decisive factor for the apparent strength value.

The flexural strength of composites decreases with increasing volume content of the pins typically owing to in-plane fiber waviness and matrix-rich zones caused by the z-pins. The decrease of flexural strength could cause the reduction of the apparent interlaminar shear strength [21].

The effect of pin diameter on the apparent interlaminar shear strength of the composite is shown in Figure 5. The apparent strength values are almost the same when the pin size was increased at fixed pin volume content. Table 4 shows the measured apparent interlaminar shear strength for each group of z-pinned Cf/Al composites and unpinned Cf/Al composites as well as their standard deviations. A possible explanation is that increasing the pin diameter may not lead to the decline of the flexural strength of z-pinned metal matrix composites. A further study is being conducted on the mechanism and effects of the steel pins' diameter on the flexural strength.

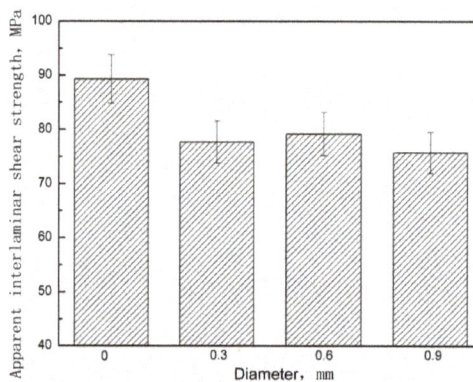

Figure 5. Effect of metal pin diameter on apparent interlaminar shear strength of z-pinned Cf/Al composites using stainless steel pins with a volume content of 1%.

Table 4. Apparent interlaminar shear strength of unpinned Cf/Al composites and z-pinned Cf/Al composites using stainless steel pins with a volume fraction of 1% measured/determined by short-beam shear test.

Specimen Number	Apparent Interlaminar Shear Strength (MPa)			
	d = 0	d = 0.3 mm	d = 0.6 mm	d = 0.9 mm
1	89.3	77.7	79.2	75.7
2	92.3	76.2	73.9	74.4
3	89.9	79.8	78.5	65.7
4	89.1	75.6	80.5	76.3
5	90.4	76.9	79.4	68.6
Average	90.2	77.2	78.3	72.1
Standard deviation	1.15	1.46	2.29	4.23

3.3. Analysis of Stress-Strain Curves of Z-Pinned Cf/Al Composites by the Short-Beam Test

Figure 6 shows the shear stress-deflection curve of a Cf/Al composite without z-pin and that of a Cf/Al composite with a z-pin volume content of 1% and 0.6 mm diameter (Other the shear stress-deflection curve of the z-pinned composite using stainless steel pins with different parameters look similar). For the unpinned composite at the beginning of loading, the shear stress increased linearly up to the maximum stress where it saw a significant and rapid decrease. The z-pinned composite had a similar elastic deformation stage to that of the unpinned composite. However, after reaching maximum shear stress, the z-pinned composite had a certain amount of deflection where the shear stress only gradually decreased with a small amount of fluctuation. The z-pinned composite in this stage retained a high shear-bearing capacity. The deformation characteristics in this post-maximum shear stress stage are similar to the tensile yield characteristics of metallic materials. Thus, this stage can be regarded as a "pseudo-yielding" stage. Following this stage, the shear stress decreased significantly to final fracture.

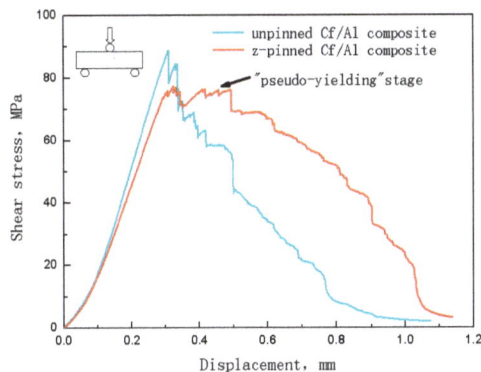

Figure 6. Typical shear stress-deflection curve of Cf/Al composites with and without z-pins.

The steel pin's geometric parameters' effect on the extent of deflection in the pseudo-yielding, herein called the yield platform, was quantitatively evaluated. This evaluation of the yield platform defines its length, Δl, as the amount of deflection between the point of maximum shear stress and the point when the stress has decreased to 90% of the maximum value. Figure 7 shows this definition of the yield platform length. The influence of the volume content and diameter of the steel z-pin on the yield platform length is shown in Figure 8. This figure shows that as the volume content of the steel pin and the diameter increased, the length of the yield platform also increased. Tables 5 and 6 show the measured yield platform length experimental values for each group of z-pinned Cf/Al composites and unpinned Cf/Al composites as well as their standard deviations. In addition, the material maintained a higher load-bearing capacity with longer yield platforms, indicating that the addition of a steel pin may also enhance the interlaminar fracture toughness of the material. This is supported by the effect of the steel z-pin diameter on the fracture work during loading as shown in Figure 9. Table 7 shows the measured fracture work length experimental values for each group of z-pinned Cf/Al composites and unpinned Cf/Al composites as well as their standard deviations. Here, it is seen that increasing the diameter of the steel pin does not increase the shear strength of the short beam; however, it did significantly increase the fracture work. This was attributed mostly to formation of a bridging zone caused z-pins spanning the crack. The crack bridging forces effectively resisted the propagation of delamination cracks and remarkably reduced the opening stress at the crack front. Thus, the fracture work and length of the yield platform could be significantly improved with z-pins. As the bridging forces are transmitted by the interface of the composite and stainless steel pin, improvement to the interlaminar shear property is controlled by the total interfacial contact area between the composite and z-pins. Hence, the length of the yield platform increases with the diameter and volume content of z-pins.

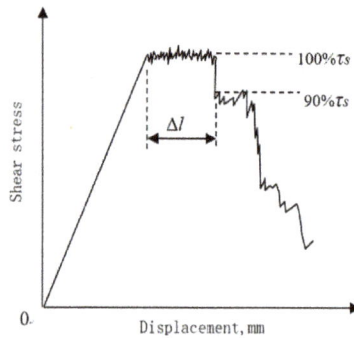

Figure 7. Schematic diagram showing definition of the yield platform length.

Table 5. Yield platform length of unpinned Cf/Al composites and z-pinned Cf/Al composites using stainless steel pins with diameters of 0.3 mm measured by short-beam shear test.

Specimen Number	Yield Platform Length (mm)			
	0 vol%	0.25 vol%	0.5 vol%	1 vol%
1	0.02	0.022	0.079	0.167
2	0.019	0.025	0.093	0.175
3	0.017	0.018	0.084	0.164
4	0.024	0.02	0.081	0.153
5	0.021	0.024	0.085	0.178
Average	0.02	0.022	0.084	0.167
Standard deviation	0.0023	0.0026	0.0048	0.0088

Table 6. Yield platform length of unpinned Cf/Al composites and z-pinned Cf/Al composites using stainless steel pins with a volume fraction of 1% measured/determined by short-beam shear test.

Specimen Number	Yield Platform Length (mm)			
	d = 0	d = 0.3 mm	d = 0.6 mm	d = 0.9 mm
1	0.02	0.167	0.223	0.286
2	0.019	0.175	0.235	0.274
3	0.017	0.164	0.218	0.285
4	0.024	0.153	0.206	0.254
5	0.021	0.178	0.241	0.296
Average	0.02	0.167	0.225	0.279
Standard deviation	0.0023	0.0088	0.0124	0.0143

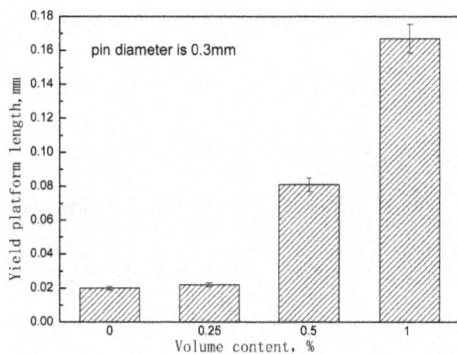

(a)

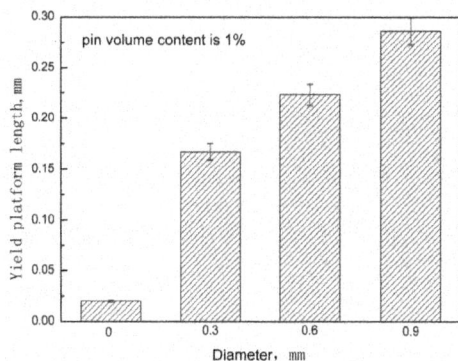

(b)

Figure 8. Effect of (**a**) metal pin content and (**b**) metal pin diameter on yield platform length.

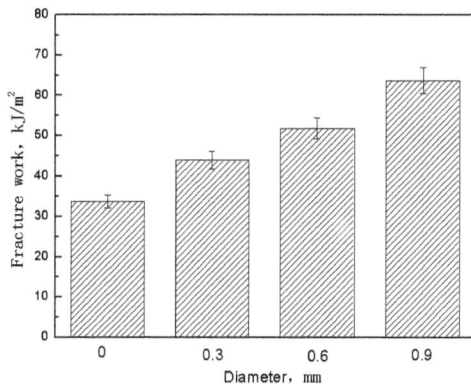

Figure 9. Effect of metal pin diameter on fracture work of z-pinned Cf/Al composites using stainless steel pins with a volume content of 1%.

Table 7. Fracture work of unpinned Cf/Al composites and z-pinned Cf/Al composites using stainless steel pins with a volume fraction of 1% measured/determined by short-beam shear test.

Specimen Number	Fracture Work (kJ/m²)			
	d = 0	d = 0.3 mm	d = 0.6 mm	d = 0.9 mm
1	33.7	44.1	51.9	63.7
2	31.9	46.6	53.4	59.2
3	36.4	47.5	54.4	62.8
4	30.6	40.5	55.1	68.5
5	32.8	41.4	50.2	60.3
Average	33.1	44	53	62.9
Standard deviation	1.95	2.76	1.77	3.24

As the z-pinned specimen deflection continued, the stress began to decrease. Compared to the unpinned composite, the pinned composite had a lower rate of decrease of the shear stress prior to failure.

3.4. Interlaminar Strengthening Mechanisms

The specimens were examined using X-ray imaging before and after testing. As well, specimens were unloaded at different deflection levels after reaching maximum shear stress to investigate the failure progression.

The unpinned composite specimen X-ray images are shown in Figure 10. The progression in this figure indicates that a delamination shear failure occurred for this type of specimen. In the early stages of failure (Figure 10b), the unpinned composite had a single delamination crack that initiated from the specimen edge and propagated to the middle of the specimen. In addition, the initially formed delamination crack at maximum shear stress showed unstable crack propagation as it progressed into the middle of the specimen.

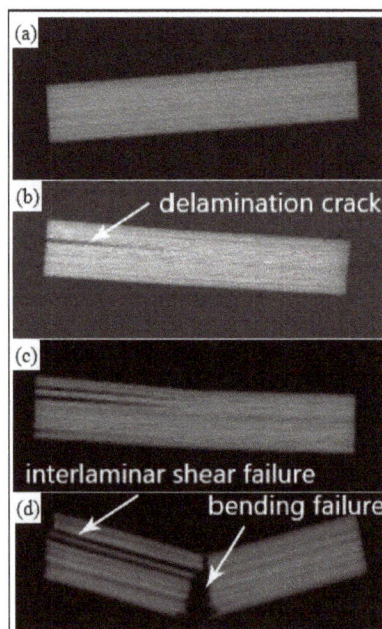

Figure 10. X-ray images of unpinned composite (**a**) before testing, (**b**) at maximum shear stress, (**c**) when load-bearing capacity dropped to 50%, (**d**) after fracture.

When the load-bearing capacity had dropped to 50% of its maximum value (Figure 10c), the unpinned composite specimen had undergone multiple delamination failures. There were no significant tensile or compressive failures observed on the upper and lower surfaces of the specimen. Thus, the failure mode was still delamination shear failure.

The final fracture of the unpinned composite specimen shown in Figure 10d demonstrates the severity of this delamination. It is clear that the delamination shear failure of the specimen becomes more severe with increasing deflection. When the tensile stress on the lower surface of the specimen exceeded the tensile strength of the composite, the specimen had a hybrid failure of shear delamination and tensile fracture.

X-ray images of the z-pinned composite with a z-pin volume content of 1% and 0.6 mm diameter are shown in Figure 11 (Other X-ray images of the z-pinned composites using stainless steel pins with different parameters look similar). The X-ray image corresponding to the end of the pseudo-yielding stage is shown in Figure 11b, which shows no visible opening fracture cracks but residual bending deformation. It is also seen that the steel pin did not plastically deform, which means that there was a good bonding at the steel pin-aluminum interface. However, the drop of shear stress here indicates that the initiation of delamination cracking in composites had occurred due to interlayer sliding displacements caused by bending deformation. It also shows that the addition of the steel pin had little effect on the initiation of delamination cracking in composites. On the other hand, metal z-pins effectively suppressed the propagation of delamination cracks. After the delamination crack was initiated, it was strongly pinned by the steel pin and could only locally expand in the section between two steel pins. The limiting of crack propagation resulted in the stable shear stress which was seen in the pseudo-yielding stage. As a result, deformation progressed with a different process compared to the unpinned composite. This reinforcement mechanism is consistent with the one of reinforcing fibers in carbon-epoxy composites.

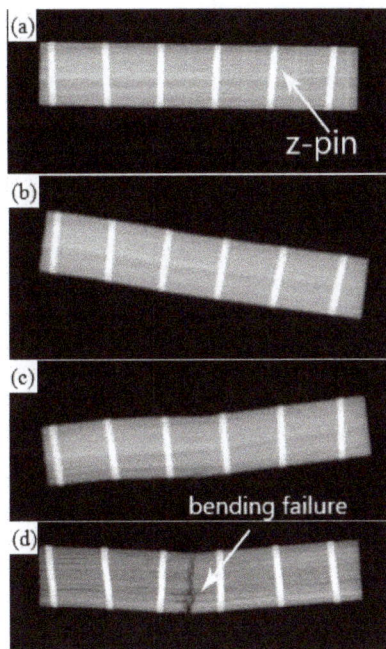

Figure 11. X-ray images of z-pinned composite with a z-pin volume content of 1% and 0.6 mm diameter (**a**) before testing (**b**) at maximum shear stress (**c**) when load-bearing capacity dropped to 50% (**d**) after fracture.

Figure 11c shows the X-ray image after the specimen reached 50% of the maximum shear stress value. In this image, the specimen still had no visible delamination damage, and the steel pin–aluminum interface remained well bonded with no observable plastic deformation of the steel pin. However, when compared with Figure 11b, the residual bending deformation in the middle of the specimen was greater. At the edge of the specimen, no deformation was observed. The lack of edge deformation demonstrates that the crack in the high-stress zone in the middle of the specimen did not extend to the stress-free zones of the edges. Residual bending deformation resulted from interlaminar movements between multiple sub-layers of composite materials. Under the action of steel pin transferring stress, the maximum shear stress of the neutral plane was distributed to each sub-layer. Consequently, the interlaminar fracture energy was dispersed and absorbed by those sub-layers. Thus, the failure mode of the specimen was transformed from delamination failure on the maximum shear stress surface to microscopic interlaminar shear failure occurring on each sublayer. As a result, the load-bearing capacity of the specimen slowly decreased with each progressive sublayer failure.

The X-ray observation of the pinned composite specimen after fracture (Figure 11d) shows that the specimen still had no obvious delamination fracture. The steel pin had an S-shaped plastic deformation. The specimen was fractured along the cross-section in the middle. There was no delamination cracking even at final fracture, for the pinned composite specimen. Thus, even if there was a layer misalignment in the specimen, the steel pin effectively inhibited the delamination fracture of the specimen due to the strong steel pin–aluminum interface. Under the condition of large shear deformation, the steel pin had seen significant plastic deformation. Figure 11b,c do not display any significant plastic deformation of the steel pin, which means at these points the pin was still behaving elastically. As a result, it is seen that the deformation resistance of the steel pin hindered the shear failure between the sub-layers of the composite material and eventually deformed as a result of this resistance. It is worthwhile mentioning that S-shaped deformation of the steel pins is caused by out-of-plane shear stress i.e., compressive stress and tensile stress as well as shear stress, which is different from the simple shear deformation due to plane stress and plane strain [22]. Due to the steel pin–aluminum interface, no delamination fracture occurred in the sub-layer in which interlayer displacement had occurred for the unpinned composite. When the tensile stress of the lower surface of the specimen exceeded its tensile strength, the specimen immediately underwent tensile fracture along the cross-section, and the stress rapidly decreased.

This is different from the effects reported for fiber z-pin deformation on shear delamination. This result is due to the difference in interfacial bonding strength and z-pin bending stiffness. In some cases, when the fiber z-pin is in the initial stage of loading the interface between the pin and the composite material is completely bonded, an S-shaped elastic deformation could occur which generates a bridging force that hinders delamination. However, the debonding or shear failure may occur at the interface as the load increases because the pin has a weak interface with the matrix. Therefore, the load causing delamination failure largely depends on the frictional pull-out force caused by interfacial friction. The contribution of deformation of the fiber z-pins is so small that many researchers have neglected the z-pins deformation when modeling the effect of fibers on the properties of the material. In this study, due to an interface reaction between the stainless steel z-pin and the aluminum matrix, there is a high degree of interfacial bonding strength. As a result, the steel z-pins can maintain a good interface with the matrix while being deformed. In addition, the steel z-pin maintains a constant bending stiffness during loading. Therefore, the deformation resistance of the steel z-pin can effectively block the interlayer shift. Thus, the z-pin bending enhances the load-bearing capacity of the material and increases the bridging resistance which prevents delamination. It should be noted that when the specimen failure appeared, both shear fracture and pull-out of the z-pins had not occurred. In other words, the carrying capacity of the z-pins was not fully utilized. This is consistent with the analysis in Section 3.2 regarding the effect of the z-pins parameters on the apparent interlaminar shear strength value.

4. Conclusions

The aim of this study was to understand the effect of steel z-pin reinforcement on the interlaminar properties of carbon fiber reinforced aluminum matrix composites (Cf/Al). The three-point beam shear test was performed with different z-pin diameters and z-pin volume contents. X-ray radiography was used to observe delamination propagation and deformation of the stainless steel pin.

The apparent interlaminar shear strength of the z-pinned composites is reduced by 1–27% due to the reduction of the flexural strength caused by the in-plane fiber waviness and aluminum-rich zones. The unpinned composites showed a combination of flexure/interlaminar shear in the failure mechanism due to the complex stress state in short beam shear test specimens. Meanwhile, bending failure was usually the dominant failure in the z-pinned composites since the steel pin is significantly effective at resisting the growth of delamination cracks. It should be emphasized that this change of failure mode is caused by the stainless steel pins improvement due to the actual interlaminar shear strength of composites.

In the shear stress and deflection curve, a yield platform similar to that in the metal tension test is observed. The length of the yield platform increases with the size and volume content of z-pins. The appearance of yield platform appears to be a direct result of the increase of crack bridging forces. The bridging force increases along with the increasing of the total size of the surface area of the composite and stainless steel pin. X-ray radiography reveals that S-shaped deformation of z-pins is the major contributor to the bridging force, which is related to the high interfacial bond strength due to the interface reaction between stainless steel and aluminum matrix.

Author Contributions: Project administration, Y.Z.; Resources, G.W.; Writing—original draft, S.W.; Writing—review & editing, Y.Z.

Funding: This research was funded by Natural Science Foundation of China (No. 51305075), the Science & Technology Innovation Foundation for Harbin Talents (Grant No. 2017RAYXJ021) and the Natural Science Foundation of Heilongjiang Province (Grant No. LC2015010).

Conflicts of Interest: The authors declare no conflict of interest.

References

1. Li, D.G.; Chen, G.Q.; Jiang, L.T.; Xiu, Z.Y.; Zhang, Y.H.; Wu, G.H. Effect of thermal cyclingon the mechanical properties of Cf/Al composites. *Mater. Sci. Eng. A* **2013**, *586*, 330–337. [CrossRef]
2. Li, Z.R.; Feng, G.J.; Wang, S.Y.; Feng, S.C. High-efficiency Joining of Cf/Al Composites and TiAl Alloys under the Heat Effect of Laser-ignited Self-propagating High-temperature Synthesis. *J. Mater. Sci. Technol.* **2016**, *32*, 1111–1116. [CrossRef]
3. Li, S.L.; Qi, L.H.; Zhang, T.; Zhou, J.Z.; Li, H.J. Microstructure and tensile behavior of 2D-Cf/AZ91D composites fabricated by liquid-solid extrusion and vacuum pressure infiltration. *J. Mater. Sci. Technol.* **2017**, *33*, 0541–546. [CrossRef]
4. Zhang, J.J.; Liu, S.C.; Lu, Y.P.; Jiang, L.; Zhang, Y.B.; Li, T.J. Semisolid-rolling and annealing process of woven carbon fibers reinforced Al-matrix composites. *J. Mater. Sci. Technol.* **2017**, *33*, 623–629. [CrossRef]
5. Zhang, Y.; Yan, L.; Miao, M.; Wang, Q.; Wu, G. Microstructure and mechanical properties of z-pinned carbon fiber reinforced aluminum alloy composites. *Mater. Des.* **2015**, *86*, 872–877. [CrossRef]
6. Mouritz, A.P. Review of z-pinned composite laminates. *Compos. Part A Appl. Sci. Manuf.* **2007**, *38*, 2383–2397. [CrossRef]
7. Yan, W.; Liu, H.Y.; Mai, Y.W. Mode II delamination toughness of z-pinned laminates. *Compos. Sci. Technol.* **2004**, *64*, 1937–1945. [CrossRef]
8. Yan, W.; Liu, H.Y.; Mai, Y.W. Numerical study of the mode I delamination toughness of z-pinned laminates. *Compos. Sci. Technol.* **2003**, *63*, 1481–1493. [CrossRef]
9. Mouritz, A.P.; Chang, P.; Isa, M.D. Z-pin composites: Aerospace structural design considerations. *J. Aero. Eng.* **2011**, *24*, 425–432. [CrossRef]
10. Isa, M.D.; Feih, S.; Mouritz, A.P. Compression fatigue properties of quasi-isotropic z-pinned carbon/epoxy laminate with barely visible impact damage. *Compos. Struct.* **2011**, *93*, 2222–2230. [CrossRef]

11. Ji, H.; Kweon, J.; Choi, J. Fatigue characteristics of stainless steel pin-reinforced composite hat joints. *Compos. Struct.* **2014**, *108*, 49–56. [CrossRef]

12. Koh, T.M.; Isa, M.D.; Feih, S.; Mouritz, A.P. Experimental assessment of the damage tolerance of z-pinned T-stiffened composite panels. *Compos. Part. B Eng.* **2013**, *44*, 620–627. [CrossRef]

13. Ko, M.G.; Kweon, J.H.; Choi, J.H. Fatigue characteristics of jagged pin-reinforced composite single-lap joints in hygrothermal environments. *Compos. Struct.* **2015**, *119*, 59–66. [CrossRef]

14. Mouritz, A.P.; Koh, T.M. Re-evaluation of mode I bridging traction modelling for z-pinned laminates based on experimental analysis. *Compos. Part. B Eng.* **2014**, *56*, 797–807. [CrossRef]

15. Mouritz, A.P. Delamination properties of z-pinned composites in hot-wet environment. *Compos. Part A. Appl. Sci. Manuf.* **2013**, *52*, 134–142. [CrossRef]

16. Pingkarawat, K.; Mouritz, A.P. Comparative study of metal and composite z-pins for delamination fracture and fatigue strengthening of composites. *Eng. Fract. Mech.* **2016**, *154*, 180–190. [CrossRef]

17. Yasaee, M.; Bigg, L.; Mohamed, G.; Hallett, S.R. Influence of Z-pin embedded length on the interlaminar traction response of multi-directional composite laminates. *Mater. Des.* **2017**, *115*, 26–36. [CrossRef]

18. Tan, K.T.; Watanabe, N.; Iwahori, Y. X-ray radiography and micro-computed tomography examination of damage characteristics in stitched composites subjected to impact loading. *Compos. Part. B Eng.* **2011**, *42*, 874–884. [CrossRef]

19. Cantwell, W.J.; Morton, J. The significance of damage and defects and their detection in composite materials: A review. *J. Strain Anal. Eng. Des.* **1992**, *27*, 29–42. [CrossRef]

20. American Society for Testing Materials. *Standard Test Method for Short-Beam Strength of Polymer Matrix Composite Materials and Their Laminates*; ASTM D 2344/D 2344M; ASTM International: West Conshohocken, PA, USA, 2000.

21. Chang, P.; Mouritz, A.P.; Cox, B.N. Flexural properties of z-pinned laminates. *Compos. Part A Appl. Sci. Manuf.* **2007**, *38*, 224–251. [CrossRef]

22. Butcher, C.; Abedini, A. Shear confusion: Identification of the appropriate equivalent strain in simple shear using the logarithmic strain measure. *Int. J. Mech. Sci.* **2017**, *134*, 273–283. [CrossRef]

materials

MDPI

Article

Severe Plastic Deformation of Fe-22Al-5Cr Alloy by Cross-Channel Extrusion with Back Pressure

Radosław Łyszkowski [1,*], Wojciech Polkowski [2] and Tomasz Czujko [1]

[1]   Faculty of Advanced Technology and Chemistry, Military University of Technology, 2 Urbanowicza,
      00-908 Warsaw, Poland; radoslaw.lyszkowski@wat.edu.pl
[2]   Foundry Research Institute, 73 Zakopiańska, 30-418 Cracow, Poland; wojciech.polkowski@iod.krakow.pl
*   Correspondence: radoslaw.lyszkowski@wat.edu.pl; Tel.: +48-261-839-028; Fax: +48-261-839-445

Received: 4 October 2018; Accepted: 2 November 2018; Published: 8 November 2018

Abstract: A new concept of the cross-channel extrusion (CCE) process under back pressure (BP) was proposed and tested experimentally. The obtained by finite element method (FEM) results showed that a triaxial compression occurred in the central zone, whereas the material was deformed by shearing in the outer zone. This led to the presence of a relatively uniformly deformed outer zone at 1 per pass and a strong deformation of the paraxial zone (3–5/pass). An increase in the BP did not substantially affect the accumulated strain but made it more uniform. The FEM results were verified using the physical modeling technique (PMT) by the extrusion of clay billet. The formation of the plane of the strongly flattened, and elongated grains were observed in the extrusion directions. With the increase in the number of passes, the shape of the resulting patterns expanded, indicating an increase in the deformation homogeneity. Finally, these investigations were verified experimentally for Fe-22Al-5Cr (at. %) alloy using of the purposely designed tooling. The effect of the CCE process is the fragmentation of the original material structure by dividing the primary grains. The complexity of the stress state leads to the rapid growth of microshear bands (MSB), grain defragmentation and the nucleation of new dynamically recrystallized grains about 200–400 nm size.

Keywords: severe plastic deformation (SPD); cross-channel extrusion (CCE); back pressure (BP); numerical simulation (FEM); physical modeling technique (PMT)

---

1. Introduction

Plastic working is one of the most popular and cost-effective methods to improve the mechanical properties of structural materials. A structural transformation occurs upon the plastic working processing, leading to a homogenization and a grain structure refinement and, thus, an increase in the mechanical strength. However, the level of strain imposed to the material is limited by its cohesion strength and is especially important in the case of materials with a lowered deformability.

In conventional processing techniques, e.g., rolling or compression, the maximum strain value is limited by the thickness/quantity of material in the deformation axis. However, severe plastic deformation (SPD) methods allow a much higher accumulated strain, due to the introduction of complex deformation schemes [1,2]. The SPD techniques are based on the domination of compression strain and a cyclically changed strain path and are actually considered the most efficient method of fabricating ultrafine grained metals and alloys [3–5]. The SPD methods are characterized by the imposition of strain values that significantly exceed those introduced in conventional processing. However, a successful implementation of the SPD techniques requires the proper selection of both the processing conditions and tool design [6,7].

Cross-channel extrusion (CCE) is a relatively seldom used method of deformation that belongs to the group of SPD processes. Unlike methods, such as high pressure torsion (HPT), where very large deformation values can be obtained in a small sample volume [8], extrusion-based methods allow one to obtain bulk materials of a certain size to ensure their practical use in the technique [9,10]. The principles of CCE processing (Figure 1) are based on the materials extrusion (by using a punch A that moves along the X-axis) with a perpendicular direction of the material flow (along the y-axis). This method allows a high accumulated strain to be obtained upon a 90° rotation of the die [11,12] without removing the material between subsequent deformation passes. Due to the continuous nature of the process and a possibility of automation, this method might be potentially adopted to industrial conditions [9,13].

Nevertheless, despite the presence of a triaxial stress state, the application of the CCE method to hardly deformable materials, such as Fe_3Al-based intermetallic alloys, is still limited due to these materials' lowered ductility [14–20]. Therefore, the main goal of the CCE process is to increase the level and extent of imposed hydrostatic strain by inhibiting a material's flow and, thereby, to produce a uniform shear deformation that prevents defect formation in the workpiece [21–24]. This assumption may be successfully accomplished by implementing a second punch (Figure 1b) that gives a back pressure (BP) and allows a controlled limitation of the material's lateral flow. To date, only a few cases of the BP effect have been reported [25–29], and its role has not been fully clarified.

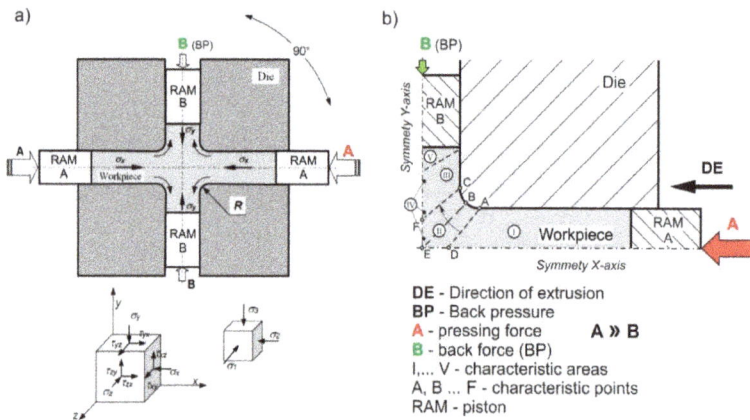

Figure 1. Scheme of (**a**) cross-channel extrusion with back pressure and (**b**) adopted analysis system.

Both the cross-channel extrusion and equal channel angular pressing (ECAP) methods belong to side extrusion-type processing. The former is described as a double axis technique, and the latter is assigned to side extrusion methods [6]. In these processes, pure shear deformation can be repeatedly imposed to a material such that an intense plastic strain is produced without any change in the cross-sectional dimensions of the workpiece. The die design in CCE methods is quite similar to that used in the ECAP-A—it may be considered a combination of four ECAP channels connected by their internal surfaces (as +). However, the main difference is that in the CCE process, a material is introduced to the die (along the X-axis) and leaves it in two opposite directions (along the y-axis) [9,10], but there is only one flow direction in the ECAP method. Consequently, a problem with filling an outer corner of the die (around a point E in Figure 1b), as is commonly observed in the ECAP method [22,25,27,28], is strongly limited in CCE processing. Moreover, this adverse effect may by additionally inhibited by using BP. Moreover, in order to alleviate deformation conditions and thereby limit the possibility of defects in hardly deformable materials [21,22], the sharp internal corner has been replaced by ABC arc.

In both methods, a shear strain is involved in a deformation mode. However, in the CCE method, the friction between a sample and die's walls, as well as the load is smaller, because a lower die's wall is replaced by a processed material that plays the same role as a movable die wall in the output channel of classical ECAP [13,28,30]. Thus, in addition to the imposed shear strain, a high hydrostatic compression occurs, which leads to a higher accumulated strain and prevents a material's cracking.

Despite the obvious advantages, the successful implementation of numerical methods requires an accurate knowledge of the material constitutive equations, process mechanics and frictional conditions. Inaccurate information regarding any of these parameters may lead to highly erroneous results. Therefore, this method requires validation, preferably based on real processes.

The physical modeling technique (PMT) is an alternative analysis method that can provide information on the plastic flow of metals, load predictions and strain distributions in metal forming processes. Using a suitable material, we can clearly observe the material flow pattern during processing, the effect of mold wall friction and a true-to-nature representation of the starting microstructure of the feedstock, all of which are possible when using this method. Usually, for modeling, materials based on the plastic mass (wax, modeling clay) [31–33], or ductile metals (Pb, Zn, Cu and other), are used [34–36]. Regardless to clearly visible differences between materials applied to PMT technique and regular constructive materials, in particular much higher plasticity, the mentioned above materials fulfill the condition of similarity and proportionality [37]. Due to this fact the PMT method allows for the qualitative and quantitative evaluation of the modeling process.

A new solution to the problem of low-ductility materials processing by CCE die has been proposed recently. Introduction of back pressure has a significant impact on the ability of the process, what was confirmed by finite element method (FEM) simulations and physical modeling of this process. The current paper focuses on checking the engineering feasibility of this idea by carrying out a laboratory experiment and investigating of macrostructure in terms of the possibility of defects occurrence.

2. Experiment Details

The results of our previous study [15,16] on the compression of Fe-22Al-5Cr intermetallic alloys carried out with GLEEBLE 3800 plastometric testing device were used to build a mathematical description of the model by FormFEM software (ITA Ltd., Ostrava, Czech republic). A rigid-plastic body model described by the following equation was used:

$$\int_V \bar{\sigma}\, \delta\dot{\bar{\varepsilon}}\, dV + \int_V \sigma_m\, \delta\dot{\varepsilon}_v\, dV + \int_V \delta\sigma_m\, \dot{\varepsilon}_v dV - \int_{S_F} F_i\, \delta u_i\, dS = 0, \tag{1}$$

where:
$\bar{\sigma}$—stress intensity V—volume
σ_m—average hydrostatic stress S—surface area
$\dot{\varepsilon}_v$—strain rate of material's volume F_i—heat flow
$\dot{\bar{\varepsilon}}$—strain rate intensity δ—material constant.

An incompressibility condition was fulfilled by the assumption of Lagrange multipliers method. A heat quantity generated upon the deformation process was calculated based on Fourier equation, and then correlated with a mechanical behavior by the following equation:

$$\int \bar{\sigma}\, \dot{\bar{\varepsilon}}\, dV = q. \tag{2}$$

where:
q—an amount of heat generated in the body deformed as a result of the deformation work.

Then, rheological parameters of the Hansel-Spittel equation for the strain rate of 0.01 s^{-1} and the assumed temperature range were taken from the conducted compression tests as follows:

$$\sigma = 1115.06e^{-0.002294T} \cdot \varepsilon^{0.120494} \cdot 0.01^{0.01026}. \tag{3}$$

where:
σ, ε—stress, strain.

Two-dimensional FEM simulations were performed to investigate the effect of BP on the deformation behavior of Fe-22Al-5Cr (at. %) intermetallic alloy during the CCE process. The assumed simulation conditions included various values of pressure and numbers of passes upon processing. The FEM models of a die and punch were developed and then limited to a representative quarter. The cross-section of the die channels was 10 mm square. The model assumed a continuity of material along the horizontal or vertical symmetry axis by attaching moving supports to all axial nodes. A movement range of the punch, heat transfer and surface conditions (e.g., coefficient of friction) between contacting parts were determined [38]. The BP was modeled by introducing a proper value of the pressure on a face plane of a sample. As a consequence, this plane was not flat, as in a real case, but possessed a lens-shaped free surface. The BP values were set as 100 MPa and 500 MPa, which corresponded to 10% and 50% of the main pressing pressure, respectively. The simulation was carried out for CCE process at room temperature.

The physical modeling technique PMT can be carried out in two ways. The first method involves the formation of an ordered structure with a characteristic pattern and then analyzing the deformation. For this purpose, a sample with alternating plates or cubes in different contrasting colors was prepared [32,39–43]. The second method, presented in this paper, is the formation of a sample of randomly distributed particles and observing the arrangement and creation of characteristic patterns [31,37]. Observing the development of the grain structure of deformed material is also possible in this much easier preparation technology.

For this purpose, extruded homogenized clay in five contrasting color rods of Ø1 mm was used. The cut rods with 1 mm long grains were mixed, pressed into a cylinder sample size of Ø8 × 36 mm. After being placed in a channel die, the sample was extruded at a rate of 5 mm/min and then thermally fixed. To evaluate the process of deformation performed by conventional methods on the longitudinal and cross section of the samples, they were subjected to macroscopic analysis.

The tool assembly used for experimental trials of the CCE process is shown in Figure 2. It consists of two equal 1.2080 (AISI D3) steel blocks that both have a cross-shaped route with a circle shape in cross-section after their assembly. The diameter of channels is 8 mm and 50 mm length. The channels in the intersection area are connected by an arc having a radius of 2 mm. Four cylinder punches are used to press material and match the four channels at each side of the die.

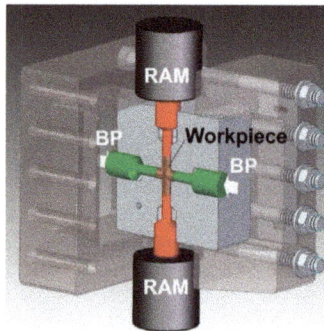

Figure 2. The die for cross-channel extrusion with back pressure (BP).

As mentioned earlier, the CCE method is predisposed for the processing of hardly deformable materials. Therefore, to verify the results of FEM model and PTM technique, tests were carried out on Fe-22Al-5Cr-0.1Zr-0.01B (at. %) alloy—called as Fe-22Al-Cr in this work. The material in as-cast state was obtained by vacuum melting of pure elements followed by casting into ceramic molds. The ingots of 8 mm diameter and 40 mm length were homogenized at 1100 °C for 10 h.

It is well known that every pass of SPD techniques already have a substantial effect on both the properties and the microstructure of the material. For this reason, one and two passes of CCE were applied to achieve a fine grain size. A samples were extruded and deformed at the speed of 2 mm/s at room temperature, without and with BP = 10 kN, which corresponds to a stress of 200 MPa. To reduce friction on the metal-tool interface, a dry graphite lubricant spray was used.

The microstructure of material in its initial and deformed state was characterized by a light optical microscopy (LOM-Eclipse MA200, Nickon, Japan), FEI Quanta™ 3D field emission gun scanning electron microscope (FEG-SEM, Hillsboro, OR, USA) coupled with X-ray energy dispersive spectroscopy (EDS-EDAX™, AMETEK, Inc., Berwyn, PA, USA) and automatic electron backscatter diffraction system (EBSD). The analysis was performed on mechanically polished longitudinal sections of samples. The effect of the applied processing on the mechanical properties of the material was evaluated in microhardness measurements, using the Vickers indenter (HMV-G, Shimadzu, Kyoto, Japan) loaded with 0.2 kg.

3. Simulation of Cross-Channel Extrusion—Results and Discussion

3.1. FEM Simulation

The FEM simulation shows that during the CCE process, a misalignment of the mesh at 45° occurs in odd passes, whereas the mesh is restored and straightened in even passes. It has been observed, that a densification of mesh lines in the axis zone of the outlet channel points to an accumulation of high strains in this area. The observed arrangement of the mesh lines is very similar to that reported for other CCE-deformed materials [9,10].

A compressive stress (−900 MPa) is formed in the inlet channel (I-area in Figure 1b) of the die, while a passing of material to the outlet channel (III-area) leads to a sudden decrease in the compressive stress (−200 MPa) (Figure 3a). An adverse stress distribution was found in the outlet channel, near the C-point. A local presence of tensile stresses with a maximum value of 600 MPa, was observed in this region. This finding points to a possible presence of plastic flow instability in this area, which may lead to a material cracking.

Replacing the sharp internal corner with the ABC arc causes that the deformation is substantial only to the volume of material delimited by two twin planes A-D and C-F, oriented approximately 45° with respect to the axis of channels (II-area), as shown in Figure 3b. In these planes, the strain rate reaches a value of 0.1–0.2 s^{-1}. A similar separation was observed by Perez and Luri [44], when a sharp internal corner of the channel die was replaced with an arc in the ECAP process. They observed that increasing the radius of 1.5 mm to 4 mm allows a more than 18% increase in the strain value to be obtained at a less than 9% increase of the processing force.

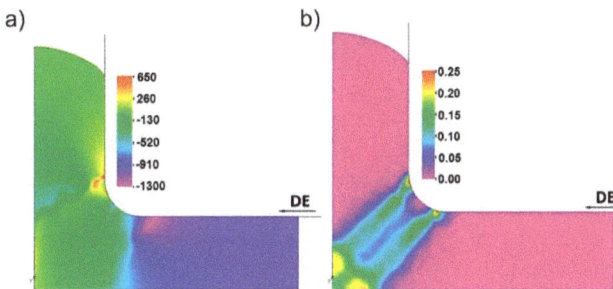

Figure 3. An example distributions of: (**a**) Stress [MPa]—corresponding to Figure 1, and strain rate [s^{-1}] (**b**) during the 1st pass without back pressure (BP = 0 MPa).

As shown in Figure 4a, the distribution of equivalent plastic strain indicates that the extruded sample deforms substantially in the central-zone of the die, in the A-D and C-F planes (Figure 1b). A triaxial stress state dominates in the II-area, where the two channels intersect, leading to a strong increase of strain, even above $\varepsilon = 5$. Upon the movement of material into the output channel, this heavily deformed region is stretched to form a strip with a thickness of approx. 2.5 mm (in one quarter of the cross section) and a degree of deformation of 2 to 3 (IV-area). As seen at the beginning of line 4 (Figure 4b), the peak of strain corresponds to an area located directly under the active punch, associated with the triaxial compression (II-area), and the remaining volume of the material deforms by shear and is significantly easier shift in the output direction (III-area). This results in a relatively uniform strain of $\varepsilon \approx 1.1$ in 3/4 of its volume (lines 4). The wave visible on the graph most likely indicates discontinuous character of deformation. As in the ECAP, in each case a portion of the material deforms in shear and then displaced into the output channel [21,23,24]. This corresponds to slight fluctuations in deformation in the plane perpendicular to the extrusion direction.

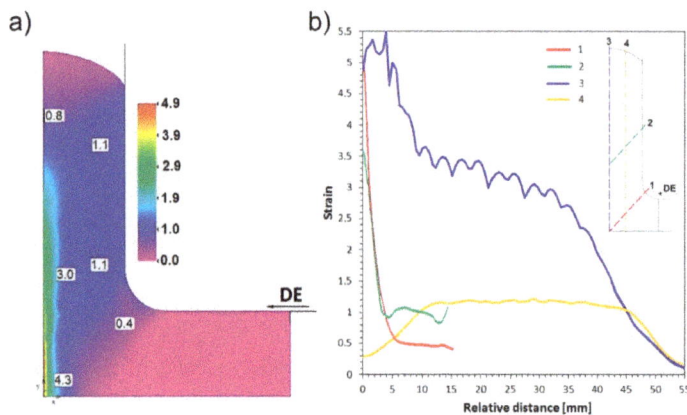

Figure 4. Equivalent plastic strain (ε): (**a**) Maps of distribution and (**b**) graph of its values at 1st pass, BP = 0 MPa.

Khan and Meredith [3] reported a formation of shear bands having several microns of width, oriented approximately 45° with respect to the pressure direction in Al 6061 that had been subjected to ECAP deformation. Zhao et al. [45] obtained similar results in a FEM simulation of the ECAP process, both in the case of a deformation behavior of the mesh grid and for the distribution of deformation. They indicate that the apparent heterogeneity of the deformation of the workpiece has been associated with friction on the bottom surface of the channel and its angle of channel refraction.

As mentioned above, this simulation was performed with a back pressure (BP) of 10% and 50% of the supposed extrusion pressure, which gives 100 MPa and 500 MPa, respectively. It should be noted that the introduction of back pressure does not substantially affect the nature of the stress distribution, but only causes its proportional increase corresponding to the variant without BP (Figure 5). The significant difference was observed for BP = 500 MPa, where behind the C-point in Figure 1b, the observed tensile stresses disappear or they change to compressive (Figure 5c). Therefore, BP = 100 MPa is an insufficient value, which should not allow a loss of flow continuity or the appearance of tensile stresses.

A strain rate distribution analysis showed a similar arrangement, as was the case for the BP absence—a substantial deformation occurs only in the A-D and C-F-diagonal planes. Regardless of the number of passes and the BP value, the instantaneous value does not exceed $\dot{\varepsilon} = 0.4 \text{ s}^{-1}$, with a mean value of 0.15–0.2 s^{-1}.

An analysis of equivalent plastic strain distribution (Figure 6a) indicates its similar nature in the CCE extruded samples to samples deformed without back pressure. For both the BP = 100 and 500 MPa samples in the first pass, a strongly deformed zone ($\varepsilon = 3.1$–3.3) is formed around the sample axis (IV-area). The rest of the material undergoes a relatively homogeneous deformation of $\varepsilon \approx 1.1$ (III-area), except for the tapered "locked" zone (V-area) located on the foreheads and central parts of the extruded sample (Figure 6b). In subsequent passes, an accumulation of strain leads to an increase in the ε value to 14–23 and 6–7 for the areas, as mentioned above. It is worth noting that the BP increasing to 500 MPa does not substantially affect the value of strain but does make it more homogeneous (Figure 6c). This is especially apparent in the IV-area, where the maximum and average value of strain were reduced (compared to ε = 21 and 16.6), with an unchanged strain value for the remaining volume of the material.

Hasani et al. [22] reported that when the BP-ECAP line-shaped flow of the material changes to a more rounded shape, the outer corner of channel matrix develops a dead-metal zone, accompanied by a twofold increase in the plastic deformation zone. It was due to the distribution of the strain rate, which reached a value of 0.35 s^{-1} for the material layers located directly at the internal corner of channel but reached only 0.2 s^{-1} in external corners. The introduction of BP (200 MPa) caused a decrease of rates and widening of the deformation zone.

Yoon et al. [24] conducted a FEM analysis of the ECAP process with and without BP. In the bottom zone of the sample, they observed the formation of a strongly deformed zone (ε~4), whereas in the remaining volume of material, the strain was lower (ε~1.1) and more homogeneous, which is in consensus with the results presented in this article. The implementation of BP reduced the gaps between the sample and the die in the outer corner of the channel, resulting in a strain increase at the sample's bottom.

Djavanroodi and Ebrahimi [25] also suggested that the use of the BP provides a complete filling of the gap in the outer corner of the channel and reduces the dead zone fraction, thereby preventing the formation of cracks and improving the uniformity of stress distribution in the material.

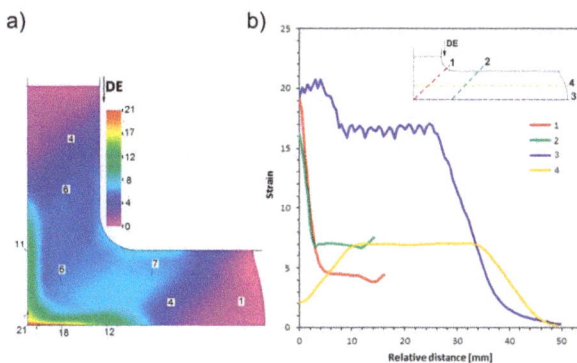

Figure 5. Equivalent plastic strain distribution (ε) at BP = 500 MPa: (**a**) First pass and (**b**) Sixth pass.

The use of BP is not limited to the ECAP process, although it is the most common case. Zangiabadi and Kazeminezhad [26] developed a new deformation method that uses tube channel pressing. They noted that BP usage leads to a greater accumulation of strain in the material (approx. 1–1.2 per pass). The strain distribution in the longitudinal direction of the pipe has a high homogeneity in the presence of BP. Furthermore, the application of BP reduces the amount of waste material. Li et al. [46] used BP to achieve a full filling by the material during its extrusion through the die with a helical deformation zone in a mutative torsion extrusion channel method (MCTE).

3.2. Physical Simulation

Figure 6 shows images of modeling clay specimens processed by the CCE. After one pass, the clay grains undergo strong deformation. They are flattened and elongated in the extrusion direction (Figure 6a). Deviation of their axis to outer surfaces with the movement from the center towards the end face of the sample was observed. A small number of sample discontinuities is observed on the samples' faces and on the outer surfaces near the C point in Figure 1b, which is compatible with FEM simulations.

The grains seen in the X-Y-section in the paraxial area underwent a strong deformation, resulting in the flattening or even blurring of their boundaries (Figure 6b). The effects of internal and the outer die wall friction are observed. This resulted in creation of a flat pattern along its axis, resembling the shape of the letter M lying on its side. In the perpendicular-section, the plane of strong deformation dissipates and a circular curvature appears as we move closer to the outer surface of the sample.

In summary, for CCE extrusion, the compressive and shearing stresses are responsible for the deformation of the material. The former affects the central part of the die (II-area) in the compression axis of the sample and causes severe flattening and elongation of the grains. The shear stresses are responsible for the deformation when the material moves towards the output channel and crosses the plane of their impact (A-D and F C-planes). They cause a disperse deformation of the remaining volume of the sample. The result is the deviation of the grains' axis on the outside, which corresponds to the shear planes. It is suggested that the basic mechanism of shear deformation is the difference in the flow route or flow path induced by the geometric character of the ECAP die. Therefore, the structure of deformed material, depending on where it is located, corresponds to these two mechanisms or reflects the cross-matrix system, which is in good agreement with the results obtained from the FEM simulation.

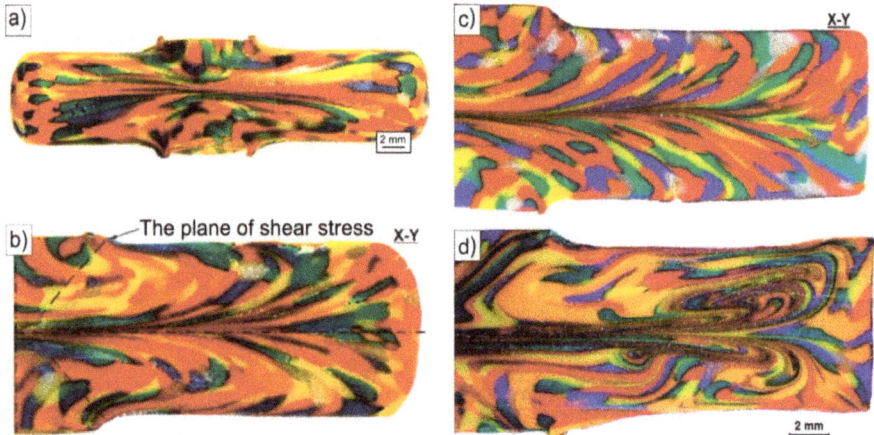

Figure 6. Modeling clay samples after: (**a**) One pass of CCE without BP and (**b**) cross-section of vertical-longitudinal samples and also (**c**) with back pressure—1 × CCE+BP, (**d**) 4 × CCE + BP.

The deformation character of the CCE process was investigated by Chou et al. [10,11]. To illustrate the deformation properties, they placed a square grid on the surface of a tin sample. After extrusion, it was deformed to form a symmetrical pattern whose lines converged at the central point of the sample and were inclined to the axis at an angle of 45°, as described above. This arrangement is similar to that formed in the ECAP by the action of shear stresses in the intersection zone of the channels. Assuming that the shape of the ECAP extruded sample by variant A corresponds to approximately 1/4 of the one obtained from the CCE [1,3], we can compare the results.

Manna et al. [32] modeled the extrusion process based on plasticine and obtained a similar pattern. Its arcuate shape with anchored edges indicates a significant effect of friction in the deformation process. Wu and Backer [43] using a plexiglass mold, obtained a more uniform pattern, due to lower friction. By introducing lubricants, we can influence on the friction in the contact zone billet-die. Han et al. [41] extruded by ECAP the billet consisted of two kinds of grains color and a size of approximately 1 mm. The evolution of the flow patterns was governed by the geometric character of the die. The simple flow line field corresponds to the geometric aspect of shear deformation in the deformation zone of the billet during the experiments. Zhan et al. [33] modeled the process of forging a compressor blade using plasticine. They observed that as the process continued, the degree of non-uniformity of deformation of each blade increased.

Sofuoglu and Gedikli [39], comparing the results of the FEM simulation of the forward extrusion and its physical modeling based on plasticine, indicated their similarity, especially regarding emerging forces. The differences might be related to the adopted method of analysis of deformed grid patterns or to the inhomogeneity of the extrusion billet. Balasundar et al. [40] modeled the impact of the tool geometry on friction and constitutive relationship of materials $\sigma = f(\varepsilon, \dot{\varepsilon}, T)$, which affect the material flow behavior, strain distribution and load requirement during the ECAP, and found their high compatibility with those of numerical simulations.

To simulate BP in the CCE process, the second set of pistons in the output channels was used. The resistance associated with their friction with the channel matrix walls reflects this pressure. The macroscopic observations of the deformed samples and of patterns formed on their outer surface did not differ substantially from those described previously. The shape of the sample reflects the exact shape of the channel die without any cracking or chipping. In the internal structure (Figure 6c), only the grains in the paraxial zone (IV-area) seemed to be more flattened and elongated with an expansion of this zone to the outer surface (area III). However, in the immediate vicinity, there are equiaxed grains. The curvature of their shape in the direction opposite to the flow is probably associated with the die wall friction.

With an increase in the number of passes, the microstructure of the CCE extruded clay underwent further development and became more complex. After four passes with BP (Figure 6d), a band structure was formed in the paraxial zone (IV-area). It consisted of highly flattened and very elongated grains, which boundaries practically blurred. Repeated of extrusion led to the curvature, looping and expansion of these bands and thus to increase of deformation homogeneity.

Manna et al. [32] modeled an extrusion process based on plasticine with as many as 15 passes and observed a significant mixing of ingredients of plasticine with a drastic reduction in the size of each section.

3.3. The Cross-Channel Extrusion of Fe-22Al-5Cr Alloy

The microstructure of the starting material is shown in Figure 7. The structure is characterized by the presence of medium sized equiaxed grains. The average grain size (d_{eq}) equals about 193 μm. X-ray phase analysis shows that the alloy has a single-phase structure based on a disordered solid solution of aluminum in α-iron with A2 lattice. However, a detailed BSE analysis proved that there are chromium or zirconium-rich precipitates on the grain boundaries GB (Figure 7b), which may be described as the $(Fe,Al)_2Zr$ Laves phase [47]. Using a static tension test, it was found that the value of yield strength is 480 MPa, tensile strength 470 MPa, and a total elongation 1.5%.

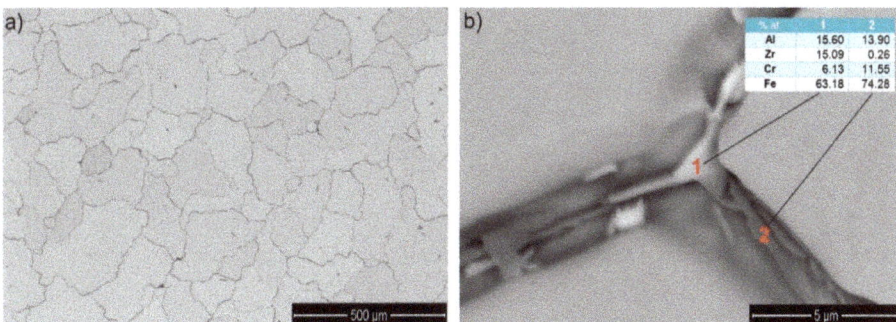

% at	1	2
Al	15.60	13.90
Zr	15.09	0.26
Cr	6.13	11.55
Fe	63.18	74.28

Figure 7. The initial microstructure of as-cast and homogenized Fe-22Al-5Cr alloy (**a**) and the chemical composition of precipitates at grain boundaries (**b**).

The macro-OIM-s after CCE process are presented in Figure 8, which shows parallel plane to the extrusion direction. The macrostructure of the samples maps the stress state caused by the cross channel system. Referring to earlier analyses, we can distinguish three types of structures (Figure 9).

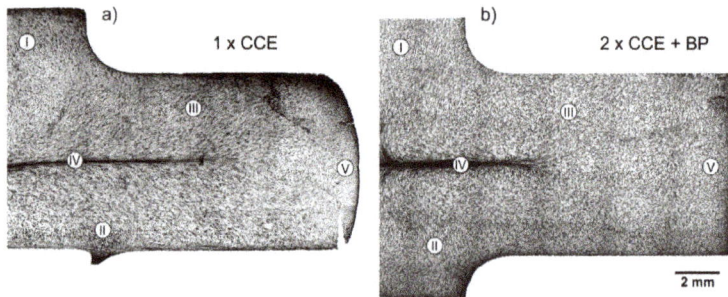

Figure 8. Cross-section of deformed alloy after: (**a**) One pass without BP and (**b**) two passes with BP = 10 kN of CCE. I–V-areas as Figure 1b.

The first, associated with shear stresses, is located on both sides of longitudinal axis of the sample and corresponds to the III-area. The structure of the material is strongly deformed with very many slip bands (Figure 9a). The grains ($d_{eq} \approx 60$ µm) are elongated and their direction of deflection towards the external surfaces is associated with the passage through the shear plane in II-area (see Figure 1b). The second, with a very strong deformation, creates a characteristic band along the X-axis of the extruded sample (Figure 9b). This corresponds to the IV-area, where we observed a band structure. The grains are strongly elongated with a blurring of the boundaries. The width of these bands, and therefore the grains is 1–5 µm. The third, associated with blocked stresses, was observed in tapered zones under the piston (I-area) or on foreheads of the sample (V-area). The grains are equi-axial and their size decreases almost twice from the initial state in this zone (Figure 9c).

In the next passage, the deformation increases and undergo propagates, which leads to a more uniform distribution. As a result of re-passing through the shear plane, the grains return to equiaxed, especially in the III-area (Figure 9d), which is consistent with previous reports [10,48,49]. The structure becomes slightly more fine-grained. The bands width in the IV-area is reduced, but as a result of the sample rotated by 90° between the first and second pass, we can observe their characteristic looping (Figure 9e), as in the case of clay modeling (Figure 6d).

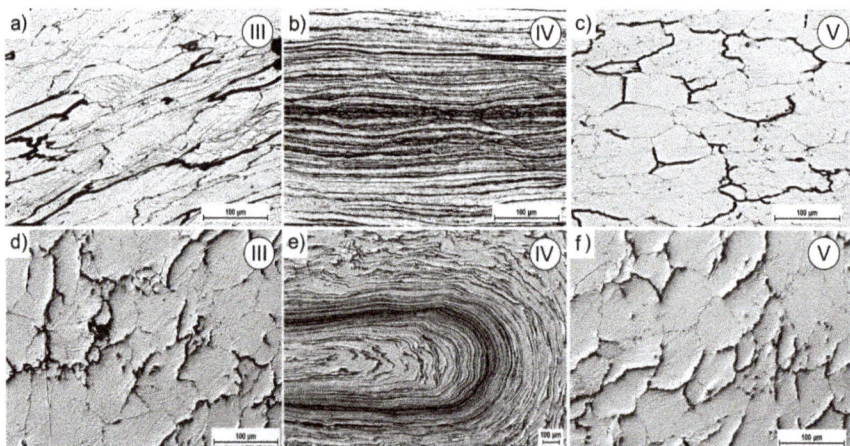

Figure 9. Microstructure of Fe-22Al-5Cr alloy from Figure 8: (**a–c**) One pass and (**d–f**) two passes of CCE.

After one pass of CCE, there were no cracks observed in the material volume except for the area (V) on the surface of the extruded sample. The BP's introduction practically eliminated cracks and improved the shape of the sample.

Typical EBSD maps of boundaries misorientation and grain size distribution of the as-CCE deformed microstructure are shown in Figure 10. As mentioned earlier, the structure of material deformed once by CCE in the area of the influence of shear stresses (III-area) consists of racked and elongated primary grains (Figure 10a). Because of a non-continuous process, their volume is traversed by numerous slip bands along where new grains are recrystallized with a size less than a few micrometers. Depending on the conditions, even whole grains may undergo refining. However, there is still a significant, almost 35% share of grains above 10 μm. Therefore, we are dealing with a typical bimodal distribution, characteristic of the processes associated with dynamic recrystallization of the material. Share of grains with a low disorientation's angle (low angle grains boundaries LAGB: $\theta < 15°$) in this process is about 40%, the highest share constitute of grains for which the angle is about $40–50°$ (high angle grains boundaries HAGB: $\theta > 15°$). Increase of deformation, as a result of repeated passes, causes a significant growth in the number of grains 0.3–2 μm, which takes place mainly at the expense of the remaining large primary grains. The actual size of the grains may be even smaller, but it would require the use of TEM testing [50].

The three-axis compressing occurring in the central part of the sample is responsible for the formation of a band structure along the X-axis (Figure 10c). Grain-bands about even a few hundred microns in length and just over a dozen microns wide are visible inside it. The changing colors inside them evidence of deformation of their lattice structure, so with a high value of the misorientation angle. However, the new, fully recrystallized, equi-axial grains with a size of 400–500 nm dominate in the structure. As a result, we are dealing with single mode the high-angular distribution of boundaries. With the next passage, as in the case of III-area, there is a general increase in the amount of fine grains in a wide range of size.

However, the material located in conical zones under the piston (I-area) or on the forehead of the extruded sample (V-area) is subject to much less impact, and thus to less deformation. This happens because it is not affected by cutting stresses or triaxial compression. Moreover, but on a much smaller scale, nucleation of new grains occurs and the size of the primary grains is reduced (Figure 10f). Thus, the present microstructural observation clearly displayed that CCE processing is able to refine the grain size of Fe-22A-5Cr alloy to UFG level.

At the same time, it seems that the introduction of BP = 200 MPa, which corresponded to 20% of the main pressing pressure, for such a hardly deformable material at RT as the Fe-22Al-5Cr alloy, it does not have a significant effect on the above-described structural changes as a result of CCE process. Undoubtedly, the application of BP leads to the suppression of a sample's damage and closure of defects, whereas its absence leads to the development of defects due to the imposed severe plastic deformation, as also Lapovok pointed out [21]. However, it seems that this impact is largely associated with material parameters [22], and therefore, it is necessary to carry out a systematic examination of the role of a BP during grain refinement.

In the described case, rebuilding of the deformed material structure occurs due to the fragmentation of its grain under the influence of bands and microshear bands (MSBs) at low temperatures below 0.5 Tm. The appearance of MSB in deformed material leads to internal structural instability, as described by Sitdikov et al. [51]. With the increasing of deformation, new ultrafine grains appear first in the MSB or at their intersections, and only then inside the grain. This is accompanied by the rotation of the crystal lattice and increase of the misorientation angle, up to formation of HAGB. Thus, the multiple deformation associated with the change in the strain path causes the material to develop a spatial network of strain-induced boundaries containing both LAGB (subgrains) and HABS [52]. A prerequisite for full refinement grain (UFG) is to create a state to achieve the maximum density of MSBs and their possible frequent intersection, which is possible to achieve only in SPD processes. Therefore, the CCE process involving multi-directional shear combined with triaxial compression speeds up the process of fragmentation of the material structure. The partial primary recrystallization it is probably caused by the inhomogeneous distributions of deformation strain, especially in III-area. Even for the relatively equiaxed grain structure after 2 passes, microstructures are also inhomogeneous. However, the resulting small grains are recrystallized grains formed during the CCE process, because they are free of dislocations, as in [53].

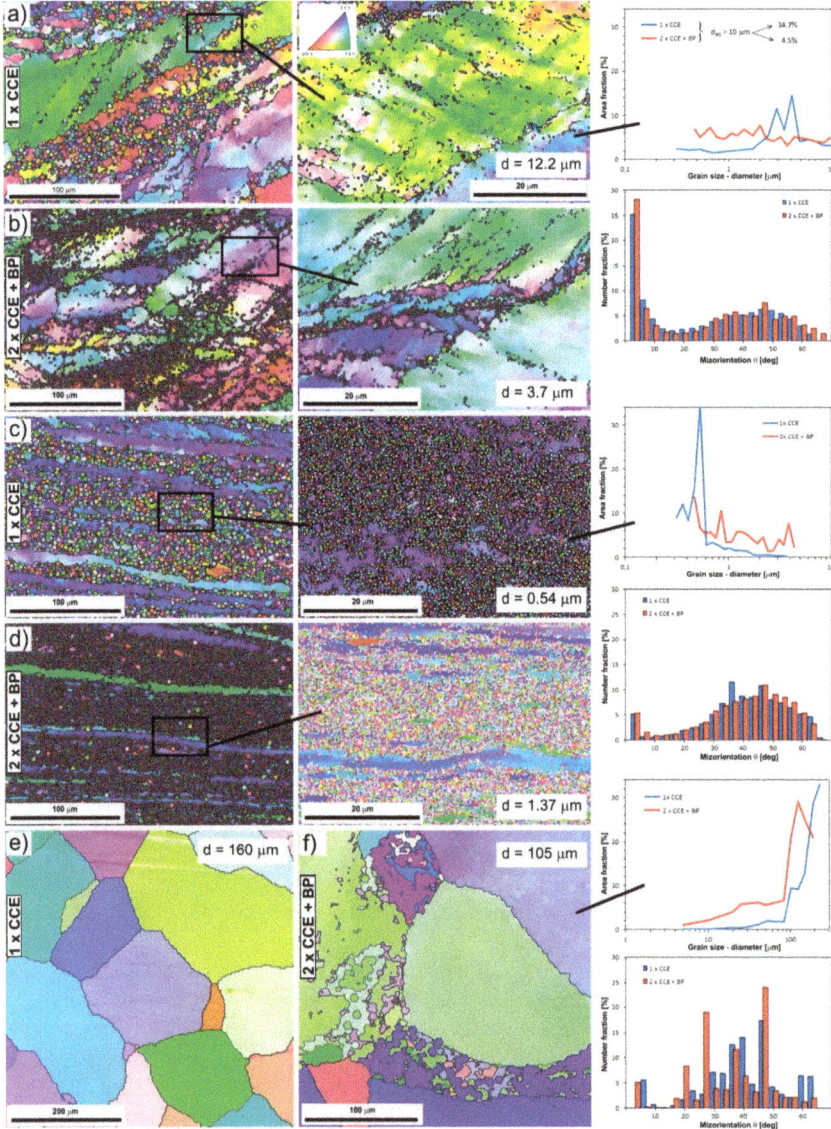

Figure 10. Electron backscatter diffraction system (EBSD) misorientation maps (OIM) obtained from samples subjected to CCE process with different areas (Figure 1b): (**a,b**) III-area of cutting, (**c,d**) IV-area of triaxial compression (**e,f**) V-area of blocked stresses, at one pass of CCE without BP and two passes with BP. Black lines of the colored areas outline correspond to the high-angular nature of grain boundaries (HAGB).

With further deformation (increasing the number of passes) most of the LAGB evolve into HAGB leading to homogeneity and an ultrafine-grain material, as results of continuous dynamic recrystallization [22,54,55].

The microhardness results along a longitudinal section of the samples show significant differences depending upon the area and conditions of deformation. The microhardness has hardly changed in I and V-area and remains at the level of 285 HV0.2 after one pass of CCE (Figure 11a,b), which is related with the lack of severe deformation mechanisms in this field. While, the transition through the plane of shear stress causes microhardness increasing to 370 HV0.2 in III-area and achieve a maximum 450 HV0.2 in IV-area. Interestingly, while the structure of the material in I and V areas has not undergone significant changes, the hardness in V-area increases after two passes. The microhardness of the tested alloy increases by an average of 10% (Figure 11b). This behavior is typical for metallic materials, especially on iron, subjected to deformation by SPD techniques [50,54–57]. Generally, the level of hardness is increased with the increase in the number of passes and the complexity of the stresses.

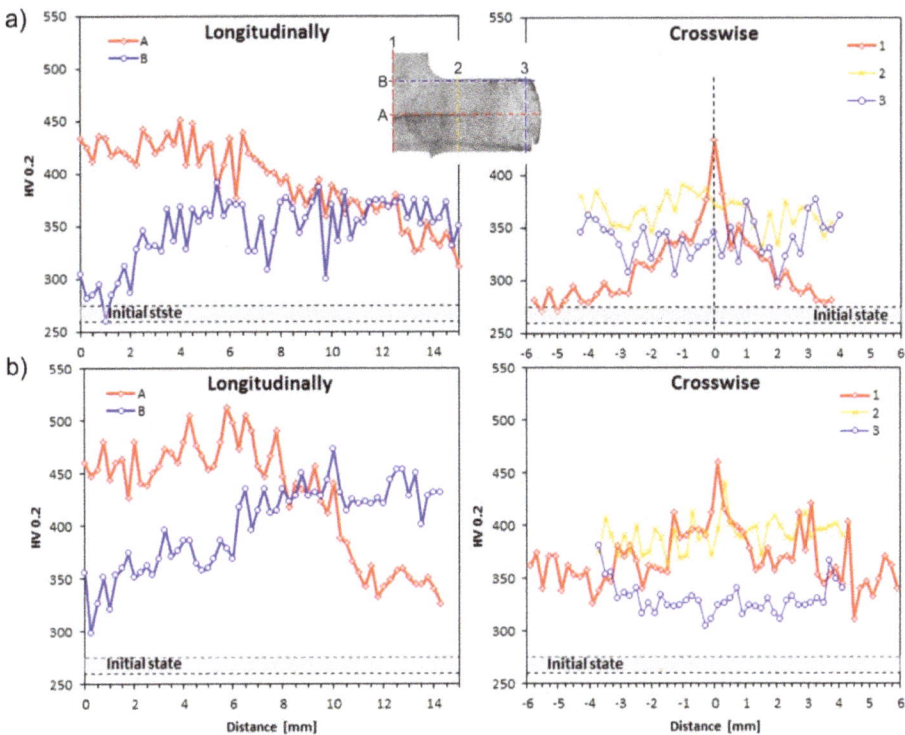

Figure 11. Microhardness of deformed alloy after: (**a**) One and (**b**) two passes of CCE.

These observations remain in full agreement with the previously described results for the extrusion of modeling clay.

4. Conclusions

To determine the behavior of the Fe-22Al-5Cr alloy during CCE process conducted with or without back pressure, the appropriate thermodynamic calculations, FEM/physical simulations and experimental analysis has been developed.

From the FEM analysis, it is observed that a movement of the material in opposite directions in the cross-channel leads to the formation of complex stress systems. A blocking of the material and its triaxial compression occur near the central point of the die. As a consequence, a strongly deformed paraxial zone ($\varepsilon \approx 5$) is formed along a channel axis perpendicular to the extrusion direction. It covers approx. 10–15% of the volume of the material. As the material is pushed to the outlet channel, the shear stress grows rapidly, which provides a relatively homogeneous deformation of approximately 1.1 in the remaining volume. The theoretical possibility of infinite accumulation of the effects of deformation in subsequent passes ($\varepsilon \approx 20/6.9$ after 6 passes) and cyclic changes in the strain paths will undoubtedly lead to a strong fragmentation of the initial structure of the material and an improvement of its mechanical properties. The introduction of BP gives a positive effect on extrusion process by reducing of tensile stress and maintaining material integrity. This prevents the material from cracking and enables the processing of materials with a limited ductility. The BP has essentially no effect on the strain rate and strain values, excluding their more homogeneous distribution.

Physical modeling investigation reveals that the structure of deformed material corresponds to two mechanisms (the compressive and shearing stresses) or reflects the cross-matrix system. The first of these leads to the formation of a band structure, while the second cause shear and looping of the structure, progressing with numbers of passes.

The CEE process carried out on Fe-22Al-5Cr (at. %) alloy confirmed the results obtained by physical modeling, which are also in good agreement with the results obtained from the FEM simulation. The effect of the CCE process is the fragmentation of the original material structure by dividing the primary grains, due to the generation under the influence of shear stresses and the microshear bands and triaxial compression of the material in its central part. The complexity of the stress state leads to the rapid growth of MSB, grain defragmentation and the nucleation of new dynamically recrystallized grains. Their size ranges from 200–400 nm.

The BP improves the homogeneity of deformation, while not substantially affecting its other structure transformations, in terms of the used parameters. It leads to the suppression of sample damage and closure of defects.

The results also indicate a need for further works related to the optimization of the shape and dimensions of each die's channel.

Author Contributions: R.Ł. conceived, designed and performed the experiments; R.Ł., W.P. and T.C. analyzed the data and contributed to writing and editing of the manuscript.

Funding: This work was supported by The National Science Centre in Poland (Grant No. 2012/05/DST8/02710) and Ministry of National Defense Republic of Poland Program—Research Grant MUT Project 13-995.

Conflicts of Interest: The authors declare no conflict of interest.

References

1. Valiev, R.Z.; Islamgaliev, R.K.; Alexandrov, I.V. Bulk nanostructured materials from severe plastic deformation. *Prog. Mater. Sci.* **2000**, *45*, 103–189. [CrossRef]
2. Korznikov, A.V.; Tram, G.; Dimitrov, O.; Korznikova, G.F.; Idrisova, S.R.; Pakieła, Z. The mechanism of nanocrystalline structure formation in Ni_3Al during SPD. *Acta Mater.* **2001**, *49*, 663–671. [CrossRef]
3. Khan, A.S.; Meredith, C.S. Thermo-mechanical response of Al 6061 with and without equal channel angular pressing (ECAP). *Int. J. Plast.* **2010**, *26*, 189–203. [CrossRef]
4. Borodachenkova, M.; Barlat, F.; Wen, W.; Bastos, A.; Grácio, J.J. A microstructure-based model for describing the material properties of Al–Zn alloys during high pressure torsion. *Int. J. Plast.* **2015**, *68*, 150–163. [CrossRef]
5. Langdon, T.G. Twenty-five years of ultrafine-grained materials: Achieving exceptional properties through grain refinement. *Acta Mater.* **2013**, *61*, 7035–7059. [CrossRef]
6. Azushima, A.; Kopp, R.; Korhonen, A.; Yang, D.Y.; Micari, F.; Lahoti, G.D.; Groche, P.; Yanagimoto, J.; Tsuji, N.; Rosochowski, A.; et al. Severe plastic deformation (SPD) processes for metals. *CIRP Ann. Manuf. Technol.* **2008**, *57*, 716–735. [CrossRef]
7. Polkowski, W.; Jóźwik, P.; Łyszkowski, R. Effect of hot differential speed rolling on microstructure and mechanical properties of Fe_3Al-based intermetallic alloy. *Int. J. Mater. Res.* **2016**, *107*, 867. [CrossRef]

8. Straumal, B.B.; Korneva, A.; Zięba, P. Phase transitions in metallic alloys driven by the high pressure torsion. *Arch. Civ. Mech. Eng.* **2014**, *14*, 242–249. [CrossRef]
9. Chou, C.Y.; Lee, S.L.; Lin, J.C. Effects of cross-channel extrusion on microstructure and mechanical properties of AA6061 aluminum alloy. *Mater. Sci. Eng. A* **2008**, *485*, 461. [CrossRef]
10. Chou, C.Y.; Lee, S.L.; Lin, J.C. Effects and deformation characteristics of cross-channel extrusion process on pure Sn and Al–7Si–0.3Mg alloy. *Mater. Chem. Phys.* **2008**, *107*, 193–199. [CrossRef]
11. Nishida, Y.; Arima, H.; Kim, J.C.; Ando, T. Rotary-die ECAP of an Al–7Si–0.35Mg alloy. *Scr. Mater.* **2001**, *45*, 261–266. [CrossRef]
12. Ma, A.; Nishida, Y.; Suzuki, K.; Shigematsu, I.; Saito, N. Characteristics of plastic deformation by rotary-die ECAP. *Scr. Mater.* **2005**, *52*, 433–437. [CrossRef]
13. Rosochowski, A.; Olejnik, L.; Richert, J.; Rosochowska, M.; Richert, M. Equal channel angular pressing with converging billets—Experiment. *Mater. Sci. Eng. A* **2013**, *560*, 358–364. [CrossRef]
14. Bystrzycki, J.; Fraczkiewicz, A.; Łyszkowski, R.; Mondon, M.; Pakieła, Z. Microstructure and tensile behavior of Fe-16Al-based alloy after severe plastic deformation. *Intermetallics* **2010**, *18*, 1338–1343. [CrossRef]
15. Łyszkowski, R.; Czujko, T.; Varin, R.A. Multi-axial forging of the Fe₃Al-base intermetallic alloy and its mechanical properties. *J. Mater. Sci.* **2017**, *52*, 2902–2914. [CrossRef]
16. Łyszkowski, R.; Bystrzycki, J. Influence of temperature and strain rate on the microstructure and flow stress of iron aluminides. *Arch. Met. Mater.* **2007**, *52/2*, 347–350.
17. Karczewski, K.; Jóźwiak, S.; Bojar, Z. Mechanisms of strength properties anomaly of Fe-Al sinters by compression tests at elevated temperature. *Arch. Met. Mater.* **2007**, *52*, 361–366.
18. Łazińska, M.; Durejko, T.; Zasada, D.; Bojar, Z. Microstructure and mechanical properties of a Fe-28%Al-5%Cr-1%Nb-2%B alloy fabricated by laser engineered net shaping. *Mater. Lett.* **2017**, *196*, 87–90. [CrossRef]
19. Siemiaszko, D.; Kowalska, B.; Jóźwik, P.; Kwiatkowska, M. The effect of oxygen partial pressure on microstructure and properties of Fe40Al alloy sintered under vacuum. *Materials* **2015**, *8*, 1513–1525. [CrossRef] [PubMed]
20. Jabłońska, M.; Kuc, D.; Bednarczyk, I. Influence of deformation parameters on the structure in selected intermetallic from Al-Fe diagram. *Solid State Phenom.* **2014**, *212*, 63–66. [CrossRef]
21. Lapovok, R. The role of back-pressure in ECAP. *J. Mater. Sci.* **2005**, *40*, 341–346. [CrossRef]
22. Hasani, A.; Lapovok, R.; Toth, L.S.; Molinari, A. Deformation field variations in equal channel angular extrusion due to back pressure. *Scr. Mater.* **2008**, *58*, 771–774. [CrossRef]
23. Kubota, M.; Wu, X.; Xu, W.; Xia, K. Mechanical properties of bulk aluminum consolidated from mechanically milled particles by back pressure equal channel angular pressing. *Mater. Sci. Eng. A* **2010**, *527*, 6533–6536. [CrossRef]
24. Yoon, S.C.; Jeong, H.G.; Lee, S.; Kim, H.S. Analysis of plastic deformation behavior during back pressure ECAP by the FEM. *Comput. Mater. Sci.* **2013**, *77*, 202–207. [CrossRef]
25. Djavanroodi, F.; Ebrahimi, M. Effect of die channel angle, friction and back pressure in ECAP using 3D FEM simulation. *Mater. Sci. Eng. A* **2010**, *527*, 1230. [CrossRef]
26. Zangiabadi, A.; Kazeminezhad, M. Computation on new deformation routes of tube channel pressing considering back pressure and friction effects. *Comput. Mater. Sci.* **2012**, *59*, 174–181. [CrossRef]
27. Ribbe, J.; Schmitz, G.; Rösner, H.; Lapovok, R.; Estrin, Y.; Wildea, G.; Divinski, S.V. Effect of back pressure during equal-channel angular pressing on deformation-induced porosity in copper. *Scr. Mater.* **2013**, *68*, 925–928. [CrossRef]
28. Nagasekhar, A.V.; Kim, H.S. Plastic deformation characteristics of cross-equal channel angular pressing. *Comput. Mater. Sci.* **2008**, *43*, 1069–1073. [CrossRef]
29. Ghazani, M.S.; Eghbali, B. Finite element simulation of cross equal channel angular pressing. *Comput. Mater. Sci.* **2013**, *74*, 124–128. [CrossRef]
30. Segal, V.M. Engineering and commercialization of equal channel angular extrusion (ECAE). *Mater. Sci. Eng. A* **2004**, *386*, 269–276. [CrossRef]
31. Kowalczyk, L. *Physical Modeling of Metal Forming Processes*; WiZP ITE: Radom, Poland, 1995; ISBN 83-86148-38-1. (In Polish)

32. Manna, R.; Agrawal, P.; Joshi, S.; Mudda, B.K.; Mukhopadhyay, N.K.; Sastry, G.V.S. Physical modeling of equal channel angular pressing using plasticine. *Scr. Mater.* **2005**, *53*, 1357–1361. [CrossRef]
33. Zhan, M.; Liu, Y.; Yang, H. Physical modeling of the forging of a blade with a damper platform using plasticine. *J. Mater. Process. Technol.* **2001**, *117*, 62–65. [CrossRef]
34. Komoria, K.; Mizuno, K. Study on plastic deformation in cone-type rotary piercing process using model piercing mill for modeling clay. *J. Mater. Process. Technol.* **2009**, *209*, 4994–5001. [CrossRef]
35. Langdon, T. The processing of ultrafine-grained materials through the application of SPD. *J. Mater. Sci.* **2007**, *42*, 3388–3397. [CrossRef]
36. Zaharia, L.; Chelariu, R.; Comaneci, R. Multiple direct extrusion—A new technique in grain refinement. *Mater. Sci. Eng. A* **2012**, *550*, 293–299. [CrossRef]
37. Hohenwarter, A. Incremental HPT as a novel SPD process: Processing features and application to copper. *Mater. Sci. Eng. A* **2015**, *626*, 80–85. [CrossRef] [PubMed]
38. Łyszkowski, R.; Šimeček, P. Numerical analysis of cross-channel pressing with extrusion at braked outflow of material. *Hut. Wiadomości Hut.* **2014**, *1*, 21–25. (In Polish)
39. Sofuoglu, H.; Gedikli, H. Physical and numerical analysis of three dimensional extrusion process. *Comput. Mater. Sci.* **2004**, *31*, 113–124. [CrossRef]
40. Balasundar, I.; Rao, M.S.; Raghu, T. Equal channel angular pressing die to extrude a variety of materials. *Mater. Des.* **2009**, *30*, 1050–1059. [CrossRef]
41. Han, W.Z.; Zhang, Z.F.; Wu, S.D.; Li, S.X. Investigation on the geometrical aspect of deformation during equal-channel angular pressing by in-situ physical modeling experiments. *Mater. Sci. Eng. A* **2008**, *476*, 224–229. [CrossRef]
42. Wong, S.F.; Hodgson, P.D.; Chong, C.J.; Thomson, P.F. Physical modelling with application to metal working, especially to hot rolling. *J. Mater. Process. Technol.* **1996**, *62*, 260–274. [CrossRef]
43. Wu, Y.; Baker, I. An experimental study of ECAP. *Scr. Mater.* **1997**, *37*, 437. [CrossRef]
44. Perez, C.J.; Luri, R. Study of ECAE process by the upper bound method considering the correct die design. *Mech. Mater.* **2008**, *40*, 617–628. [CrossRef]
45. Zhao, W.J.; Ding, H.; Ren, Y.P.; Hao, S.M.; Wang, J.; Wang, J.T. Finite element simulation of deformation behavior of pure aluminum during equal channel angular pressing. *Mater. Sci. Eng. A* **2005**, *410–411*, 348–352. [CrossRef]
46. Li, J.; Li, F.; Li, P.; Ma, Z.; Wang, C.; Wang, L. Micro-structural evolution in metals subjected to simple shear by a particular severe plastic deformation method. *J. Mater. Eng. Perform.* **2015**, *24*, 2944–2956. [CrossRef]
47. Kratochvíl, P.; Dobes, F.; Pesicka, J.; Málek, P.; Bursík, J.; Vodicková, V.; Hanus, P. Microstructure and high temperature mechanical properties of Zr-alloyed Fe$_3$Al-type aluminides: The effect of carbon. *Mater. Sci. Eng. A* **2012**, *548*, 175–182. [CrossRef]
48. Sepahi-Boroujeni, S.; Fereshteh-Saniee, F. Expansion equal channel angular extrusion, as a novel severe plastic deformation technique. *J. Mater. Sci.* **2015**, *50*, 3908–3919. [CrossRef]
49. Mogucheva, A.; Babich, E.; Ovsyannikov, B.; Kaibyshev, R. Microstructural evolution in a 5024 aluminum alloy processed by ECAP with and without back pressure. *Mater. Sci. Eng. A* **2013**, *560*, 178–192. [CrossRef]
50. Tang, H.Y.; Hao, T.; Wang, X.P.; Luo, G.N.; Liu, C.S.; Fang, Q.F. Structure and mechanical behavior of Fe–Cr alloy processed by equal-channel angular pressing. *J. Alloys Compd.* **2015**, *640*, 141–146. [CrossRef]
51. Sitdikova, O.; Sakaia, T.; Miura, H.; Hama, C. Temperature effect on fine-grained structure formation in high-strength Al alloy 7475 during hot severe deformation. *Mater. Sci. Eng. A* **2009**, *516*, 180–188. [CrossRef]
52. Tikhonova, M.; Kaibyshev, R.; Fang, X.; Wang, W.; Belyakov, A. Grain boundary assembles developed in an austenitic stainless steel during large strain warm working. *Mater. Charact.* **2012**, *70*, 14–20. [CrossRef]
53. Liang, N.; Zhao, Y.; Li, Y.; Topping, T.; Zhu, Y.; Valiev, R.Z.; Lavernia, E.J. Influence of microstructure on thermal stability of ultrafine-grained Cu processed by ECAP. *J. Mater. Sci.* **2018**, *53*, 13173. [CrossRef]
54. Muñoz, J.A.; Higuera, O.F.; Cabrera, J.M. Microstructural and mechanical study in the plastic zone of ARMCO iron processed by ECAP. *Mater. Sci. Eng. A* **2017**, *697*, 24–36. [CrossRef]
55. Wang, H.; Li, W.; Hao, T.; Jiang, W.; Fang, Q.; Wang, X.; Zhang, T.O.; Zhang, J.; Wang, K.; Wang, L. Mechanical property and damping capacity of ultrafine-grained Fe-13Cr-2Al-1Si alloy produced by equal channel angular pressing. *Mater. Sci. Eng. A* **2017**, *695*, 193–198. [CrossRef]

56. Polkowski, W.; Jóźwik, P.; Bojar, Z. Electron Backscatter Diffraction Study on Microstructure, Texture, and Strain Evolution in Armco Iron Severely Deformed by the Differential Speed Rolling Method. *Met. Mater. Trans. A* **2015**, *46*, 2216–2226. [CrossRef]

57. Jabłońska, M.; Śmiglewicz, A. A study of mechanical properties of high manganese steels after different rolling conditions. *Metalurgija* **2015**, *54*, 619–622.

materials

MDPI

Article

Microstructure and Flexural Properties of Z-Pinned Carbon Fiber-Reinforced Aluminum Matrix Composites

Sian Wang [1], Yunhe Zhang [1,*], Pibo Sun [2], Yanhong Cui [1] and Gaohui Wu [3]

[1] College of Mechanical and Electrical Engineering, Northeast Forestry University, Harbin 150040, China; daowsa@nefu.edu.cn (S.W.); cyh_1995@126.com (Y.C.)
[2] Information Technology Department, Qingdao Vocational and Technical College of Hotel Management, Qingdao 266100, China; sunpibo@126.com
[3] School of Materials Science and Engineering, Harbin Institute of Technology, Harbin 150080, China; wugh@hit.edu.cn
* Correspondence: yunhe.zhang@nefu.edu.cn; Tel.: +86-451-8219-2843 or +86-15663526798; Fax: +86-451-8219-2843

Received: 5 December 2018; Accepted: 28 December 2018; Published: 7 January 2019

Abstract: Z-pinning can significantly improve the interlaminar shear properties of carbon fiber-reinforced aluminum matrix composites (Cf/Al). However, the effect of the metal z-pin on the in-plane properties of Cf/Al is unclear. This study examines the effect of the z-pin on the flexural strength and failure mechanism of Cf/Al composites with different volume contents and diameters of the z-pins. The introduction of a z-pin leads to the formation of a brittle phase at the z-pin/matrix interface and microstructural damage such as aluminum-rich pockets and carbon fiber waviness, thereby resulting in a reduction of the flexural strength. The three-point flexural test results show that the adding of a metal z-pin results in reducing the Cf/Al composites' flexural strength by 2–25%. SEM imaging of the fracture surfaces revealed that a higher degree of interfacial reaction led to more cracks on the surface of the z-pin. This crack-susceptible interface layer between the z-pin and the matrix is likely the primary cause of the reduction of the flexural strength.

Keywords: metal–matrix composites (MMCs); carbon fiber; mechanical properties; z-pin reinforcement; laminate

1. Introduction

Carbon fiber-reinforced aluminum matrix composites (Cf/Al) have high specific stiffness, high specific strength, good electrical conductivity and good fatigue properties, and thus have great potential in the aerospace and automotive industries [1–3]. While the traditional laminated plate structures have excellent in-plane properties, the lack of reinforcement in the thickness direction can cause delamination failure under thermal stress during the material forming process and the cutting force during machining processing [4]. Z-pinning is an effective technique for improving delamination resistance by inserting rods with high strength and high modulus such as titanium, steel, or fibrous carbon composite in the thickness direction of a composite [5]. Z-pins can enhance the interlaminar strength and impact damage tolerance of composites as well as the ultimate failure load and fatigue life of composite joints based on crack bridging forces [6–9]. Zhang et al. reported that the interlaminar shear strength of the Cf/Al composites could be increased by as much as 230% using the stainless steel z-pin [9].

Z-pinning increases the interlaminar mechanical properties but reduces the in-plane mechanical properties of the laminated composites [10–12]. Hoffmann and Scharr reported that z-pinning reduced the tensile strength of carbon fiber/epoxy laminates by 24–47% and the fatigue strength by 6–11% [10].

Knopp et al. found that using z-pins with a density of 2% and a diameter of 0.28 mm can reduce the flexural strength of the laminate by 27–31% [11]. Li et al. tested the compressive properties of laminates at room temperature and in dry, hot, and humid environments, and found that the z-pin reduced the compressive modulus by 17.2% and the compressive strength by 13.9% [12]. The reduction in the in-plane properties is attributed to the damage of the microstructure caused by the z-pin, including the waviness of in-plane fiber around the z-pin, fiber breakage, resin-rich pockets, and swelling of the z-pinned laminates [13,14]. The damage of these microstructures increases with increasing z-pin diameter and content, and thus the in-plane performance decreases with increasing z-pin diameter and content [10,14].

While the effect of z-pins on the in-plane mechanical properties of polymer matrix composites has been recognized, less is known about those of metal matrix composites. The interfacial layer composed of $FeAl_3$ is formed by the interdiffusion of Fe and Al at the interface during the fabrication of z-pinned Cf/Al composites [15,16]. This strong interface can effectively transfer the interlaminar load from the matrix to the z-pin, resulting in improved interlaminar mechanical properties, but the brittle $FeAl_3$ may adversely affect the in-plane mechanical properties of the composite [17]. Therefore, the degree of the interfacial reaction may be a key factor influencing the flexural performance of z-pinned Cf/Al.

Cf/Al composites are often subjected to flexural stress in typical applications. Thus, it is important to understand how the flexural strength changes at different z-pin parameters. This understanding could allow for better prediction and help maximize the performance of z-pinned Cf/Al. This work is an extension of earlier works by Zhang et al. [7,9] on the interlaminar shear strength and failure mechanism of z-pinned Cf/Al composites. The purpose of this paper is to study the influence of metal z-pin parameters on the flexural strength of z-pinned Cf/Al materials. Since the interface reaction is important for the mechanical properties of metal–matrix composites, the influence of the interface reaction degree on the strength and the failure mechanism of z-pinned Cf/Al is also discussed.

2. Experimental

2.1. Preparation of Z-Pinned and Unpinned Cf/Al Composites

The metal z-pin-reinforced Cf/Al composites were prepared using the pressure infiltration method. The matrix, in-plane reinforcement, and interlaminar reinforcement were 5A06Al, M40, and AISI321, respectively. The basic properties of M40 carbon fibers and AISI321 are listed in Table 1. The chemical compositions of 5A06 Al alloy and AISI321 are listed in Tables 2 and 3, respectively. The flow chart of the manufacturing process of investigated composites is shown in Figure 1. In order to fabricate the preforms of carbon fibers, the carbon fibers were first unidirectionally wound by a winding machine (3FW250 × 1500, Harbin Composite Equipment Company, Harbin, China) to the desired shape. Then, the AISI321 z-pin was inserted into preforms of carbon fibers. The preforms of carbon fibers with z-pins were preheated to 500 ± 10 °C. The 5A06Al alloy was melted at 780 ± 20 °C and then infiltrated into the preforms under a pressure of 0.5 MPa. The pressure was carried on for 2 h to obtain Cf/Al composites. An unpinned composite was manufactured as a control martial under the same procedure to determine the damage to flexural performance caused by the z-pin. The volume fraction of the carbon fiber was 55%, which was measured by Archimedes method.

Table 1. Basic properties of in-plane reinforcement and interlaminar reinforcement.

Materials	Tensile Strength (MPa)	Elastic Modulus (GPa)	Elongation to Fracture (%)	Density (g/cm^3)
M40	4410	377	1.2	1.76
AISI321	1905	198	2	7.85

Table 2. Chemical composition of 5A06 Al alloy (wt %).

Material	Mg	Mn	Si	Fe	Zn	Cu	Ti	Al
5A06 Al	5.8–6.8	0.5–0.8	0.4	0.4	0.2	0.1	0.02–0.1	Bal.

Table 3. Chemical composition of AISI321 (wt %).

Material	Cr	Ni	Ti	Mn	Si	C	S	P	Fe
AISI321	17–19	8–11	0.5-0.8	<2.0	<1.0	<0.12	<0.03	<0.035	Bal.

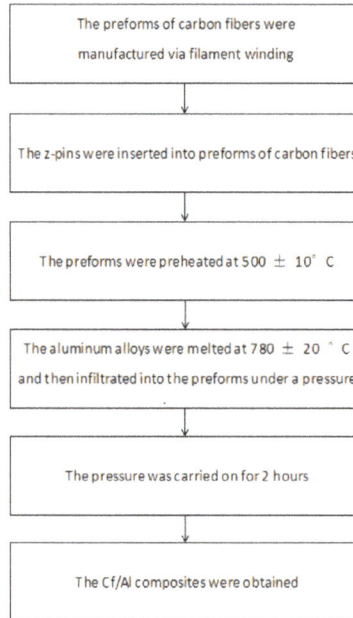

The preforms of carbon fibers were manufactured via filament winding

The z-pins were inserted into preforms of carbon fibers

The preforms were preheated at $500 \pm 10°$ C

The aluminum alloys were melted at 780 ± 20 ° C and then infiltrated into the preforms under a pressure

The pressure was carried on for 2 hours

The Cf/Al composites were obtained

Figure 1. The flow chart of the manufacturing process of investigated composites.

2.2. Characterization Technique

Several types of z-pinned Cf/Al composite samples and unpinned Cf/Al composite samples were made for a three-point flexural test. The composites were reinforced by AISI321 stainless steel z-pins with a diameter of 0.3 mm, volume contents of 0.25%, 0.5%, and 1.0% in order to investigate the influence of the pin content on the flexural properties. Moreover, the composites were reinforced by AISI321 stainless steel z-pins with a volume content of 1% and diameters of 0.3 mm, 0.6 mm, and 0.9 mm in order to investigate the influence of the pin content on the flexural properties. The composites were machined into three-point flexural samples with dimensions of 60 mm × 10 mm × 2 mm and the z-pins aligned in parallel rows along the samples. The scheme and photo of the composite structure with the arrangement of the z-pins are shown in Figure 2. The row spacing between the z-pins of several types of z-pinned Cf/Al composite samples is listed in Table 4. The z-pin spacing can be determined using

$$S = \frac{D}{2}\sqrt{\frac{\pi}{\rho \sin A}} \qquad (1)$$

where S, D, ρ, A are z-pin spacing, diameter, volume content and angle, respectively. In this work, all z-pins were inserted vertically, with a z-pin angle of 90°. It can be seen that when fixing the z-pin volume content and angle, the z-pin spacing increases with the diameter. Fixing the z-pin diameter and angle, the z-pin spacing increases with the volume content. The three-point flexural test was performed on an Instron-5569 electronic universal tensile test machine according to GB/T232-2010 with a beam speed of 0.5 mm/min. A schematic of the test set-up is shown in Figure 3. The diameters of the cylindrical supports and cylindrical head were 5 mm and 10 mm, respectively. The support span was 40 mm (the span-to-thickness ratio was kept 20:1). The flexural strengths of unpinned and z-pinned composites were calculated using

$$\sigma = \frac{3PL}{2bh^2} \tag{2}$$

where σ is the flexural strength, P is the maximum load during the test, L is the support span, and b and h are the width and thickness of the sample, respectively. The microstructure of the composites was observed by a ZEISS 40MAT optic microscopy (OM, Carl Zeiss Microscopy, Jena, Germany). The fracture morphologies of samples were examined using a S-4700 scanning electron microscope (SEM, Royal Dutch Philips Electronics Ltd., Amsterdam, Netherlands).

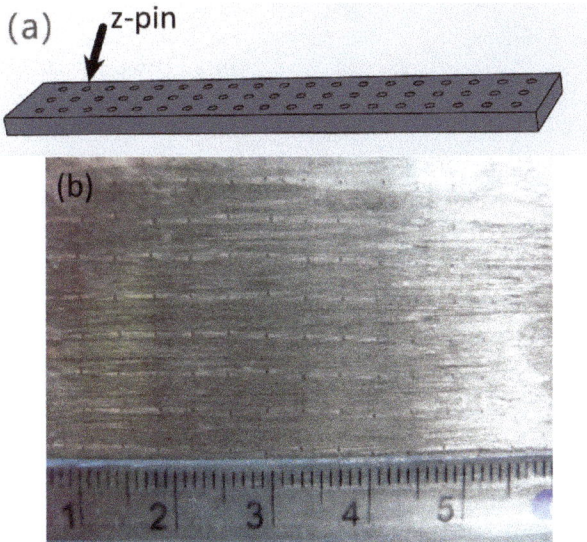

Figure 2. The (**a**) scheme and (**b**) photo of the z-pinned composite structure.

Table 4. The z-pin spacing of several types of z-pinned Cf/Al composite samples.

Z-Pin Volume Content (%)	Z-Pin Diameter (mm)	Z-Pin Angle (°)	Z-Pin Spacing (mm)
0.25	0.3	90	5.3
0.5	0.3	90	3.8
1	0.3	90	2.7
1	0.6	90	5.3
1	0.9	90	8.0

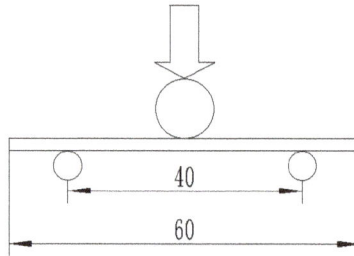

Figure 3. Scheme of the three-point flexural testing.

3. Results and Discussion

3.1. Microstructure

Figure 4 shows the microstructure of the z-pinned Cf/Al composites. A distinct interfacial reaction layer was formed between AISI 321 steel and Cf/Al, indicating that the two had strong interfacial bonds. Zhang et al. [9] determined that the interfacial product formed during the sample preparation of z-pinned Cf/Al composite is $FeAl_3$. As the metal pin diameter increased, the thickness of the interfacial reaction layer decreased. This means that the interfacial reaction of Cf/Al composites with thin metal z-pins was stronger than that of the Cf/Al composites with thick metal z-pins. This is because the interface layer was formed by the diffusion of Al and Fe atoms at a high temperature. When the aluminum liquid with high temperature was in contact with the z-pin during the preparation process, the heat of the aluminum liquid was transferred to the metal z-pin and caused the metal pin to heat up. The amplitude of the z-pin temperature increase depends on the quantity of heat absorbed. Larger z-pins require more heat to achieve the same increase in temperature. In other words, the thick z-pin takes more time to complete the heat exchange. During the sample preparation, the time limited the contact between the aluminum liquid and the z-pin, so the atomic diffusion was insufficient, and the thickness of the interface layer was small. A thin z-pin reached higher temperatures quicker. When the higher temperature is reached quickly, there is a longer time for atomic diffusion, and consequently, a higher amount of interfacial reaction occurs.

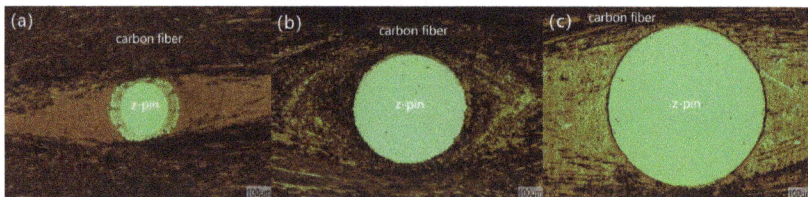

Figure 4. Microstructure of z-pinned Cf/Al composites. (**a**) φ0.3 mm metal pin, (**b**) φ0.6 mm metal pin, and (**c**) φ0.9 mm metal pin.

A distinct spindle region was formed around the z-pin. This region forms because the sample preparation process deform the fibers around the metal z-pin. During the waviness of the fiber, voids are formed, and these voids are filled with liquid aluminum in subsequent steps to form aluminum-rich zones characterized by fiber waviness, and aluminum enrichment. It is clear from the observations that larger z-pin diameters have a greater degree of carbon fiber waviness, and a larger area of aluminum-rich zones surrounding the z-pin. The mechanical damage to the fiber most likely occurred during z-pin insertion, although this is difficult to observe through the SEM.

3.2. Flexural Strength

Figure 5 shows the stress-deflection curves of the unpinned Cf/Al composite and the z-pinned Cf/Al composites with a diameter of 0.3 mm and volume fractions of 0.25%, 0.5%, and 1%. It suggests that the addition of the z-pin did not change the fracture behavior but did significantly reduce the flexural strength. The effect of the z-pin on the flexural strength is shown in Figure 6. It was found that the flexural strength of the composite decreased by 2%, 3%, and 25% after the z-pin was added. As the volume fraction of the z-pin increased, the flexural strength of the composite decreased, which is consistent with experimental results for z-pin-reinforced polymer matrix composites [11,14]. This is due to increased microstructural damage resulting from the increased volume fraction of the z-pin. By increasing the z-pin content, the degree of buckling of the carbon fiber increased and the overall area of the aluminum-rich zones also increased, leading to the dilution of the volume fraction of in-plane fiber. In addition, increasing the number of metal z-pins may increase the fiber breakage, and result in the loss of strength, although this was difficult to determine quantitatively.

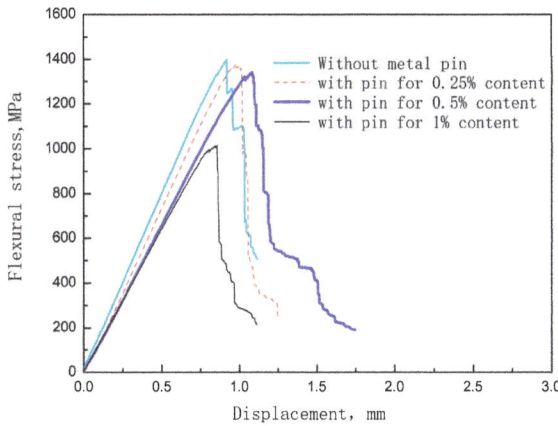

Figure 5. Typical Cf/Al stress-deflection results for samples without a z-pin and those with varying z-pin contents and a 0.3 mm diameter.

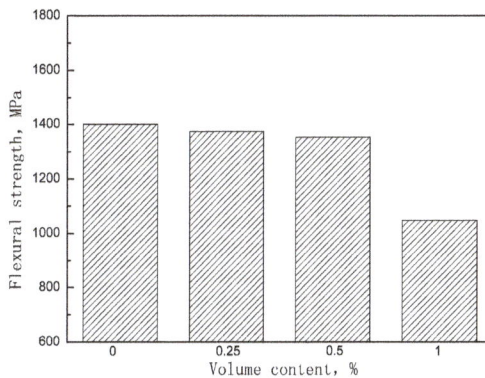

Figure 6. Effect of metal z-pin volume content on the flexural strength of the Cf/Al composite.

Many studies have reported that when matrix-enriched areas become connected and a channel is formed due to the small z-pin's spacing, crack expansion is more likely to occur, and the in-plane properties are further reduced [5,8]. As the z-pin content increases to 1%, small spacing between z-pins may be another reason for sudden drop in the strength of z-pinned Cf/Al composites.

Figure 7 shows the stress-deflection curves of the unpinned Cf/Al composite and the z-pinned Cf/Al composites with a volume fraction of 1% and diameters of 0.3 mm, 0.6 mm, and 0.9 mm respectively. The influence of the z-pin on the flexural strength is summarized in Figure 8 according to the stress-deflection curves. It can be seen that the addition of z-pins of diameters 0.3 mm, 0.6 mm, and 0.9 mm results in a 25%, 18%, and 13% reduction in flexural strength of the composites, respectively. This indicates that the knockdown of flexural strength decreased with the increase of the z-pin diameter, which is inconsistent with the experimental results of z-pinned polymer matrix composites [14]. For z-pinned polymer matrix composites, the deflection of carbon fibers and the dilution of carbon fiber content due to matrix-rich zones are the main factor for a reduction of the in-plane mechanical properties [12,13]. Thus, larger diameter z-pins lead to the greater deflection of carbon fibers and a larger area of matrix-rich zones, causing the lower flexural strength of the z-pinned polymer matrix composites. For the z-pinned Cf/Al composite, increasing the diameter causes the problems described above as well as other changes. Thin z-pin-enhanced Cf/Al has a more severe interface reaction than thick z-pin enhanced Cf/Al. Thin z-pin enhanced Cf/Al has a thicker interface diffusion layer due to the presence of a larger number of brittle reactants of FeAl$_3$, and this layer becomes harder, more brittle, and more susceptible to fracture [9,16,17]. Thus, the degree of interfacial reaction is the main factor controlling Cf/Al flexural strength, and its effect is greater than that of damage to the microstructure. The thin metal z-pin-reinforced material has lower strength since it has minor microstructural damage but a high reaction degree. In addition, the thick z-pins have stronger fracture resistance, which results in the higher interlaminar strength of Cf/Al composites. Thus, when Cf/Al is required to have high flexural performance and delamination resistance, it is recommended to use relatively large metal z-pins (approximately 1 mm) for enhancement.

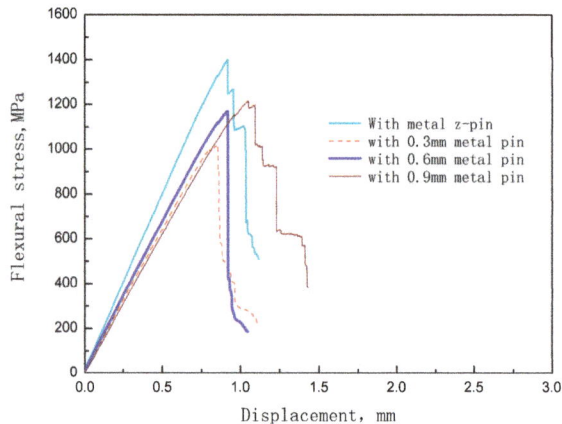

Figure 7. Typical Cf/Al stress-deflection results for samples without a z-pin and those with a z-pin content of 1% and different diameters.

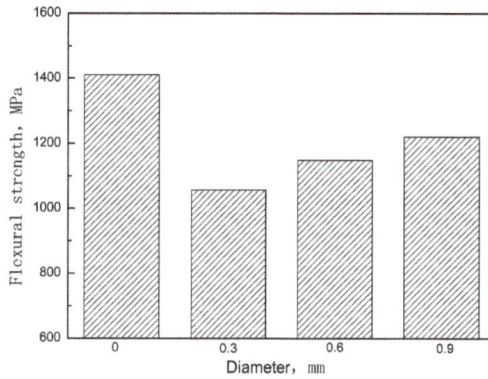

Figure 8. Effect of the diameter of the metal z-pins on the flexural strength of the Cf/Al composite.

3.3. Fracture Surface

Figure 9 presents the fracture photograph of the unpinned Cf/Al composite and z-pinned Cf/Al composites with a metal pin volume fraction of 1% and diameters of 0.3 mm, 0.6 mm, and 0.9 mm. These images show that there are a large number of fibers pulled out from the fracture surface of the z-pinned Cf/Al composite (Figure 9a,c,e). This fracture feature is the same as that of the unpinned Cf/Al composite (Figure 9f). This similarity of fracture features indicates that the addition of the metal pin does not change the fracture mode of Cf/Al composites and that carbon fiber is still the main load carrier. These results are consistent with the results of the stress-deflection curve discussed in Section 3.2.

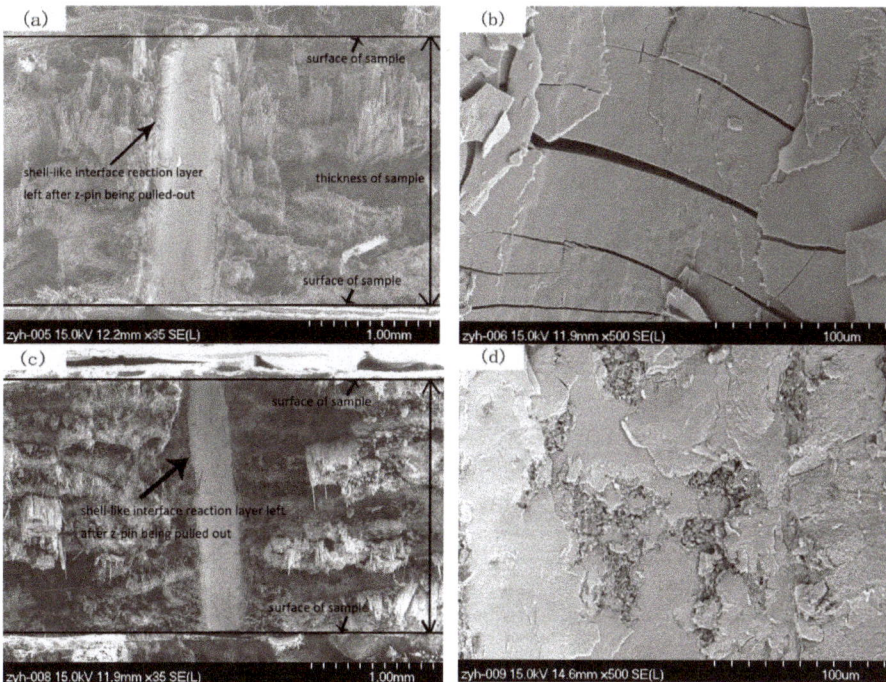

Figure 9. *Cont.*

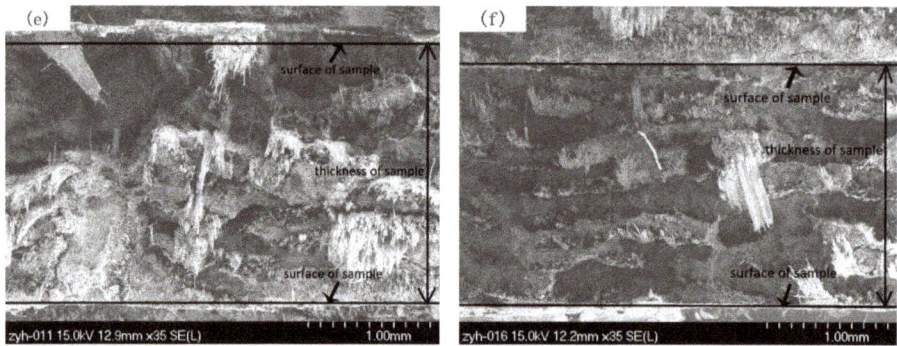

Figure 9. SEM images of the fracture surface of Cf/Al composites (**a**)–(**b**) with φ0.3 mm z-pin, (**c**)–(**d**) with φ0.6 mm z-pin, and (**e**) with φ0.9 mm z-pin, (**f**) without z-pin.

At the fracture surface of the z-pinned Cf/Al composites with diameters of 0.3 mm and 0.6 mm, a continuous shell-like interface reaction layer $FeAl_3$ between the metal pin and aluminum was found (Figure 9a,c). The cracking of the metal pin/aluminum interface was along the $FeAl_3$ interfacial reaction layer. There was no significant reaction layer with $FeAl_3$ for the z-pinned Cf/Al fracture with a diameter of 0.9 mm (Figure 9e). This shows that the 0.3 mm and 0.6 mm metal z-pin reinforced Cf/Al composites were fractured along the cross-section containing the metal pin, while the 0.9 mm metal z-pin-reinforced Cf/Al composite did not fracture along the cross-section containing the metal pin.

Further high-magnification observations on the shell-like $FeAl_3$ reaction layers revealed that the thick shell-like $FeAl_3$ reaction layer of the z-pinned Cf/Al composite with a diameter of 0.3 mm was covered with microcracks (Figure 9b), whereas, less microcracks were observed in the thin reaction layer of the z-pinned Cf/Al composite with a diameter of 0.6 mm (Figure 9d). This shows that larger thicknesses of the reaction layer led to more of the brittle phase of $FeAl_3$ and also resulted in more interface cracking. In addition, the thickness of the reaction layer not only affected the content of the brittle phase but also affected the crack growth. When the interface layer is thin, the crack size is also small, and it is more difficult to induce fiber breakage. That is, cracks are more likely to initiate and propagate when the reaction is severe. It is worth noting that the effects of interfacial reactions are not all harmful. Neither too heavy nor too small interfacial reactions are perfect. Achieving an appropriate amount of interface reaction is conducive to maintaining a good interface bonding strength and an effective load transfer from the matrix to reinforcement, therefore improving the interlaminar mechanical properties of composites. Nevertheless, in this work the interfacial reaction is relatively serious. After melted aluminum alloy was infiltrated into the preforms, the degree of interfacial reaction was dependent on the temperature of the preform and the melted aluminum alloy, and the diffusion time apart from contact area between the z-pin and the aluminum during the preparation process. Thus it is possible to decrease the temperature of the preform and melted aluminum alloy to optimize the z-pin/matrix interface, thereby minimizing the knockdown of flexural strength.

Based on the above analysis, the failure process of z-pinned Cf/Al composites was inferred. For 0.3 mm and 0.6 mm reinforced Cf/Al composites with a thick $FeAl_3$ reaction layer, under stress in the fiber direction, the fracture occurred in the brittle $FeAl_3$ phase first, which led to cracking of the interface. However, as carbon fiber was the main load-bearing phase, the composite did not fail. With increased loading, the cracks produced at the interface gradually extended into the carbon fibers, leading to the final failure of the sample, whereas, for the 0.9 mm z-pin-reinforced Cf/Al with a thin $FeAl_3$ reaction layer, no significant interface cracks occurred, or at least interface cracks did not propagate into the fibers, due to good interfacial bonding. With further loading, the buckling fibers in the vicinity of the metal pin broke and eventually led to the overall fracture of the composite.

4. Conclusions

In the current work, the effect of the metal z-pin on the flexural strength of Cf/Al was investigated. Three-point flexural tests showed that the flexural strength of Cf/Al composites was reduced by 2–25% due to the introduction of the z-pin. The fracture surfaces revealed that the fracture mode of Cf/Al composites was not changed by the z-pins, and that carbon fibers contributed to the flexural strength of Cf/Al composites. The reduced flexural strength was attributed to the microstructural damage caused by the z-pin, such as waviness of the in-plane fiber, fiber breakage, aluminum-rich regions and formation of the brittle phase caused by the interfacial reaction between the metal pin and the matrix. The flexural strength declined with the increasing z-pin volume content due to greater damage to the microstructure, which is consistent with the results of polymer–matrix composites. The thick metal z-pin-reinforced Cf/Al composite caused more damage to the microstructure, but also had a lower interface reaction degree and less brittle phase, and thereby showed higher flexural strength than the thin metal z-pin reinforced Cf/Al composite. The study also showed that the brittle phase caused by the interfacial reaction is the main factor for the decline in the flexural strength of z-pinned Cf/Al composites. Thus it is expected that adjusting the process parameters, for example by decreasing the temperature of the preform and melted aluminum alloy, could possibly reduce the degree of interface reaction and thereby maximize the flexural strength of z-pinned Cf/Al composites.

Author Contributions: Resources, G.W.; writing—original draft preparation, S.W.; writing—review and editing, Y.Z., P.S. and Y.C.; project administration, Y.Z.

Funding: This research was funded by National Natural Science Foundation of China, grant number 51305075, the Science and Technology Innovation Foundation for Harbin Talents, grant number 2017RAYXJ021 and the Heilongjiang Province Scientific Research Foundation for Postdoctoral Scholar, grant number LBH-Q18004.

Conflicts of Interest: The authors declare no conflict of interest.

References

1. Wang, C.C.; Chen, G.Q.; Wang, X.; Zhang, Y.H.; Yang, Y.S. Effect of Mg content on the thermodynamics of interface reaction in Cf/Al composite. *Metall. Mater. Trans. A* **2012**, *43*, 2514–2519. [CrossRef]
2. Daoud, A. Microstructure and tensile properties of 2014 Al alloy reinforced with continuous carbon fibers manufactured by gas pressure infiltration. *Mater. Sci. Eng. A* **2005**, *391*, 114–120. [CrossRef]
3. Zhang, Y.H.; Wu, G.H. Comparative study on the interface and mechanical properties of T700/Al and M40/Al composites. *Rare metals* **2010**, *29*, 102–107. [CrossRef]
4. Su, J.; Wu, G.H.; Li, Y.; Gou, H.S.; Chen, G.H.; Xiu, Z.Y. Effects of anomalies on fracture processes of graphite fiber reinforced aluminum composite. *Mater. Des.* **2011**, *32*, 1582–1589. [CrossRef]
5. Mouritz, A.P. Review of z-pinned composite laminates. *Compos. Part A Appl. Sci. Manuf.* **2007**, *38*, 2383–2397. [CrossRef]
6. Ranatunga, V.; Clay, S.B. Cohesive modeling of damage growth in z-pinned laminates under mode-I loading. *J. Compos. Mater.* **2012**, *47*, 3269–3283. [CrossRef]
7. Wang, S.; Zhang, Y.; Wu, G. Interlaminar shear properties of z-pinned carbon fiber reinforced aluminum matrix composites by short-beam shear test. *Materials* **2018**, *11*, 1874. [CrossRef] [PubMed]
8. Ko, M.G.; Kweon, J.H.; Choi, J.H. Fatigue characteristics of jagged pin-reinforced composite single-lap joints in hygrothermal environments. *Compos. Struct.* **2015**, *119*, 59–66. [CrossRef]
9. Zhang, Y.; Yan, L.; Miao, M.; Wang, Q.; Wu, G. Microstructure and mechanical properties of z-pinned carbon fiber reinforced aluminum alloy composites. *Mater. Des.* **2015**, *86*, 872–877. [CrossRef]
10. Hoffmann, J.; Scharr, G. Mechanical properties of composite laminates reinforced with rectangular z-pins in monotonic and cyclic tension. *Compos. Part A* **2018**, *109*, 163–170. [CrossRef]
11. Knopp, A.; Düsterhoft, C.; Reichel, M.; Scharr, G. Flexural properties of z-pinned composite laminates in seawater environment. *J. Mater. Sci.* **2014**, *49*, 8343–8354. [CrossRef]
12. Li, C.; Yan, Y.; Wang, P.; Qi, D.; Wen, Y. Study on compressive properties of z-pinned laminates in RTD and hygrothermal environment. *Chin. J. Aeronaut.* **2012**, *25*, 64–70. [CrossRef]

Materials **2019**, *12*, 174

13. Steeves, C.A.; Fleck, N.A. In-plane properties of composite laminates with through-thickness pin reinforcement. *Int. J. Solids Struct.* **2006**, *43*, 3197–3212. [CrossRef]
14. Chang, P.; Mouritz, A.P.; Cox, B.N. Flexural properties of z-pinned laminates. *Compos. Part A Appl. Sci. Manuf.* **2007**, *38*, 224–251. [CrossRef]
15. Elrefaey, A.; Takahashi, M.; Ikeuchi, K. Preliminary investigation of friction stir welding aluminum/copper lap joints. *Weld. World* **2005**, *49*, 93–101. [CrossRef]
16. Sun, X.J.; Tao, J.; Guo, X.Z. Bonding properties of interface in Fe/Al clad tube prepared by explosive welding. Trans. *Trans. Nonferrous Met. Soc. China* **2011**, *21*, 2175–2180. [CrossRef]
17. Kimapong, K.; Watanabe, T. Lap joint of A5083 aluminum alloy and SS400 steel by friction stir welding. *Mater. Trans.* **2005**, *46*, 835–841. [CrossRef]

![materials logo] *materials*

MDPI

Article

Structural and Mechanical Properties of Ti–Co Alloys Treated by High Pressure Torsion

Boris B. Straumal [1,2,3,*], **Anna Korneva** [4], **Askar R. Kilmametov** [1,2], **Lidia Lityńska-Dobrzyńska** [4], **Alena S. Gornakova** [1], **Robert Chulist** [4], **Mikhail I. Karpov** [1] and **Paweł Zięba** [4]

1   Institute of Solid State Physics and Chernogolovka Scientific Center, Russian Academy of Sciences, Chernogolovka 142432, Russia; Askar.Kilmametov@kit.edu (A.R.K.); alenahas@issp.ac.ru (A.S.G.); Karpov@issp.ac.ru (M.I.K.)
2   Karlsruhe Institute of Technology (KIT), Institute of Nanotechnology, 76344 Eggenstein-Leopoldshafen, Germany
3   National University of Science and Technology «MISIS», Moscow 119049, Russia
4   Institute of Metallurgy and Materials Science, Polish Academy of Sciences, 30-059 Krakow, Poland; a.korniewa@imim.pl (A.K.); L.Litynska@imim.pl (L.L.-D.); R.Chulist@imim.pl (R.C.); p.zieba@imim.pl (P.Z.)
*   Correspondence: straumal@issp.ac.ru

Received: 26 December 2018; Accepted: 25 January 2019; Published: 29 January 2019

Abstract: The microstructure and properties of titanium-based alloys can be tailored using severe plastic deformation. The structure and microhardness of Ti–4 wt.% Co alloy have been studied after preliminary annealing and following high pressure torsion (HPT). The Ti–4 wt.% Co alloy has been annealed at 400, 500, and 600 °C, i.e., below the temperature of eutectoid transformation in the Ti–4 wt.% Co system. The amount of Co dissolved in α-Ti increased with increasing annealing temperature. HPT led to the transformation of α-Ti in ω-Ti. After HPT, the amount of ω-phase in the sample annealed at 400 °C was about 8085%, i.e., higher than in pure titanium (about 40%). However, with increasing temperature of pre-annealing, the portion of ω-phase decreased (60–65% at 500 °C and about 5% at 600 °C). The microhardness of all investigated samples increased with increasing temperature of pre-annealing.

Keywords: titanium alloys; high pressure torsion; microhardness

---

1. Introduction

Titanium and its alloys possess low density, high strength, as well as high corrosion resistance in the broad temperature interval. Titanium alloys were found to have broad applications in the aircraft, building, and medicinal industries. Due to their outstanding biocompatibility, Ti-alloys are increasingly applied in orthopaedic and dental implants. But titanium alloys don't only have advantages. Unfortunately, the high melting temperature, high elastic modulus, and high affinity for oxygen can limit their application as biomaterials [1,2]. Fortunately, the structure and properties of titanium alloys can be tailored using various combinations of thermal and mechanical treatments. One of the promising new options is the so-called severe plastic deformation (SPD). SPD permits to rich the extremely high strains in a material without its failure. Disadvantages of titanium and its alloys can also be improved by the addition of alloying elements like niobium, zirconium, hafnium, molybdenum, cobalt, and chromium [2,3].

At the focus of this work will be Ti–Co alloys subjected to high pressure torsion (HPT), being one of the SPD modes. The Ti–Co alloys are broadly used as implant alloys in dentistry and medicine for many years [4–8]. Thus, the Ti-based alloys with cobalt addition show higher strength [9,10] and have lower melting temperature, which can alleviate many casting problems. The addition of cobalt improves the corrosion resistance of titanium [11] and its mechanical properties [12]. The ternary (Ti–Co)-based alloys also found broad applications [13–18]. The Ti–Co alloys are frequently used as coatings on other titanium alloys like Ti6Al4V [19–23]. Such surface modifications permit improvement of the endurance of Ti6Al4V alloy due to the formation of hard Ti-Co intermetallic particles. The Ti–Co thin films were used also as diffusion barriers, or as an element of integrated circuits [24,25].

SPD not only refines the grains of metallic alloys (including those of titanium) [26–28]. SPD also drives bulk and grain-boundary phase transformations [29–33]. In titanium these are the transitions between the low-temperature α-phase, high-temperature β-phase, and high-pressure ω-phase [34–39]. The high-pressure ω-phase appears in Ti-based alloys during HPT and then retains after pressure release [26,27,40,41]. Previously, it has been studied how ω-phase transforms during HPT from the mixture of α- and β-phases [26,42,43]. It has been observed that β-to-ω transformation goes along quite easily [26,43,44]. It is martensitic, follows a special orientation relationship, and does not need intensive mass transfer. However, the HPT-driven β-to-ω transformation in Ti-4 wt.% Co alloy proceeds less easily in comparison to the Ti-4 wt.% Fe one [44]. Most probably, the reason is the less favorable coincidence of lattice constants between β- and ω-phases in the Ti-4 wt.% Co alloy. The α-to-ω transformation in Ti-based alloys encounter more troubles than β-to-ω [26,43]. Mainly it is because the orientation relationship between α and ω phases is less favorable [26,34–39,44]. How would the high-pressure ω-phase form in the case of only α-phase and intermetallic precipitates existing in a sample before HPT? In order to answer this question, we studied the properties of Ti–4 wt.% Co alloy where the HPT of the α + β mixture had already been investigated [19,43–45]. We annealed the Ti–4 wt.% Co alloy for extremely long durations below eutectoid temperature in order to produce the α-Ti solid solution with a different (and equilibrium) concentration of cobalt, as well as a different amount of possible coarse Ti_2Co precipitates.

2. Experimental

For the preparation of Ti–4 wt.% Co alloys, pure titanium (99.98%) and cobalt (99.99%) were been used. The concentration of 4 wt.% Co was on the left side of the point of eutectoid $\beta \rightarrow \alpha + Ti_2Co$ transformation (8.5 wt.% Co, see Figure 1). The alloy was melted in the argon atmosphere with the aid of an induction furnace and cast into ingots cylindrical with a diameter of 10 mm. The resulting ingots were spark erosion cut into 0.7 mm thick disks. The resulted slices were chemically etched and put into the ampoules. The residual pressure in the sealed quartz ampoules was about 4×10^{-4} Pa. The annealing temperatures were 400, 500, and 600 °C, i.e., below the temperature of eutectoid transformation in the Ti–Co system. We annealed the ampoules during a very long period (for 5685, 5685, and 2774 h, respectively) in order to reach the equilibrium cobalt content in the αTi-based solid solution. The ampoules with samples inside were quenched in cold water after annealing. Then, the ampoules were broken and disks were treated at room temperature with the aid of HPT in a Bridgman anvil type unit using a custom built computer-controlled device (W. Klement GmbH, Lang, Austria) with 5 plunger rotations. The strain rate was 1 rpm, the pressure was 7 GPa, and the thickness of the samples after HPT was 0.35 mm.

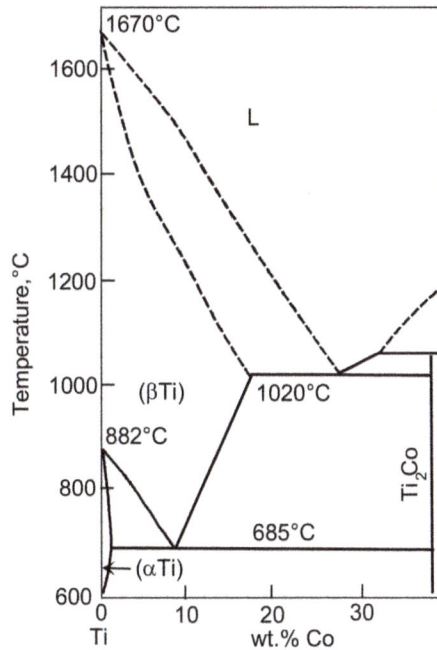

Figure 1. The Co-rich part of the Ti–Co phase diagram [46].

Measurements of the microhardness were performed using the PMT-3 unit with the load of 20 g. The samples were carefully polished before measurements with 1 μm diamond paste. We measured microhardness at least 10 times for each sample at the distance from the disk center of about half of its radius [45]. The Siemens D-500 X-ray diffractometer with Cu-Kα radiation was used for the investigations of X-ray diffraction (XRD). The software PowderCell for Windows Version 2.4.08.03.2000 (Werner Kraus & Gert Nolze, BAM Berlin) allowed us to calculate the lattice parameters and to perform the phase analysis. The FEI E-SEM XL30 SEM (Hillsboro, OR, USA) equipped with EDAX Genesis energy-dispersive X-ray spectrometer (EDS) permitted us to conduct the scanning electron microscopy (SEM) investigations. The TECNAI G2 FEG super TWIN (200 kV) TEM (Hillsboro, OR, USA) operating at an accelerating voltage of 200 kV was used for the transmission electron microscopy (TEM) studies. The TEM instrument was equipped with an energy dispersive X-ray (EDS) spectrometer manufactured by EDAX. We prepared thin foils for TEM using an electrolyte D2 manufactured by Struers company (Cleveland, OH, USA).

3. Results and Discussion

Figure 1 shows part of the Ti–Co phase diagram (with low cobalt content) [46]. The sample composition of Ti–4 wt.% Co is on the left side of the point of eutectoid β → α + Ti₂Co transformation (8.5 wt.% Co). The annealing temperatures were 400, 500, and 600 °C, and they are located below the temperature of eutectoid transformation T_e = 685 °C. The maximum solubility of cobalt in α-Ti is about 1.2 wt.% Co and is reached at T_e = 685 °C.

In Figure 2, the X-ray diffraction patterns for Ti–4 wt.% Co alloy preliminary annealed at 400, 500, 600 °C (lower patterns), and for the same samples, but after following HPT (upper patterns) are shown. After annealing, all samples contained α-Ti and intermetallic compound Ti₂Co. According to the phase diagram (Figure 1), the Ti₂Co phase is daltonide and its composition does not change with the temperature. Therefore, the position of Ti₂Co peaks in the X-ray diffraction patterns are the same for all three annealing temperatures. However, the amount of Ti₂Co phase slightly decreased with increasing annealing temperature (see Table 1). This is because the total amount of cobalt in the alloy remained constant, and the cobalt solubility in the α-Ti based solid solution increased when the annealing temperature approached the eutectoid one. According to the phase diagram (Figure 1), the solubility of cobalt in α-Ti-based solid solution increased with a temperature below T_e = 685 °C. The increase of Co content in α-Ti decreases the lattice parameter [44]. Indeed, we can see in lower patterns in Figure 2 that the α-Ti peaks in the sample annealed at 600 °C are shifted to the right in comparison to samples annealed at lower temperatures (it means the decrease of lattice parameter) with increasing temperature. Thus, the amount of cobalt dissolved in α-Ti increases with increasing temperature. Since the total amount of cobalt remains the same (4 wt.% Co), the cobalt atoms for α-Ti were taken from Ti_2Co precipitates, and their amount slightly decreased with increasing temperature (see Table 1). After HPT, all peaks in XRD patterns were broadened and their intensity decreased. It marked the usual for SPD strong grain refinement. Moreover, the ω-Ti phase appeared in all samples. A certain amount of α-Ti phase remained. The intermetallic phase Ti_2Co was also present. The lattice parameters for α-Ti and ω-Ti before and after HPT were given in Table 1. The lattice parameters of ω-Ti are less sensitive to the temperature of pre-annealing than those of α-Ti. The lattice parameter a of α-Ti phase increased after HPT in all studied samples. The lattice parameter c of α-Ti phase also increased in the sample pre-annealed at 600 °C and slightly decreased in samples pre-annealed at 400 and 500 °C

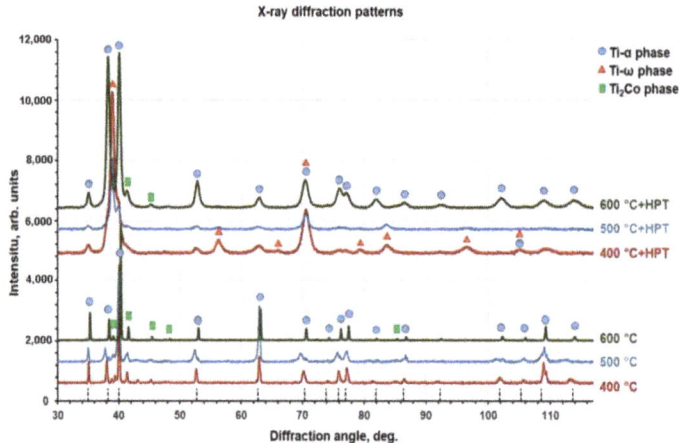

Figure 2. X-ray diffraction patterns for Ti–4 wt.% Co alloy after annealing at 400, 500 and 600 °C (lower patterns) and after high pressure torsion (HPT) with preliminary heat treatment (upper patterns). Vertical dotted lines show the positions of the reflections for pure α-Ti.

After HPT, the amount of ω-phase in the sample annealed at 400 °C was about 80–85%, i.e., higher than in pure titanium (about 40% [26]). However, with increasing temperature of pre-annealing the portion of ω-phase decreased (60–65% at 500 °C and about 5% at 600 °C). Earlier we observed that both Ti–Fe and Ti–Co alloys annealed above eutectoid temperature contain after HPT more ω-phase than the same HPT-treated alloys annealed before HPT below eutectoid temperature [44]. Also, the addition of aluminum to the binary Ti–V alloys completely suppressed the formation of (ωTi) phase after HPT [47]. The decrease of the amount of ω-phase with increasing temperature of pre-annealing can be indirectly driven by the change of the amount and morphology of intermetallic precipitates (see Table 1 and Figures 3–5).

Figures 3–5 show the microstructure of Ti–4 wt.% Co alloy after annealing at different temperatures and HPT. Figures 3a, 4a and 5a show SEM micrographs. Figures 3b, 4b and 5b show bright field and Figures 3c, 4c and 5c dark field TEM micrographs after annealing and following HPT. Figures 3d, 4d and 5d show selected area electron diffraction patterns (SAED). The part of SAED used for DF images is shown by the circle. The main input to the DF images give the ω-100 ring. Therefore, the grains appearing bright in the DF images mainly represent the ω-phase. Particularly, it is visible how some ω-grains are elongated in the rotation direction of the HPT anvil. After HPT the grains of α-Ti and ω-Ti phases are very fine. Figures 3c,d, 4c,d, and 5c,d witness that the grain size of α-Ti and ω-Ti phases after HPT increased with increasing temperature of preliminary annealing (about 70 nm for 400 °C, 100 nm for 500 °C and 150 nm for 600 °C). It can be seen in SEM micrographs that the morphology of Ti_2Co particles (they appear bright) is different after different temperatures of pre-annealing and HPT. With increasing temperature of pre-annealing, the Ti_2Co particles (also after HPT) become bigger. It seems that the hard and coarse Ti_2Co particles were less refined by HPT than the smaller ones. SAED-patterns witness that the samples contained at least two finely dispersed phases.

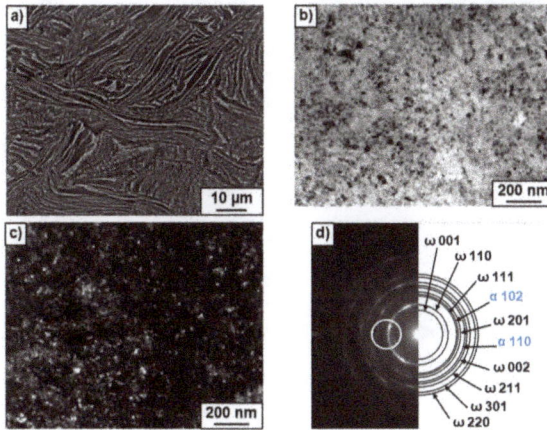

Figure 3. Microstructure of Ti–4 wt.% Co alloy after annealing at 400 °C and high pressure torsion (HPT). (**a**) Scanning electron microscopy (SEM) micrograph. (**b**) Bright field and (**c**) dark field transmission electron microscopy (TEM) micrographs after annealing at 400 °C and following HPT. (**d**) Selected area electron diffraction pattern.

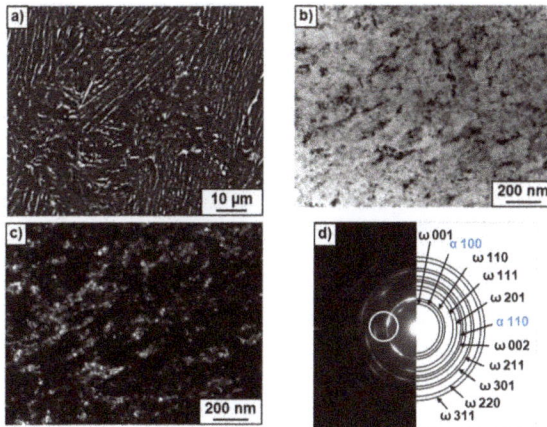

Figure 4. Microstructure of Ti–4 wt.% Co alloy after annealing at 500 °C and high pressure torsion (HPT). (**a**) Scanning electron microscopy (SEM) micrograph. (**b**) Bright field and (**c**) dark field transmission electron microscopy (TEM) micrographs after annealing at 400 °C and following HPT. (**d**) Selected area electron diffraction pattern.

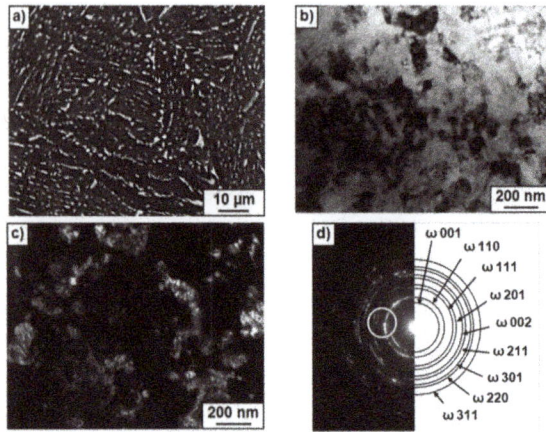

Figure 5. Microstructure of Ti–4 wt.% Co alloy after annealing at 600 °C and and high pressure torsion (HPT). (**a**) Scanning electron microscopy (SEM) micrograph. (**b**) Bright field and (**c**) dark field transmission electron microscopy (TEM) micrographs after annealing at 400 °C and following HPT. (**d**) Selected area electron diffraction pattern.

HPT changes of the lattice parameters *a* and *c* for α-Ti phase (Table 1). These changes are equivalent to the decrease of cobalt content in α-Ti phase [44]. This behavior is very similar to the recently observed "purification" of α-Ti phase in the Ti–Fe alloys after HPT [27].

Table 1. Phases, lattice parameter and their amount in studied titanium alloys after annealing and after following HPT.

Sample	Lattice Parameter, nm	Lattice Parameter, nm	Lattice Parameter, nm	Volume
-	Before HPT	After HPT	After HPT	Fraction, %
-	α-Ti	α-Ti	ω-Ti	ω-Ti
Ti–4 wt. %Co 600 °C, 2774 h	$a = 0.2941, c = 0.4689, c/a = 1.594$	$a = 0.2957, c = 0.4703, c/a = 1.590$	$a = 0.4622, c = 0.2833$	5
Ti–4 wt. %Co 500 °C, 5685 h	$a = 0.2954, c = 0.4759, c/a = 1.611$	$a = 0.2966, c = 0.4718, c/a = 1.591$	$a = 0.4622, c = 0.2833$	65
Ti–4 wt. %Co 400 °C, 5685 h	$a = 0.2953, c = 0.4729, c/a = 1.602$	$a = 0.2963, c = 0.4725, c/a = 1.595$	$a = 0.4622, c = 0.2833$	80
Pure Ti	$a = 0.2953, c = 0.4694, c/a = 1.588$	$a = 0.2959, c = 0.4690, c/a = 1.585$	$a = 0.4627, c = 0.2830$	40

Where can we move the cobalt atoms from the α-Ti phase during HPT? The first possibility is that they migrate into newly formed ω-Ti phase. It is known, for example, that the solubility of iron in ω-Ti is much higher than in the α-Ti phase [26]. We can suppose that a similar law is true for solubility of cobalt in ω-Ti and α-Ti. X-ray microanalysis in SEM mode indeed demonstrated that the areas predominately filled with ω-phase contained more cobalt than the areas predominately filled with α-phase. The second possibility is that the cobalt atoms are used to form the fine precipitates of Ti$_2$Co phase. They are visible in the dark-field TEM images and contribute into SAED patterns. The third possibility are the grain boundaries (GBs) that additionally appear in the samples afterward HPT. In all materials subjected to HPT, the grain size decreased at least one order of magnitude [28,48–56]. In our case, the grains after HPT became almost a thousand times smaller. As a result, the GB area in the volume unit strongly increased. Cobalt atoms segregate in these new GBs. They are taken from the bulk solid solution. Due to this phenomenon of GB segregation, the overall (apparent) solubility of a second component strongly increased in nanograined materials [57]. Thus, the third reason for the "cleaning" of α-Ti phase during HPT is that the cobalt atoms are used to form the GB segregation. Similar HPT effect exists also in steels [58–60]. In steels also, only a few carbon atoms can be diluted in the α-Fe lattice. However, the GBs "help" to dissolve a large amount of carbon without formation of carbides [58–60]. During the HPT-driven "cleaning" of α-Ti phase in our experiments the cobalt atoms migrate from the volume solid solution to the GBs. Such HPT-driven atomic migration in Ti-alloys proceed very quickly [26,27]. The estimated equivalent diffusion coefficients of this diffusion-like mass transfer are several orders of magnitude higher than the coefficient of conventional volume diffusion extrapolated to the HPT temperature of 300 K [26]. This acceleration is especially astonishing because the applied pressure always decreases the rate of diffusion-controlled processes [61,62].

One can find in the published papers the mechanical properties of different Ti- phases [63–65]. The elastic moduli of α-Ti and β-Ti were determined in a Ti–4 wt.% V–6 wt.% Al alloy [63]. It appeared that was the elastic modulus of α-Ti in this coarse-grained alloy is 22% higher than that of the β-Ti. Moreover, the shear modulus of β-Ti in the samples annealed between 600 and 975 °C decreased with increasing temperature of annealing [64]. It was also shown theoretically that the specific energy of the α/β interphase boundary decreased with increasing temperature about two times [65].

In Figure 6, the microhardness values were given for the HPT-treated samples after preliminary annealing. The microhardness measured in the middle of the radius increased from 210 to 250 HV with increasing temperature of pre-annealing (Figure 6a). We can suppose that this increase is due to the decrease of the portion of ω-phase in the samples. The hardness is influenced also by the hard Ti$_2$Co intermetallic particles. The increase of the pre-annealing temperature slightly decreased the amount of Ti$_2$Co particles (Table 1). They become larger (compare Figures 3a, 4a and 5a). These two facts would be the reason for the certain softening. On the other hand, the concentration of cobalt in α-Ti increased with the increasing temperature of pre-annealing (even after HPT). It can lead to a certain solid-solution hardening of the α-phase. The resulted influence of these three factors leads to the increase of microhardness. The increase of microhardness with an increase of the annealing temperature has been observed recently in Ti–V and Ti–V–Al alloys [47]. However, this similarity is superficial because the factors leading to the increase of microhardness in the Ti–V and Ti–V–Al alloys are most probably different. First, these alloys do not contain any intermetallic precipitates. Second, the low amount of ω-phase is present after HPT only in binary Ti–V alloys [47].

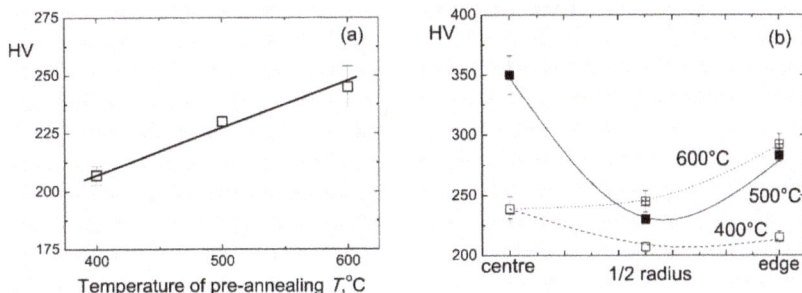

Figure 6. Dependence of microhardness of Ti–4 wt.% Co alloy after HPT on the temperature of preliminary annnealing (**a**) and on the position in the sample (**b**).

4. Conclusions

High pressure torsion leads to the phase transformations in the studied Ti–4 wt.% Co alloy. The samples were annealed below eutectoid temperature in order to produce the mixture of α-Ti phase with different cobalt concentrations and Ti$_2$Co intermetallic precipitates. Thus, the initial phases before HPT were different from the previously studied α + β mixture. After HPT, the ω-Ti phase appeared in the samples. Its portion decreased with increasing temperature of pre-annealing. The microhardness of all investigated samples increased with increasing temperature of pre-annealing.

Author Contributions: Investigation, A.K., A.R.K., L.L.-D., A.S.G. and R.C.; Supervision, M.I.K. and P.Z.; Writing—original draft, B.B.S.

Funding: The work was partially supported by German Research Foundation (grant numbers FA999/1, IV 98/5, HA 1344/32, RA 1050/20), Russian Foundation for Basic Research (grant numbers 16-53-12007, 16-03-00285, 18-03-00067), National Science Center of Poland (grant OPUS 2017/27/B/ST8/010192), Ministry of Education and Science of the Russian Federation in the framework of the Program to Increase the Competitiveness of NUST "MISiS" and Karlsruhe Nano Micro Facility

Acknowledgments: The authors are deeply grateful to M.I. Egorkin (Material Science Laboratory, ISSP RAS) for the production of titanium alloys.

Conflicts of Interest: The authors declare no conflict of interest

References

1. Long, M.; Rack, H.J. Titanium alloys in total joint replacement-a materials science perspective. *Biomaterials* **1998**, *19*, 1621–1639. [CrossRef]
2. Niinomi, M. Recent metallic materials for biomedical applications. *Metall. Mater. Trans. A* **2002**, *33*, 477–486. [CrossRef]
3. Matsuno, H.; Yokoyama, A.; Watari, F.; Uo, M.; Kawasaki, T. Biocompatibility and osteogenesis of refractory metal implants, titanium, hafnium, niobium, tantalum and rhenium. *Biomaterials* **2001**, *22*, 1253–1262. [CrossRef]
4. Wang, K. The use of titanium for medical applications in the USA. *Mater. Sci. Eng. A* **1996**, *213*, 134–137. [CrossRef]
5. Fasching, A.; Norwich, D.; Geiser, T.; Paul, G.W. An evaluation of a NiTiCo alloy and its suitability for medical device applications. *J. Mater. Eng. Perform.* **2011**, *20*, 641–645. [CrossRef]
6. Wang, R.; Welsch, G. Evaluation of an experimental Ti-Co alloy for dental restorations. *J. Biomed. Mater. Res. B* **2013**, *101*, 1419–1427. [CrossRef] [PubMed]
7. Niinomi, M. Mechanical biocompatibilities of titanium alloys for biomedical applications. *J. Mech. Behav. Biomed.* **2008**, *1*, 30–42. [CrossRef] [PubMed]
8. Mutlu, I. Synthesis and characterization of Ti-Co alloy foam for biomedical applications. *Trans. Nonferr. Met. Soc. China* **2016**, *26*, 126–137. [CrossRef]
9. Liu, X.T.; Chen, S.Y.; Tsoi, J.K.H.; Matinlinna, J.P. Binary titanium alloys as dental implant materials-a review. *Regenerat. Biomater.* **2017**, *4*, 315–323. [CrossRef]
10. Ohnsorge, J.; Holm, R. Surface investigations of oxide layers on cobalt-chromium alloyed orthopedic implants using ESCA technique. *Med. Prog. Technol.* **1978**, *5*, 171–177.
11. Onagawa, J. Preparation of high corrosion resistant titanium alloys by spark plasma sintering. *J. Jpn. Inst. Met.* **1999**, *63*, 1149–1152. [CrossRef]
12. Liu, Q.; Yang, W.Y.; Chen, G.L. On superplasticity of two phase alpha-titanium-intermetallic Ti–(Co, Ni)–Al alloy. *Acta Metall. Mater.* **1995**, *43*, 3571–3582. [CrossRef]
13. Hu, K.; Huang, X.M.; Lu, J.; Liu, H.S.; Cai, G.M.; Jin, Z.P. Measurement of phase equilibria in Ti-Co-Pt ternary system. *Calphad* **2018**, *60*, 191–199. [CrossRef]
14. Zeng, Y.; Zhu, L.L.; Cai, G.M.; Liu, H.S.; Huang, J.W.; Jin, Z.P. Investigation of phase equilibria in the Ti-Co-Zr ternary system. *Calphad* **2017**, *56*, 260–269. [CrossRef]
15. Fartushna, I.; Bulanova, M.; Ayral, R.M.; Tedenac, J.C.; Zhludenko, N.; Meleshevich, K.; Romanenko, Y. Phase equilibria in the crystallization interval in the Ti-Co-Sn system at above 50 at % Ti. *J. Alloys Compd.* **2016**, *673*, 433–440. [CrossRef]
16. Samal, S.; Agarwal, S.; Gautam, P.; Biswas, K. Microstructural evolution in novel suction cast multicomponent Ti-Fe-Co alloys. *Metall. Mater. Trans. A* **2015**, *46A*, 851–868. [CrossRef]
17. Xue, Y.; Wang, H.M. Microstructure and properties of Ti-Co-Si ternary intermetallic alloys. *J. Alloys Compd.* **2008**, *464*, 138–145. [CrossRef]
18. Lin, J.H.C.; Chen, Y.F.; Ju, C.P. Effect of nickel addition on microstructure and properties of Ti-Co-Ni alloys. *Biomaterials* **1995**, *18*, 1401–1407. [CrossRef]
19. Fatoba, O.S.; Adesina, O.S.; Popoola, A.P.I. Evaluation of microstructure, microhardness, and electrochemical properties of laser-deposited Ti-Co coatings on Ti-6Al-4V alloy. *Int. J. Adv. Manuf. Technol.* **2018**, *97*, 2341–2350. [CrossRef]

20. Adesina, O.S.; Popoola, A.P.I.; Pityana, S.L.; Oloruntoba, D.T. Microstructural and tribological behavior of in situ synthesized Ti/Co coatings on Ti-6Al-4V alloy using laser surface cladding technique. *Int. J. Adv. Manuf. Technol.* **2018**, *95*, 1265–1280. [CrossRef]

21. Adesina, O.S.; Popoola, A.P.I. A study on the influence of laser power on microstructural evolution and tribological functionality of metallic coatings deposited on Ti-6Al-4V alloy. *Tribology* **2017**, *11*, 145–155. [CrossRef]

22. Adesina, O.S.; Mthisi, A.; Popoola, A.P.I. The effect of laser based synthesized Ti–Co coating on microstructure and mechanical properties of Ti6A14V alloy. *Procedia Manuf.* **2016**, *7*, 46–52. [CrossRef]

23. Langelier, B.C.; Esmaeili, S. In-situ laser-fabrication and characterization of TiC-containing Ti-Co composite on pure Ti substrate. *J. Alloys Compd.* **2009**, *482*, 246–252. [CrossRef]

24. Gromov, D.G.; Mochalov, A.I.; Pugachevich, V.P.; Kirilenko, E.P.; Trifonov, A.Y. Study of phase separation in Ti-Co-N thin films on silicon substrate. *Appl. Phys. A* **1997**, *64*, 517–521. [CrossRef]

25. Gromov, D.G.; Mochalov, A.I.; Pugachevich, V.P. CoSi2 formation in contact systems based on Ti-Co alloy with low cobalt content. *Appl. Phys. A* **1995**, *61*, 565–567.

26. Kilmametov, A.; Ivanisenko, Y.; Mazilkin, A.A.; Straumal, B.B.; Gornakova, A.S.; Fabrichnaya, O.B.; Kriegel, M.J.; Rafaja, D.; Hahn, H. The $\alpha\rightarrow\omega$ and $\beta\rightarrow\omega$ phase transformations in Ti-Fe alloys under high-pressure torsion. *Acta Mater.* **2018**, *144*, 337–351. [CrossRef]

27. Kilmametov, A.; Ivanisenko, Y.; Straumal, B.; Mazilkin, A.A.; Gornakova, A.S.; Kriegel, M.J.; Fabrichnaya, O.B.; Rafaja, D.; Hahn, H. Transformations of α' martensite in Ti-Fe alloys under high pressure torsion. *Scr. Mater.* **2017**, *136*, 46–49. [CrossRef]

28. Valiev, R.Z.; Islamgaliev, R.K.; Alexandrov, I. Bulk nanostructured materials from severe plastic deformation. *Prog. Mater. Sci.* **2000**, *45*, 103–189. [CrossRef]

29. Sauvage, X.; Wetscher, F.; Pareige, P. Mechanical alloying of Cu and Fe induced by severe plastic deformation of a Cu-Fe composite. *Acta Mater.* **2005**, *53*, 2127–2135. [CrossRef]

30. Straumal, B.B.; Sauvage, X.; Baretzky, B.; Mazilkin, A.A.; Valiev, R.Z. Grain boundary films in Al–Zn alloys after high pressure torsion. *Scr. Mater.* **2014**, *70*, 59–62. [CrossRef]

31. Straumal, B.; Korneva, A.; Zięba, P. Phase transitions in metallic alloys driven by the high pressure torsion. *Arch. Civ. Mech. Eng.* **2014**, *14*, 242–249. [CrossRef]

32. Straumal, B.B.; Kilmametov, A.R.; Ivanisenko, Y.; Mazilkin, A.A.; Kogtenkova, O.A.; Kurmanaeva, L.; Korneva, A.; Zięba, P.; Baretzky, B. Phase transitions induced by severe plastic deformation: Steady-state and equifinality. *Int. J. Mater. Res.* **2015**, *106*, 657–664. [CrossRef]

33. Straumal, B.B.; Kilmametov, A.R.; Korneva, A.; Mazilkin, A.A.; Straumal, P.B.; Zięba, P.; Baretzky, B. Phase transitions in Cu-based alloys under high pressure torsion. *J. Alloys Compd.* **2017**, *707*, 20–26. [CrossRef]

34. Donachie, M.J., Jr. *Titanium: A Technical Guide*, 2nd ed.; ASM International: Materials Park, OH, USA, 2000.

35. Errandonea, D.; Meng, Y.; Somayazulu, M.; Häusermann, D. Pressure-induced $\alpha\rightarrow\omega$ transition in titanium metal: A systematic study of the effects of uniaxial stress. *Physica B* **2005**, *355*, 116–125. [CrossRef]

36. Trinkle, D.R.; Hennig, R.G.; Srinivasan, S.G.; Hatch, D.M.; Jones, M.D.; Stokes, H.T.; Albers, R.C.; Wilkins, J.W. New mechanism for the α to ω martensitic transformation in pure titanium. *Phys. Rev. Lett.* **2003**, *91*, 025701. [CrossRef] [PubMed]

37. Sikka, S.K.; Vohra, Y.K.; Chidambaram, R. Omega phase in materials. *Prog. Mater. Sci.* **1982**, *27*, 245–310. [CrossRef]

38. Banerjee, S.; Mukhopadhyay, P. *Phase Transformations: Examples from Titanium and Zirconium Alloy*; Elsevier: Amsterdam, The Netherlands, 2010.

39. Hickman, B.S. The formation of omega phase in titanium and zirconium alloys: A review. *J. Mater. Sci.* **1969**, *4*, 554–563. [CrossRef]

40. Kriegel, M.J.; Kilmametov, A.; Rudolph, M.; Straumal, B.B.; Gornakova, A.S.; Stöcker, H.; Ivanisenko, Y.; Fabrichnaya, O.; Hahn, H.; Rafaja, D. Transformation pathway upon heating of Ti–Fe alloys deformed by high-pressure torsion. *Adv. Eng. Mater.* **2018**, *20*, 1700933. [CrossRef]

41. Kriegel, M.J.; Kilmametov, A.; Klemm, V.; Schimpf, C.; Straumal, B.B.; Gornakova, A.S.; Ivanisenko, Y.; Fabrichnaya, O.; Hahn, H.; Rafaja, D. Thermal stability of athermal ω-Ti(Fe) produced upon quenching of β-Ti(Fe). *Adv. Eng. Mater.* **2018**, *20*, 201800158. [CrossRef]

42. Straumal, B.B.; Kilmametov, A.R.; Ivanisenko, Y.; Gornakova, A.S.; Mazilkin, A.A.; Kriegel, M.J.; Fabrichnaya, O.B.; Baretzky, B.; Hahn, H. Phase transformations in Ti-Fe alloys induced by high pressure torsion. *Adv. Eng. Mater.* **2015**, *17*, 1835–1841. [CrossRef]

43. Kilmametov, A.R.; Ivanisenko, Y.; Straumal, B.B.; Gornakova, A.S.; Mazilkin, A.A.; Hahn, H. The $\alpha \to \omega$ transformation in titanium-cobalt alloys under high-pressure torsion. *Metals* **2018**, *8*, 1. [CrossRef]

44. Straumal, B.B.; Kilmametov, A.R.; Ivanisenko, Y.; Mazilkin, A.A.; Valiev, R.Z.; Afonikova, N.S.; Gornakova, A.S.; Hahn, H. Diffusive and displacive phase transitions in Ti-Fe and Ti-Co alloys under high pressure torsion. *J. Alloys Compd.* **2018**, *735*, 2281–2286. [CrossRef]

45. Permyakova, I.E.; Glezer, A.M.; Grigorovich, K.V. Deformation behavior of amorphous Co-Fe-Cr-Si-B alloys in the initial stages of severe plastic deformation. *Bull. Russ. Acad. Sci. Phys.* **2014**, *78*, 996–1000. [CrossRef]

46. *Binary Alloy Phase Diagrams*; Massalski, T.B. (Ed.) American Society for Metals: Metals Park, OH, USA, 1991.

47. Gornakova, A.S.; Straumal, A.B.; Khodos, I.I.; Gnesin, I.B.; Mazilkin, A.A.; Afonikova, N.S.; Straumal, B.B. Effect of composition, annealing temperature and high pressure torsion on structure and hardness of Ti-V and Ti-V-Al alloys. *J. Appl. Phys.* **2019**, *125*. [CrossRef]

48. Lojkowski, W.; Djahanbakhsh, M.; Burkle, G.; Gierlotka, S.; Zielinski, W.; Fecht, H.J. Nanostructure formation on the surface of railway tracks. *Mater. Sci. Eng. A* **2001**, *303*, 197–208. [CrossRef]

49. Cepeda-Jiménez, C.M.; García-Infanta, J.M.; Zhilyaev, A.P.; Ruano, O.A.; Carreño, F. Influence of the thermal treatment on the deformation-induced precipitation of a hypoeutectic Al-7 wt% Si casting alloy deformed by high-pressure torsion. *J. Alloys Compd.* **2011**, *509*, 636–643. [CrossRef]

50. Ivanisenko, Y.; Lojkowski, W.; Valiev, R.Z.; Fecht, H.J. The mechanism of formation of nanostructure and dissolution of cementite in a pearlitic steel during high pressure torsion. *Acta Mater.* **2003**, *51*, 5555–5570. [CrossRef]

51. Sagaradze, V.V.; Shabashov, V.A. Deformation-induced anomalous phase transformations in nanocrystalline FCC Fe-Ni based alloys. *Nanostruct. Mater.* **1997**, *9*, 681–684. [CrossRef]

52. Ohsaki, S.; Kato, S.; Tsuji, N.; Ohkubo, T.; Hono, K. Bulk mechanical alloying of Cu-Ag and Cu/Zr two-phase microstructures by accumulative roll-bonding process. *Acta Mater.* **2007**, *55*, 2885–2895. [CrossRef]

53. Sergueeva, A.V.; Song, C.; Valiev, R.Z.; Mukherjee, A.K. Structure and properties of amorphous and nanocrystalline NiTi prepared by severe plastic deformation and annealing. *Mater. Sci. Eng. A* **2003**, *339*, 159–165. [CrossRef]

54. Prokoshkin, S.D.; Khmelevskaya, I.Y.; Dobatkin, S.V.; Trubitsyna, I.B.; Tatyanin, E.V.; Stolyarov, V.V.; Prokofiev, E.A. Alloy composition, deformation temperature, pressure and post-deformation annealing effects in severely deformed Ti-Ni based shape memory alloys. *Acta Mater.* **2005**, *53*, 2703–2714. [CrossRef]

55. Sauvage, X.; Renaud, L.; Deconihout, B.; Blavette, D.; Ping, D.H.; Hono, K. Solid state amorphization in cold drawn Cu/Nb wires. *Acta Mater.* **2001**, *49*, 389–394. [CrossRef]

56. Miyazaki, T.; Terada, D.; Miyajima, Y.; Suryanarayana, C.; Murao, R.; Yokoyama, Y.; Sugiyama, K.; Umemoto, M.; Todaka, T.; Tsuji, N. Synthesis of non-equilibrium phases in immiscible metals mechanically mixed by high pressure torsion. *J. Mater. Sci.* **2011**, *46*, 4296–4301. [CrossRef]

57. Straumal, B.B.; Protasova, S.G.; Mazilkin, A.A.; Goering, E.; Schütz, G.; Straumal, P.B.; Baretzky, B. Ferromagnetic behaviour of ZnO: Role of grain boundaries. *Beilstein J. Nanotechnol.* **2016**, *7*, 1936–1947. [CrossRef] [PubMed]

58. Ivanisenko, Y.; Sauvage, X.; Mazilkin, A.; Kilmametov, A.; Beach, J.A.; Straumal, B.B. Bulk nanocrystalline ferrite stabilized through grain boundary carbon segregation. *Adv. Eng. Mater.* **2018**, *20*, 1800443. [CrossRef]

59. Straumal, B.B.; Mazilkin, A.A.; Protasova, S.G.; Dobatkin, S.V.; Rodin, A.O.; Baretzky, B.; Goll, D.; Schütz, G. Fe–C nanograined alloys obtained by high pressure torsion: Structure and magnetic properties. *Mater. Sci. Eng. A* **2009**, *503*, 185–189. [CrossRef]

60. Straumal, B.B.; Dobatkin, S.V.; Rodin, A.O.; Protasova, S.G.; Mazilkin, A.A.; Goll, D.; Baretzky, B. Structure and properties of nanograined Fe–C alloys after severe plastic deformation. *Adv. Eng. Mater.* **2011**, *13*, 463–469. [CrossRef]

61. Molodov, D.A.; Straumal, B.B.; Shvindlerman, L.S. The effect of pressure on migration of the [001] tilt grain boundaries in the tin bicrystals. *Scr. Metall.* **1984**, *18*, 207–211. [CrossRef]

62. Straumal, B.B.; Klinger, L.M.; Shvindlerman, L.S. The influence of pressure on indium diffusion along single tin–germanium interphase boundaries. *Scr. Metall.* **1983**, *17*, 275–279. [CrossRef]

63. Trofimov, E.A.; Lutfullin, R.Y.; Kashaev, R.M. Elastic modulus of TI-6AL-4V titanium alloy. *Lett. Mater.* **2015**, *5*, 67–69. [CrossRef]

64. Elmer, J.W.; Palmer, T.A.; Babu, S.S.; Specht, E.D. In situ observations of lattice expansion and transformation rates of α and rβ phases in Ti-6Al-4V. *Mater. Sci. Eng. A* **2005**, *391*, 104–113. [CrossRef]

65. Murzinova, M.A.; Zherebtsov, S.V.; Salishchev, G.A. Dependence of the specific energy of the β/α interface in the VT6 titanium alloy on the heating temperature in the interval 600–975 °C. *JETP* **2016**, *122*, 705–715. [CrossRef]

materials

MDPI

Article

Dissolution of Ag Precipitates in the Cu–8wt.%Ag Alloy Deformed by High Pressure Torsion

Anna Korneva [1],*, Boris Straumal [2,3,4], Askar Kilmametov [2,3], Robert Chulist [1], Grzegorz Cios [5], Brigitte Baretzky [3] and Paweł Zięba [1]

[1] Institute of Metallurgy and Materials Science, Polish Academy of Sciences, 25 Reymonta Street, 30-059 Krakow, Poland; r.chulist@imim.pl (R.C.); p.zieba@imim.pl (P.Z.)
[2] Institute of Solid State Physics and Scientific Center of RAS, Russian Academy of Sciences, Ac. OssipyanStr. 2, 142432 Chernogolovka, Russia; straumal@mf.mpg.de (B.S.); askar.kilmametov@kit.edu (A.K.)
[3] Karlsruhe Institute of Technology (KIT), Institute of Nanotechnology, Hermann-von-Helmholtz-Platz 1, 76344 Eggenstein-Leopoldshafen, Germany; brigitte.baretzky@kit.edu
[4] National University of Science and Technology «MISIS», Leniskijprosp. 4, 119049 Moscow, Russia
[5] AGH University of Science and Technology, Academic Centre for Materials and Nanotechnology, 30 Mickiewicza Av, 30-059 Krakow, Poland; grzegorz.cios@agh.edu.pl
*   Correspondence: a.korniewa@imim.pl; Tel.: +48-12-295-2867

Received: 10 January 2019; Accepted: 29 January 2019; Published: 1 February 2019

Abstract: The aim of this work was to study the influence of severe plastic deformation (SPD) on the dissolution of silver particles in Cu–8wt.%Ag alloys. In order to obtain different morphologies of silver particles, samples were annealed at 400, 500 and 600 °C. Subsequently, the material was subjected to high pressure torsion (HPT) at room temperature. By means of scanning and transmission electron microscopy, as well as X-ray diffraction techniques, it was found that during SPD, the dissolution of second phase was strongly affected by the morphology and volume fraction of the precipitates in the initial state. Small, heterogeneous precipitates of irregular shape dissolved more easily than those of large size, round-shaped and uniform composition. It was also found that HPT led to the increase of solubility limit of silver in the copper matrix as the result of dissolution of the second phase. This unusual phase transition is discussed with respect to diffusion activation energy and mixing enthalpy of the alloying elements.

Keywords: Cu–Ag alloy; high-pressure torsion; ultrafine microstructure; phase dissolution; microhardness

---

1. Introduction

Over the last decade, Cu–Ag alloys have attracted attention due to their high strength and high conductivity [1–3]. The mechanical and electrical properties of these alloys with low Ag content mainly depend on the volume fraction and distribution of Ag precipitates. Small Ag precipitates act as obstacles against the dislocation movement, increasing both strength and conductivity when Ag is extracted from the Cu matrix. It was found that applying cold drawing to the Cu–6wt.%Ag alloy resulted in a significant increase of strengthening effect from Ag precipitates (from 100 to 560 MPa) at the drawing strain equal to 6 [3]. In this process, fine Ag precipitates are produced which are elongated and evolve into filamentary structure along the drawing direction. It can be assumed that applying severe plastic deformation (SPD) with higher strain should further increase the strength of the material conserving its high conductivity.

It is well known that SPD allows producing bulk ultra-fine-grained materials with extraordinary mechanical properties, such as exceptionally high strength with considerable ductility at room temperatures or large super-plasticity at elevated temperatures [4–6]. Recently, it has been shown that the presence of the second phase (precipitates) additionally influences the grain refinement when the material is subjected to SPD [7,8]. This effect depends on morphology of the second phase (size, volume fraction, distribution). For instance, it has been shown in [8] that an extensive grain refinement of the Al–5.4%Mg–0.5%Mn–0.1%Zr alloy under equal-channel angular pressing (ECAP) was facilitated by a dispersion of Al_6Mn particles with an average size of 25 nm. They precipitated during the homogenization annealing at intermediate temperature. On the contrary, the formation of coarse recrystallized grains took place in this material after ECAP, when coarse Al_6Mn particles with a plate-like shape were present in the initial state. On the other hand, the second phase is also affected by SPD, because SPD frequently induces phase transformations [9–11]. For example, some phases precipitate or dissolve after SPD at room temperature [12–14], which usually should happen after heat treatment at higher temperatures. W. Huang [15] reported that the dissolution of the second phase occurring in SPD is an important phenomenon of phase transition. The dissolution of the second phase induced by SPD also depends on distribution and morphology of the particles. The influence of dissolved precipitated phases on the grain refinement and improvement of mechanical properties of severely deformed materials was reported in [16,17], however there is no systematic research about the effect of SPD on the behavior of particles, with respect to their morphology and properties. Therefore, the aim of this work is to study the influence of high-pressure torsion (HPT) on dissolution of the second phase and microhardness of the Cu–8wt.%Ag alloy characterized by different morphology of Ag precipitates.

2. Materials and Methods

The Cu–8wt.%Ag alloy was manufactured by induction melting in vacuum from the high-purity 5N5 Cu and 4N5 Ag. The resulting ingots with the diameter of 10 mm were cut by means of spark erosion into 0.7 mm thick disks. Three disks of the Cu–8wt.%Ag alloy were sealed separately into quartz ampoules (HMB Quarzglass, Helmenhorst, Germany) with the residual pressure of 4×10^{-4} Pa for annealing in a SUOL resistance furnace (Tula-Term, Tula, Russia) at 400 °C (1200 h), 500 °C (710 h) and 600 °C (86 h). After annealing, the samples were quenched in water (ampoules were broken). Afterwards, discs were subjected to HPT in a Bridgman anvil chamber (W. Klement GmbH, Lang, Austria) at a pressure of 5 GPa, five revolutions at the speed of 1 rpm at room temperature. In our experiment three discs for each state were deformed and then examined with X-ray diffraction (XRD, Siemens, München, Germany) technique. Since the obtained XRD-curves were identical for each state, one deformed disc was used for further microstructural investigations using scanning electron microscopy (SEM, FEI, Hillsboro, OR, USA) and transmission electron microscopy (TEM, TECNAI, Hillsboro, OR, USA) techniques. The samples for the structure studies were cut at the distance of 3 mm from the center of the deformed discs. For the metallographic investigations, the samples were ground with SiC grinding paper, and sequentially polished with 6, 3 and 1 μm diamond pastes. The detailed microstructure observation and electron backscatter diffraction (EBSD) analysis were carried out on a FEI Quanta 3D FEGSEM scanning electron microscope (SEM, FEI, Hillsboro, OR, USA) equipped with a field emission gun (FEG) and an energy-dispersive X-ray spectrometer (EDX) manufactured by EDAX (Mahwah, NJ, USA). The SEM images were taken using backscattered electron signal (BSE mode) in order to obtain the composition contrast between different phases. The EBSD maps combined with element maps were carried out on a sample plane perpendicular to the normal direction. The working voltage and current were set to 20 KV and 8 nA, respectively. The mappings were carried out in the beam-scanning mode with a step size of 50 nm. Crystallographic and microstructural features were analyzed with the TSL OIM software (version Analysis 7 EDAX, Mahwah, NJ, USA). In order to visualize orientation maps the so-called inverse pole figure (IPF) color coding was used. The details of the resulting microstructure components, especially in nanoscale, were revealed using a TECNAI G2 FEG super TWIN (200 kV) transmission electron microscope (TEM, TECNAI, Hillsboro, OR, USA) equipped with an energy dispersive X-ray (EDS) spectrometer manufactured by EDAX TECNAI (Mahwah, NJ; USA). Thin foils of the alloy for TEM observation were prepared by a twin-jet polishing technique using D2 electrolyte of Struers company (Ballerup, Denmark) with parameters recommended by the manufacturer. The X-ray diffraction patterns were obtained in the Bragg–Brentano geometry on a Philips X′Pert powder diffractometer with the use of Cu-$K\alpha$ radiation. Lattice parameters were evaluated by the Fityk software [18] using a Rietveld-like whole profile refinement. Relative amounts of Cu- and Ag-rich solid solutions were estimated from the integrated intensities. Pure polycrystalline copper was used as reference. The phases in the alloys were identified by comparing with the X′PertHighScorePANalytical phase database [19]. The hardness measurements were performed using an G200 nanoindenter (KLA-Tencor, Milpitas, CA, USA) with XP head with the indentation load of 1.96 mN at the distance of 3 mm from the center of the deformed disc. There were 100 microhardness measurements taken on each sample with 10×10 points with the measurement step of 3 μm. After that, each measurement was categorized as copper matrix or silver particle by means of SEM observations. The indents lying on the boundaries between matrix or particles were discarded. A minimum of ten measurements for both phases were taken into account. Low force was chosen to measure hardness of particles and matrix because the particles were small. The same force for both particles and matrix were used in order to compare them. The size of the indenter imprint was about 0.5 and 0.26 μm before and after HPT, respectively.

3. Results and Discussion

The Cu–Ag equilibrium phase diagram is presented in Figure 1a [20]. The vertical line shows the chemical composition of the examined alloy, while the dots show the annealing temperatures, i.e., 400, 500 and 600 °C. These temperatures correspond to the ($\alpha + \beta$) state, where the α-phase is the solid solution of Ag in Cu whereas the β-phase is the solid solution of Cu in Ag.

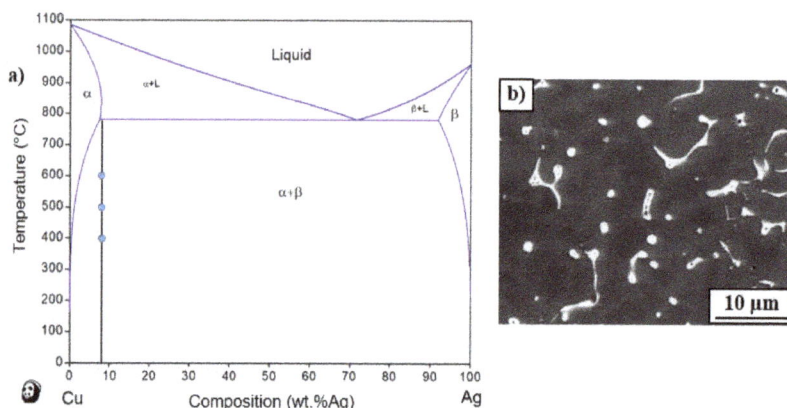

Figure 1. (a) phase diagram of Cu–Ag system [20] with vertical line showing chemical composition of the examined alloys, while dots show annealing temperatures of examined samples; (b) SEM image of the as-cast Cu–8wt.%Ag alloy.

The SEM microstructure observations show that the as-cast Cu–8wt.%Ag alloy contains the precipitates of ($\alpha + \beta$) eutectic with irregular shape and small rounded particles of the β-phase of about 1 μm size) uniformly distributed in the α-phase (Cu-matrix), Figure 1b. The eutectic and β particles were enriched with silver (from 11 to 55 wt.%). The chemical composition of as-cast Cu-matrix was inhomogeneous: the Ag content changed from 3.6 ± 0.5 to 6.7 ± 0.3 wt.%. Annealing the as-cast alloy at 400 °C resulted in the preservation of ($\alpha + \beta$) eutectic and β-phase particles with the average size of about 1.1 μm. The β-phase particles were observed within the grains and at the grain boundaries of the Cu-matrix (Figure 2a). Annealing at 400 °C resulted also in the discontinuous precipitation (DP) of a duplex structure containing a new fine β-phase and the solute depleted initial α-phase (see SEM image in Figure 2a and TEM images in Figure 3a,b). The DP phenomenon was frequently observed in Cu-based alloys [21,22] including Cu–Ag alloys with low Ag content [3]. A similar microstructure of DP was observed in the as-cast Cu–6wt.%Ag alloy after ageing at 450 °C for 32 h [3]. The DP process was controlled by the diffusion at the moving reaction front between the supersaturated Cu-matrix and forming (α + β) lamellae. The moving reaction front corresponded to the high-angle grain boundary. The selected area electron diffraction (SAED) pattern (Figure 3c) of the bright field image presented in Figure 3b confirms that the observed duplex structure is related to fine β precipitates with the thickness of about 50 nm in the depleted initial Cu-matrix. The measurement of chemical composition by means of EDS in TEM showed that the silver content in the primary β particles remained after casting, (see Figure 3a) was about 87 wt.%Ag, in the fine β precipitates around 30 wt.%Ag, while in the Cu-matrix it was 0.9 wt.%Ag.

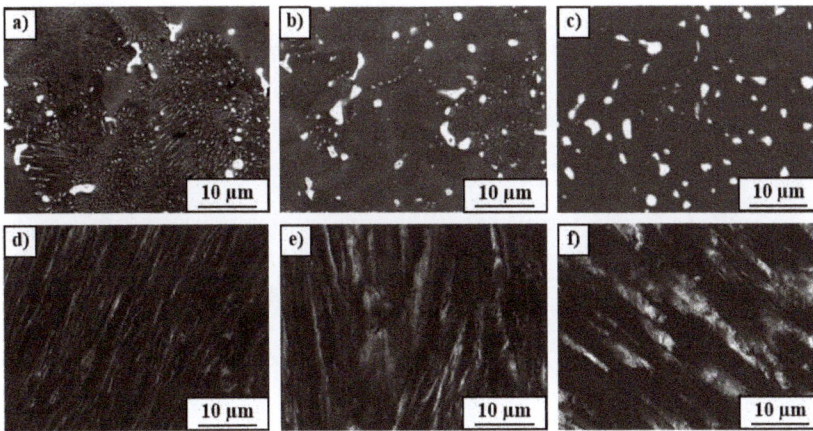

Figure 2. SEM images of the Cu–8wt.%Ag alloy after annealing at (**a**) 400 °C; (**b**) 500 °C; (**c**) 600 °C and after HPT deformation: (**d**) 400 °C + HPT; (**e**) 500 °C + HPT; (**f**) 600 °C + HPT.

Figure 3. (**a**,**b**) Bright field TEM images of the Cu–8wt.%Ag alloy after annealing at 400 °C and (**c**) SAED pattern taken from image (**b**).

The microstructure of the Cu–8wt.%Ag alloy after annealing at 500 °C (Figure 2b) also contained the eutectic, primary β-phase particles and fine β-phase lamellae from the discontinuous precipitation, however the DP volume fraction is much lower than that in the sample annealed at 400 °C. The microstructure of the Cu–8wt.%Ag alloy after annealing at 600 °C differs significantly from that described above; only one type of homogeneous β-phase particles with the average sizes of about 1.4 µm were observed (Figure 2c). It is well known that discontinuous precipitation basically occurs at low temperatures when grain boundary diffusion dominates over bulk diffusion. Generally, the increase of temperature causes the increase of the bulk diffusion coefficient in comparison with the grain boundary one. This is because the activation enthalpy of bulk diffusion is nearly two times higher than that of grain boundary diffusion. Therefore, the increase of annealing temperature resulted in the reduction of discontinuous precipitation at 500 °C and its complete disappearance at 600 °C. The increase of bulk diffusion taking place through the volume of crystal grains at 600 °C explains also the transformation of (α + β) eutectic precipitates into the homogeneous β-phase particles. The measurement of volume fraction of the β-phase on the basis of X-ray diffraction (XRD) patterns showed that with increasing annealing temperature, the volume fraction of the β-phase decreases from about 8.6 to 6.3% (Table 1). This trend correlates with the lever rule for the calculation of the relative number of phases in a two-phase mixture in a binary alloy system. It should be noted that the amount of the β-phase measured in XRD patterns includes all β precipitates in the material; namely, primary ones, those from the eutectic and from the discontinuous precipitation. It should be also noted that according to the equilibrium Cu–Ag phase diagram, the higher the temperature of annealing, the higher the solubility of Ag in the Cu-matrix. The measurement of chemical composition by means of EDS/SEM confirmed that with increasing annealing temperature, the silver concentration in the Cu-matrix (α-phase) and in the coarse β-phase particles grew (Table 2).

Table 1. The volume fraction (%) of β-phase precipitates in the Cu–8wt.%Ag alloy after annealing at different temperatures *T* (°C) measured on the bases of XRD data.

T °C	Volume Fraction %		
	Before HPT	After HPT	Increment Δ
400	8.6	2.6	6.0
500	6.6	3.5	3.1
600	6.3	3.9	2.4

Table 2. Distribution of silver (wt.%) in the Cu–8wt.%Ag alloy before and after HPT process measured by EDS/SEM.

T °C	Before HPT		After HPT	
	Cu-Matrix	Coarse Ag Precipitates	Cu-Matrix	Coarse Ag Precipitates
400	4.0 ± 0.4	32.5 ± 0.6	5.6 ± 0.2	14.1 ± 0.7
500	4.7 ± 0.5	44.1 ± 0.9	5.4 ± 0.2	33.8 ± 0.7
600	5.0 ± 0.4	52.3 ± 1.0	5.1 ± 0.2	39.2 ± 0.8

The observation of microstructure of the Cu–8wt.%Ag alloy after HPT by means of TEM technique showed that HPT resulted in a strong grain refinement of Cu-matrix (down to about 400 nm) with high density of dislocations within the grains (see bright and dark field TEM images in Figure 4). The selected area electron diffraction (SAED) patterns (Figure 4c,f,i) also showed a strong grain refinement; a large number of spots formed almost continuous diffraction rings which indicates the occurrence of large quantity of small grains with different crystallographic orientation. The maximum number of spots in the diffraction rings were observed in the deformed sample after prior annealing at 600 °C. It seems that the higher temperature of preliminary annealing, the stronger grain refinement of Cu-matrix. However, the EBSD measurements and XRD diffraction analysis with much better statistics showed the reverse dependencies of grain refinement on temperature of preliminary annealing (see this in the subsequent part of manuscript). A close examination of diffraction rings (Figure 4c,f,i) also showed the presence of small Ag particles in all deformed samples. Most likely, they are related to the fragmentation and refinement of β-phase precipitates (enriched by Ag) due to the HPT deformation. The analysis of their behavior under HPT was additionally carried out using SEM. The SEM observations of deformed samples of the Cu–8wt.%Ag alloy showed blurring of the (α + β) eutectic and β-phase precipitates along slip lines formed during the HPT deformation (Figure 2d–f).

The lower the temperature of preliminary annealing, the more blurry precipitate boundaries could be observed. Heterogeneity of chemical composition and irregular shape of eutectic and the presence of fine discontinuous precipitation may be the reason that the β-phase observed in the initial samples annealed at 400 and 500 °C dissolved easier and had more blurry boundaries after the HPT. Moreover, the primary precipitates and eutectics in the samples annealed at 400 and 500 °C showed slightly lower hardness (Table 3) than precipitates that appeared after annealing at 600 °C, which contained more Ag content (Table 2) and were characterized by a rounded shape and uniform composition.

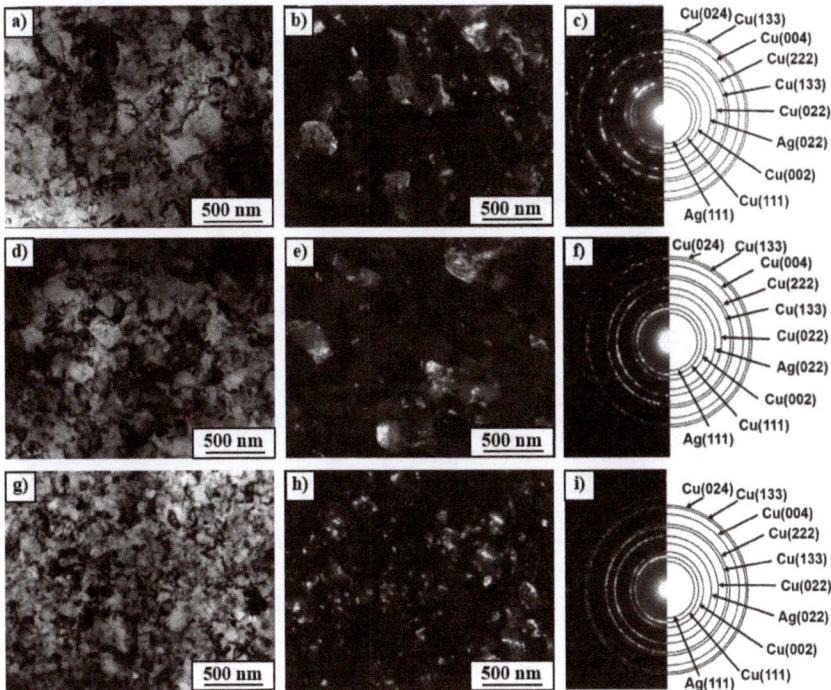

Figure 4. TEM images of the Cu–8wt.%Ag alloy after annealing and HPT: (a–c) 400 °C + HPT; (d–f) 500 °C + HPT; (g–i) 600 °C + HPT; (a,d,g) bright field images; (b,e,h) dark field images; (c,f,i) SAED patterns taken from (b,e,h).

Figure 5 presents XRD patterns of the Cu–8wt.%Ag alloy after annealing and after the HPT deformation. The Ag and Cu peaks were observed in all the samples after annealing. The Cu peaks (α-phase) observed at higher diffraction angles (137 and 144°) were split, which indicated low internal stresses and larger grain sizes in the annealed samples. The significant broadening of Cu peaks was observed after HPT suggesting a strong grain refinement and significant micro-distortion of the crystal lattice. A slight shift of Cu peaks to lower diffraction angles after HPT was observed as well. This corresponds to the increase of Cu lattice parameters due to migration of Ag atoms from the β-phase (Ag particles) into the Cu-matrix. At the same time the (111) reflection of the β-phase became smaller and more broadened (in the samples annealed at 500 and 600 °C prior to deformation), while the remaining β-phase peaks disappeared or became blurred and undetectable. The disappearance of most β-phase reflections after HPT indicates the partial dissolution of this phase.

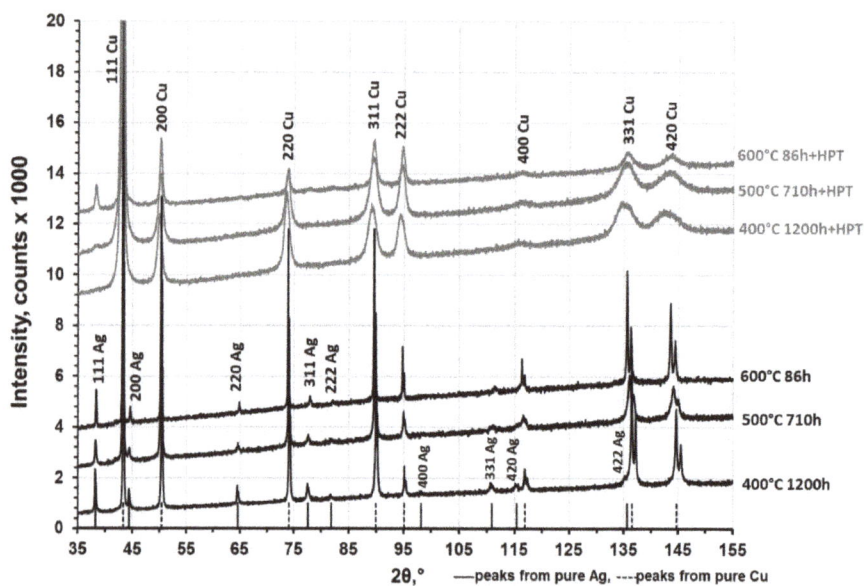

Figure 5. X-ray diffraction patterns of the Cu–8wt.%Ag alloy after annealing at different temperatures before and after HPT.

Figure 6a shows an enlarged part of the X-ray diffraction patterns of the deformed samples preliminarily annealed at 400 and 600 °C. It is easy to note that maximum broadening and maximum shift of (111) Cu peak is observed in the sample annealed at 400 °C. Additionally the (111) Ag peak disappeared after HPT in the sample. It confirms that the maximum grain refinement and maximum dissolution of β-phase takes place there.

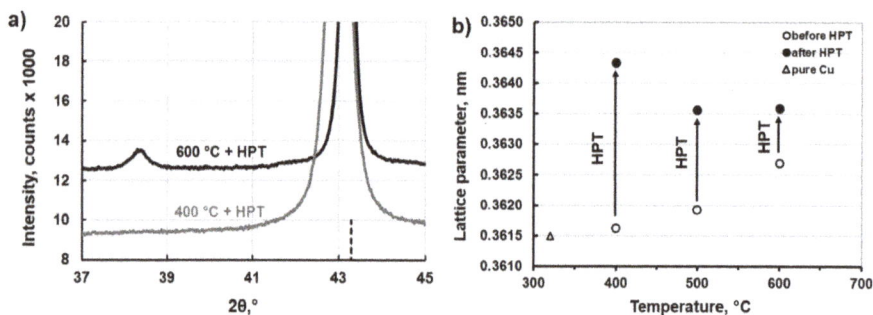

Figure 6. (a) Enlarged part of X-ray diffraction patterns of samples of the Cu–8wt.%Ag alloy after HPT and preliminary annealed at 400 and 600 °C. Vertical dotted line shows the (111) peak position of pure copper; (b) Changes of lattice parameter of Cu-matrix induced by HPT in samples preliminarily annealed at 400, 500 and 600 °C.

The changes of lattice parameter of Cu-matrix before and after HPT are graphically presented in Figure 6b. Generally, doping copper with silver leads to an increase in the lattice parameter of copper [23,24]. One can see from Figure 6b that along with the increase of annealing temperature, the lattice parameter increases, because the solubility of Ag in Cu increases alongside the solvus line on the phase Cu–Ag diagram. HPT resulted in the increase of the lattice parameters values, wherein their maximum increase is observed in the sample preliminary annealed at 400 °C. These results show a good correlation with the data presented in Table 1, i.e., the largest amount of dissolved β-phase under action of HPT process is observed in the sample pre-annealed at 400 °C, while the smallest one in the sample pre-annealed at 600 °C. Therefore, it can be concluded that the lower the annealing temperature, the greater the enrichment of matrix in Ag after HPT. It is a result of partial dissolution of the primary β-phase, (α + β) eutectic and probably the effect of complete dissolution of the fine discontinuous precipitates. It should be also noted, that the lattice parameter of Cu-matrix obtained after HPT in the samples pre-annealed at 400, 500 and 600 °C reached 0.36433, 0.36356 and 0.36359 ± 0.00001 nm, respectively. According to the [24] database, a slightly smaller lattice parameter of 0.36421 nm in the sample pre-annealed at 400 °C, corresponds to the Cu–5.82 at.%Ag alloy obtained after homogenizing at 760 °C for 48 days and 780 °C for 70 days. The composition of 5.82 at.%Ag corresponds to 9.5 wt.%Ag, while 8 wt.%Ag is the maximum amount of Ag which can be dissolved in Cu at eutectic temperature 780 °C. In other words, HPT led to a greater dissolution of Ag than the possible maximum solubility of Ag in the Cu–8wt.%Ag alloy obtained after prolonged annealing at 780 °C. This effect of increasing solubility limit of Ag induced by HPT was not observed in the other deformed samples (pre-annealed at 500 and 600 °C) of the examined alloy.

The phenomenon of anomalous phase transitions, i.e., dissolution of the second phase under the influence of SPD, was observed in many works [9,25] and systematically studied in Fe-based alloys [26,27]. For example, it was shown that the C, N and Niwas transferred from the second phase particles into the iron solid solution as a result of SPD, wherein the concentration of these elements could exceed the equilibrium saturation concentrations [26]. The authors presented the following scheme of such a dissolution. At the first stage of the SPD the particles were refined, at a more developed stage of deformation the contribution of "non-crystallographic" deformation mechanisms increased, accompanied by an intense generation of vacancies. Then the separation and drift of atoms from particles in the field of edge dislocations became more efficient. Such phase transitions are also characterized by anomalous rapid diffusion, which can be explained by the generation of excess vacancies and grain boundary diffusion in ultra-fine-grained materials with a large fraction of high angle grain boundaries, and it happened despite the fact that the applied pressure additionally slowed down the diffusion [28,29]. It should be noted that the HPT-induced mass transfer led not only to the increase of the solubility limit, but also to obtaining nanocomposites from mixtures of micro-powders even for immiscible systems [30].

Figure 7 shows SEM/EBSD orientation maps (a–c) and corresponding chemical composition maps (e–g) of the deformed alloy. The character of silver distribution in the form of elongated coarse β-phase particles well correlates with the microstructure observed in Figure 2d–f. The average grain sizes of deformed Cu-matrix (measured on the basis of orientation topography maps) reached 0.35 ± 0.12, 0.44 ± 0.16 and 0.48 ± 0.17 μm, respectively for the samples preliminary annealed at 400, 500 and 600 °C. It seems that the presence of the discontinuous precipitates (with the maximum fraction in the sample annealed at 400 °C) in the form of fine duplex structure additionally leads to the grain refinement of the microstructure during HPT. The statement that the presence of fine dispersed particles promotes the microstructure refinement during SPD has been confirmed in works [7,8]. For example, the nanosize Al₃(Sc,Zr) precipitates play an important role in the grain refinement of the Al–0.2Sc–0.1Zr alloy subjected to the process of accumulative continuous extrusion forming [8]. The grain size of the alloy dramatically refined from 100 μm to 800 nm through continuous dynamic recrystallization (CDRX). The nanosize particles promoted grain refinement through three mechanisms: the precipitates (i) facilitated retention of high dislocation density in the alloy by enabling the generation of dislocation and pinning dislocation slip, which increased the driving force for CDRX; (ii) promoted the formation of deformation bands, providing sites for activation of CDRX, and (iii) activated CDRX near the grain boundary. The texture of HPT processed samples of the Cu–8wt.%Ag alloy did not show significant differences in intensity and component position. All samples yielded a typical shear texture with A,

B, C components of fcc metals. It strongly suggested that the discussed particles did not change the deformation mechanism during the HPT process.

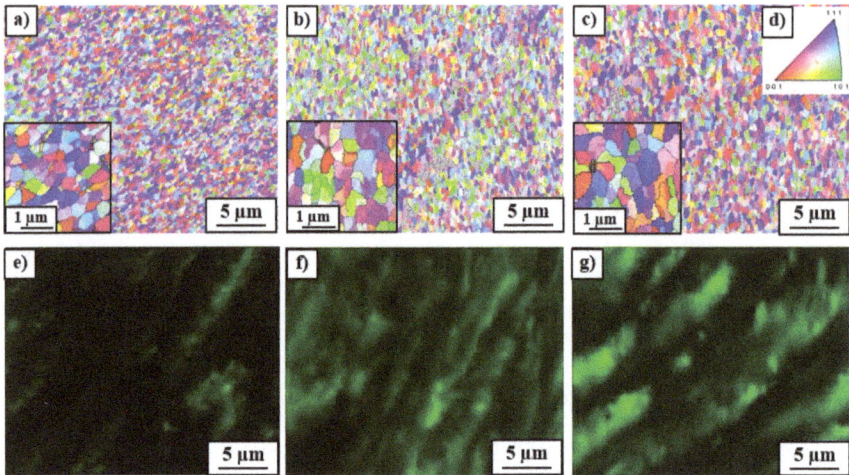

Figure 7. (a–c) SEM/EBSD orientation maps and (e–g) corresponding Ag distribution (green color) in Cu-matrix (black color) of the Cu–8wt.%Ag alloy after HPT, preliminary annealed at (a,e) 400 °C, (b,f) 500 °C and (c,g) 600 °C together with (d) standard unit triangle for fcc α-phase. The inserts inside (a–c) display enlarged areas of the EBSD maps with high angle grain boundaries (HAGBs) larger than 10°.

The results of microhardness measurements are presented in Table 3. The Cu and Ag-solid solutions have the same fcc lattice and the hardness values of Cu-matrix and Ag precipitates are also similar reaching about 134 H_V before HPT. After HPT only, the microhardness of the sample pre-annealed at 600 °C was measured, in which the Cu-matrix did not include fine Ag particles resulting from discontinuous precipitation. HPT led to the increase of microhardness of both phases to the value of about 310 H_V. The increase of microhardness was associated with an increase in the crystal structure defect density and with the increase of high-angle grain boundary fraction as a result of grain refinement induced by HPT.

Table 3. Microhardness of Cu-matrix and coarse Ag precipitates in the Cu–8wt.%Ag alloy before and after HPT process.

Treatment	Microhardness H_V	
	Cu-Matrix	Ag Precipitates
400 °C	132 ± 8	134 ± 6
500 °C	131 ± 5	128 ± 10
600 °C	136 ± 8	142 ± 8
600 °C + HPT	311 ± 20	310 ± 32

Based on results obtained in Ref. [25], the effect of SPD on phase transformations was established to depend strongly on diffusion activation energy and mixing enthalpy of the alloying elements. The higher the activation enthalpy of diffusion, the lower the diffusion relaxation of crystallographic defects formed by SPD. This, in turn, promoted grain refinement and phase transitions (such as dissolution of the second phase or the decomposition of supersaturated solid solution). For example, it was shown that the higher the activation enthalpy of diffusion of the second component (like

Ag, Co, Hf, Cr in the Cu-based alloys), the higher the as-called effective temperature (T_{eff}), which corresponded to a new position of the deformed alloy in the equilibrium phase diagram [25]. In other words, the phases forming during SPD at ambient temperature can also appear after long annealing at a certain elevated temperature T_{eff} with subsequent quenching. However, if the real annealing at the elevated temperature led to grain growth, the SPD resulted in the grain refinement of microstructure.

The effect of SPD on phase transformation of compounds with a positive or negative enthalpy of mixing was different. Due to strong interatomic bonds such as compounds with a negative enthalpy of mixing (such as Cu–Sn, Cu–In), the process of dissolution of second phase practically did not occur [31,32] compared with compounds of positive mixing enthalpy (Cu–Ag, Cu–Ni, Cu–Co). For example, in as-cast Cu–36wt.%Sn alloy which was subjected to HPT at room temperature [31], the microstructure of the alloy before HPT, contained alternating coarse-grained or even single-crystalline plates of the hard intermetallic ζ ($Cu_{10}Sn_3$) and ε (Cu_3Sn) phases. After HPT, neither dissolution of phases, nor phase transformations were observed. Only slight grain refinement took place inside the ζ and ε plates, however, the shape of alternating plates remained unchanged.

4. Conclusions

The microstructure of the Cu–8wt.%Ag alloy after annealing at 400 and 500 °C contained coarse (α + β) eutectic, coarse β-phase precipitates and fine (α + β) duplex structure formed due to the discontinuous precipitation. The volume fraction of discontinuous precipitates after annealing at 500 °C was smaller than that after annealing at 400 °C. Annealing of the alloy at 600 °C resulted in the transformation of eutectic precipitates into coarse homogeneous β-phase particles. The higher the temperature of annealing, the lower volume fraction of β-phase precipitates and the higher content of Ag in the Cu-matrix (α-phase).

Applying HPT to the Cu–8wt.%Ag alloy resulted in: (1) strong grain refinement of the Cu-matrix (down to about 400 nm); (2) partial dissolution of coarse (α + β) eutectic and coarse β-phase particles, their fragmentation and refinement; (3) dissolution of fine discontinuous precipitates; and (4) an increase of solubility limit of Ag in the Cu-matrix in the sample pre-annealed at 400 °C.

The maximum HPT effect on grain refinement of the Cu-matrix and the dissolution of β-phase was observed in the sample preliminary annealed at 400 °C due to the fact that the initial state of this sample was characterized by: (1) the largest volume fraction of dispersed duplex structure; (2) the presence of a slightly softer (α + β) eutectic in comparison with coarse and homogeneous Ag particles; and (3) the lowest solubility of Ag in Cu-matrix. Specifically, small, heterogeneous precipitates of the irregular shape dissolved more easily than those of large sizes, rounded shape and uniform composition.

Author Contributions: Conceptualization, A.K. (Anna Korneva) and B.S.; Data Curation, A.K. (Anna Korneva) and A.K. (Askar Kilmametov); Formal Analysis, A.K. (Anna Korneva) and A.K. (Askar Kilmametov); Investigation, A.K. (Anna Korneva), A.K. (Askar Kilmametov), R.C. and G.C.; Methodology, A.K. (Anna Korneva); Supervision, A.K. (Anna Korneva); Writing—Original Draft Preparation, A.K. (Anna Korneva); Writing—Review and Editing, A.K. (Anna Korneva), P.Z., B.S. and R.C.; Funding Acquisition, P.Z., B.S. and B.B.

Funding: This research was funded by the National Science Centre of Poland (grant OPUS number 2014/13/B/ST8/04247) and Deutsche Forschungsgemeinsgaft (AB 1768). All the research was performed within the Accredited Testing Laboratories possessing the certificate No. AB 120 issued by the Polish Centre of Accreditation. This means that all the research is performed according to European standard PN-ISO/IEC 17025:2005 as well as the EA-2/15. B.S. acknowledges the support of Ministry of Education and Science of the Russian Federation in the framework of the Program to Increase the Competitiveness of NUST "MISiS".

References

1. Sakai, Y.; Schneider-Muntau, H.J. Ultra-high strength high conductivity Cu-Ag alloy wires. *Acta Mater.* **1997**, *45*, 1017–1023. [CrossRef]

2. Tian, Y.Z.; Wu, S.D.; Zhang, Z.F.; Figueiredo, R.B.; Gao, N.; Langdon, T.G. Microstructural evolution and mechanical properties of a two-phase Cu–Ag alloy processed by high-pressure torsion to ultrahigh strains. *Acta Mater.* **2011**, *59*, 2783–2796. [CrossRef]

3. Bao, G.; Xu, Y.; Huang, L.; Lu, X.; Zhang, L.; Fang, Y.; Meng, L.; Liu, J. Strengthening Effect of Ag Precipitates in Cu–Ag Alloys: A Quantitative Approach. *Mater. Res. Lett.* **2016**, *1*, 37–42. [CrossRef]

4. Valiev, R.Z. Approach to nanostructured solids through the studies of submicron grained policrystals. *Nanostruct. Mater.* **1995**, *6*, 73–82. [CrossRef]

5. Krasilnikov, N.; Lojkowski, W.; Pakiela, Z.; Valiev, R. Tensile strength and ductility of ultra-fine-grained nickel processed by severe plastic deformation. *Mater. Sci. Eng. A* **2005**, *397*, 330–337. [CrossRef]

6. Valiev, R.Z.; Islamgaliev, R.K.; Alexandrov, I.V. Bulk nanostructured materials from severe plastic deformation. *Prog. Mater. Sci.* **2000**, *45*, 103–189. [CrossRef]

7. Nikulin, I.; Kipelova, A.; Malopheyev, S.; Kaibyshev, R. Effect of second phase particles on grain refinement during equal-channel angular pressing of an Al–Mg–Mn alloy. *Acta Mater.* **2012**, *60*, 487–497. [CrossRef]

8. Shen, Y.F.; Guan, R.G.; Zhao, Z.Y.; Misra, R.D.K. Ultrafine-grained Al–0.2Sc–0.1Zr alloy: The mechanistic contribution of nano-sized precipitates on grain refinement during the novel process of accumulative continuous extrusion. *Acta Mater.* **2015**, *100*, 247–255. [CrossRef]

9. Straumal, B.B.; Kilmametov, A.R.; Ivanisenko, Y.; Mazilkin, A.A.; Kogtenkova, O.A.; Kurmanaeva, L.; Korneva, A.; Zięba, P.; Baretzky, B. Phase transitions induced by severe plastic deformation: Steady-state and equifinality. *Int. J. Mater. Res.* **2015**, *106*, 657–664. [CrossRef]

10. Sauvage, X.; Chbihi, A.; Quelennec, X. Severe plastic deformation and phase transformations. *J. Phys. Conf. Ser.* **2010**, *240*, 012003. [CrossRef]

11. Glezer, A.M.; Sundeev, R.V. General view of severe plastic deformation in solid state. *Mater. Lett.* **2015**, *139*, 455–457. [CrossRef]

12. Sauvage, X.; Ivanisenko, Y. The role of carbon segregation on nanocrystallisation of pearlitic steels processed by severe plastic deformation. *J. Mater. Sci.* **2007**, *42*, 1615–1621. [CrossRef]

13. Straumal, B.B.; Protasova, S.G.; Mazilkin, A.A.; Rabkin, E.; Goll, D.; Schütz, G.; Baretzky, B.; Valiev, R.Z. Deformation-driven formation of equilibrium phases in the Cu–Ni alloys. *J. Mater. Sci.* **2011**, *47*, 360–367. [CrossRef]

14. Cepeda-Jiménez, C.M.; García-Infanta, J.M.; Zhilyaev, A.P.; Ruano, O.A.; Carreño, F. Influence of the thermal treatment on the deformation-induced precipitation of a hypoeutectic Al–7 wt% Si casting alloy deformed by high-pressure torsion. *J. Alloy. Comp.* **2011**, *509*, 636–643. [CrossRef]

15. Huang, W.; Liu, Z.; Xia, L.; Xia, P.; Zeng, S. Severe plastic deformation-induced dissolution of θ'' particles in Al–Cu binary alloy and subsequent nature aging behavior. *Mater. Sci. Eng. A* **2012**, *556*, 801–806. [CrossRef]

16. Hu, N.; Xu, X.C. Influence of dissolved precipitated phases on mechanical properties of severely deformed Al-4 wt% Cu alloys. *Mater. Sci. Forum* **2010**, *667–669*, 1021–1026. [CrossRef]

17. Gutierrez-Urrutia, I.; Munoz-Morris, M.A.; Morris, D.G. Influence of course second phase particles and fine precipitates on microstructural refinement during severe plastic deformation by ECAP and on structural stability during annealing. In Proceedings of the 4th International Symposium on Ultrafine Grained Materials IV, San Antonio, TX, USA, 12–16 March 2006; pp. 269–282.

18. Wojdyr, M.J. Fityk: A general-purpose peak fitting program. *J. Appl. Crystallogr.* **2010**, *43*, 1126–1128. [CrossRef]

19. Smith, D. *ICDD Grant_in_Aid*; Penn State University: University Park, PA, USA, 1978; pp. 22–212.

20. Dinsdale, A.T.; Kroupa, A.; Vízdal, J.; Vrestal, J.; Watson, A.; Zemanova, A. *COST531 Database for Lead-Free Solders*, Ver. 3.0. 2008; unpublished research.

21. Zięba, P.; Gust, W. Analytical electron microscopy of discontinuous solid state reactions. *Int. Mat. Rev.* **1998**, *43*, 70–97. [CrossRef]

22. Manna, I.; Pabi, S.K.; Gust, W. Discontinuous Reactions in Solids. *Int. Mater. Rev.* **2001**, *46*, 53–91. [CrossRef]

23. Landolt-Börnstein. *Phase Equilibria, Crystallographic and Thermodynamic Data of Binary Alloy*; New Series IV/12A, Supplement to IV/5A; Springer: Berlin/Heidelberg, Germany, 2006; pp. 1–4.

24. Subramanian, R.; Perepezko, J.H. The Ag-Cu (Silver-Copper) System. *J. Phase Equilib.* **1993**, *1*, 62–75. [CrossRef]

25. Straumal, B.B.; Kilmametov, A.R.; Korneva, A.; Mazilkin, A.A.; Straumal, P.B.; Zięba, P.; Baretzky, B. Phase transitions in Cu-based alloys under high pressure torsion. *J. Alloy. Comp.* **2017**, *707*, 20–26. [CrossRef]

26. Faizov, I.A.; Raab, G.I.; Faizova, S.N.; Aksenov, D.A.; Zaripov, N.G.; Gunderov, D.V.; Golubev, O.V. Rastvorenie chastits vtorykh faz splava Cu-Cr-Zr v usloviyakh ravnokanal'nogo uglovogo pressovaniya. *Fizika* **2016**, *21*, 1387–1390. (In Russian)
27. Shabashov, V.A. Neravnovesnye diffuzionnye fazovye prevrashcheniya i nanostrukturirovanie pri intensivnoy kholodnoy deformatsii. *Voprosy Materialovedeniya* **2008**, *3*, 169–179. (In Russian)
28. Molodov, D.A.; Straumal, B.B.; Shvindlerman, L.S. The effect of pressure on migration of the [001] tilt grain boundaries in the tin bicrystals. *Scr. Metall.* **1984**, *18*, 207–211. [CrossRef]
29. Straumal, B.B.; Klinger, L.M.; Shvindlerman, L.S. The influence of pressure on indium diffusion along single tin–germanium interphase boundaries. *Scr. Metall.* **1983**, *17*, 275–279. [CrossRef]
30. Kilmametov, A.; Kulagin, R.; Mazilkin, A.; Seils, S.; Boll, T.; Heilmaier, M.; Hahn, H. High-pressure torsion driven mechanical alloying of CoCrFeMnNi high entropy alloy. *Scr. Mater.* **2019**, *158*, 29–33. [CrossRef]
31. Korneva, A.; Straumal, B.; Chulist, R.; Kilmametov, A.; Cios, G.; Bała, P.; Schell, N.; Zięba, P. Grain refinement of intermetallic compounds in the Cu–Sn system under high pressure torsion. *Mater. Lett.* **2016**, *179*, 12–15. [CrossRef]
32. Korneva, A.; Straumal, B.; Kogtenkova, O.; Ivanisenko, Y.; Wierzbicka-Miernik, A.; Kilmametov, A.; Zieba, P. Microstructure evolution of Cu—22% In alloy subjected to the high pressure torsion. *IOP Conf. Ser. Mater. Sci. Eng.* **2014**, *63*, 012093. [CrossRef]

![materials logo]

materials

MDPI

Article

The Influence of Post-Weld Heat Treatment on the Microstructure and Fatigue Properties of Sc-Modified AA2519 Friction Stir-Welded Joint

Robert Kosturek *, Lucjan Śnieżek, Marcin Wachowski and Janusz Torzewski

Faculty of Mechanical Engineering, Military University of Technology, 2 gen. W. Urbanowicza str., 00-908 Warsaw, Poland; lucjan.sniezek@wat.edu.pl (L.Ś.); marcin.wachowski@wat.edu.pl (M.W.); janusz.torzewski@wat.edu.pl (J.T.)
* Correspondence: robert.kosturek@wat.edu.pl; Tel.: +48-261-839-245

Received: 4 January 2019; Accepted: 12 February 2019; Published: 15 February 2019

Abstract: The aim of this research was to investigate the influence of post-weld heat treatment (PWHT, precipitation hardening) on the microstructure and fatigue properties of an AA2519 joint obtained in a friction stir-welding process. The welding process was performed with three sets of parameters. One part of the obtained joints was investigated in the as-welded state and the second part of joints was subjected to the post-weld heat treatment (precipitation hardening) and then investigated. In order to establish the influence of the heat treatment on the microstructure of obtained joints both light and scanning electron microscopy observations were performed. Additionally, microhardness analysis for each sample was carried out. Fatigue properties of the samples in the as-welded state and the samples after post-weld heat treatment were established in a low-cycle fatigue test with constant true strain amplitude equal to $\varepsilon = 0.25\%$ and cycle asymmetry coefficient R = 0.1. Hysteresis loops together with changes of stress and plastic strain versus number of cycles are presented in this paper. The fatigue fracture in tested samples was analyzed with the use of scanning electron microscope. Our results show that post-weld heat treatment of AA2519 friction stir-welded joints significantly decreases their fatigue life.

Keywords: friction stir welding; heat treatment; AA2519; microstructure; fatigue; fractography

1. Introduction

High-strength aluminum alloys are very interesting engineering materials due to their high specific strength, forming abilities, good mechanical properties at low temperature and corrosion resistance [1–3]. Some of these alloys also present good ballistic resistance, which makes them especially attractive to the military, as well as, the space industry [1,4]. AA2519 is a heat treatable aluminum-copper alloy with copper content within the 5.3–6.4% range, used in the military for advanced amphibious assault vehicles (AAAV) [5]. The precipitation hardening of this alloy is realized by a two-step heat treatment—the solution treatment (annealing in 530 °C/2 h and cooling in cold water) and artificial aging (165 °C/10 h) [6–8]. After this process AA2519 alloy is strengthened by θ' precipitates, semi-coherent metastable Al_2Cu phase with body-centered tetragonal crystal structure, which increases its mechanical properties significantly [7]. Additionally the modification of AA2519 alloy used in this research consists of the addition of scandium and has been developed by the Institute of Non-Ferrous Metals, Light Metals Division in Skawina (Poland). Adding scandium to aluminum alloys affects their properties significantly by increasing their mechanical properties, recrystallization temperature and causes grain refinement due to presence of Al_3Sc precipitates [9–12]. Despite the advantages of AA2519 alloy, the high concentration of copper causes problems with its welding, especially hot cracking during solidification of the weld [13,14]. One of the most promising technologies for joining aluminum alloys (including AA2519) is friction stir-welding [15–22]. This solid-state welding process is based on friction between workpieces and the rotating tool, which generates heat leading to plasticizing of the material to be welded [15–17]. The movement of the rotating tool along the edges of two workpieces causes a mixing of the plasticized material, and as a result the creation of a joint between them [15,17,21]. The process of welding significantly influences the microstructure of joined alloy especially dissolution and coarsening of strengthening precipitates which causes a decrease of joint mechanical properties compared to the base material [21–24]. An approach worth considering to avoid this disadvantageous change in distribution of strengthening precipitates is to join the alloy in a non-strengthened state and then subject it to the post-welded heat treatment (PWHT) [15,23–25]. Most of the research concerned with friction stir welding of high-strength aluminum alloys is focused on basic mechanical properties of the joint established in tensile tests, omitting investigation of fatigue properties which are far more important in terms of application for construction of machines—particularly in the aerospace industry [2,18,19,21,22]. It has been reported that PWHT can increase the mechanical properties of high-strength aluminum alloys established during tensile tests, such as tensile strength and elongation [15,24]. Although the increase of joint efficiency of heat-treated welds is significant, the response of such a joint on the cyclic load is rarely the subject of investigations. The literature does not contain a research concerned with the influence of post-weld heat treatment of an AA2519 friction stir-welded joint on its fatigue properties. The present work is aimed at investigating the microstructure of AA2519 friction stir welded (FSW) joints in the as-welded state and after PWHT and to explore its influence on the low-cycle fatigue properties of welds.

2. Materials and Methods

The workpieces to be joined were 5 mm thick sheets made of AA2519-O alloy with chemical composition as presented in Table 1.

Table 1. Chemical composition of AA2519 alloy to be welded.

Si	Fe	Cu	Mg	Zn	Ti	V	Zr	Sc	Al
0.06	0.08	5.77	0.18	0.01	0.04	0.12	0.2	0.36	Base

The friction stir welding process has been performed by using ESAB FSW Legio 4UT machine (Military University of Technology, Warsaw, Poland) with the set of welding parameters presented in Table 2.

Table 2. Welding parameters and state for each sample with designation.

Sample Designation	Tool Rotation Speed (rpm)	Tool Traverse Speed (mm/min)	State of the Sample
400-100-HT0	400	100	As-welded
400-100-HT1	400	100	After PWHT
800-100-HT0	800	100	As-welded
800-100-HT1	800	100	After PWHT
800-200-HT0	800	200	As-welded
800-200-HT1	800	200	After PWHT

For each welding process axial force was equal to 17 kN and the tilt angle of Triflute type tool was set to 2°. From the welded workpiece there were cut two types of samples, designated as HT0 and HT1. The HT0 samples were investigated in the as-welded state, and HT1 samples were subjected to the post-welded precipitation hardening process and then investigated. The precipitation hardening process has been performed by solution treatment in 530 °C for 2 h and cooling in cold water, and then aging in 165 °C for 10 h. The welded joints were sectioned perpendicular to the welding direction where metallurgical examinations and hardness measurements were carried out. The very important matter in the friction stir welding process is that the material flow around the tool is not symmetrical. A friction stir-welded joint has its advancing side (AS, where the direction of the tool rotation is accordant with the welding direction) and retreating side (RS, where the direction of the tool rotation is opposed to the welding direction). In this study every cross-section of the joint subjected to the microstructure observations or microhardness analysis has a retreating side on its right side. In order to investigate the joint microstructure, the samples were examined using a digital light microscope (Olympus LEXT OLS 4100, Military University of Technology, Warsaw) and scanning electron microscope (Jeol JSM-6610, Military University of Technology, Warsaw,) with energy-dispersive x-ray spectroscopy (EDX) detector. The samples were etched by using Kroll reagent (20 mL H_2O + 5 mL HNO_3 + 3 drops of HF) with etching time equal to 15 s. The Vickers microhardness measurements of the polished cross sections were performed across the welds by applying a load of 0.1 kg. The top left corner of each sample has been set as "point zero" for the measurements. For each joint, microhardness distributions were prepared for the upper, middle and lower part of the cross section: 0.7 mm, 2.8 mm and 4.2 mm from the top respectively. Fatigue properties were established on an Instron 8802 Servohydraulic Fatigue Testing System (Military University of Technology, Warsaw, Poland) with constant true strain amplitude ε = 0.25%, cycle asymmetry coefficient R = 0.1 and frequency equal to f = 1 Hz. The strain during testing has been measured using a 2520–603 dynamic extensometer. The scheme of sample for fatigue testing is presented in Figure 1.

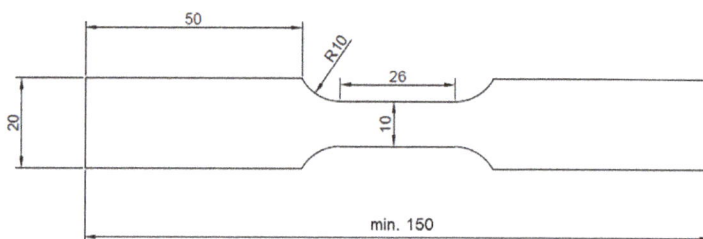

Figure 1. Scheme of sample for fatigue testing. All dimension are in mm.

3. Results

The joints obtained in the as-welded state (HT0 samples) have a typical for friction stir-welding process macrostructure consisting of a stir zone localized in the center of the joint, a thermo-mechanically affected zone, and a heat affected zone (Figure 2).

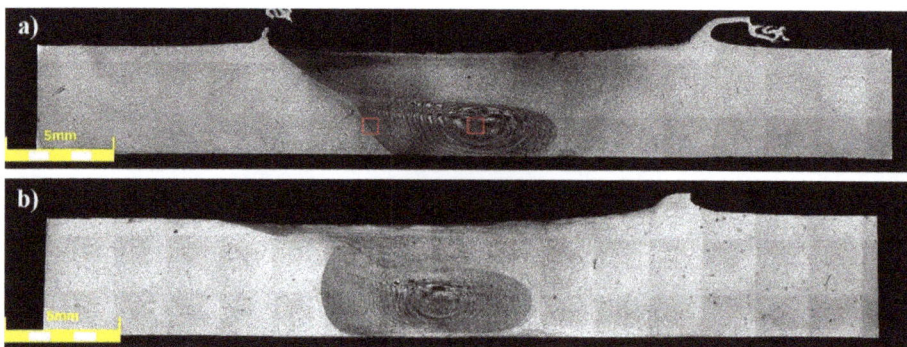

Figure 2. *Cont.*

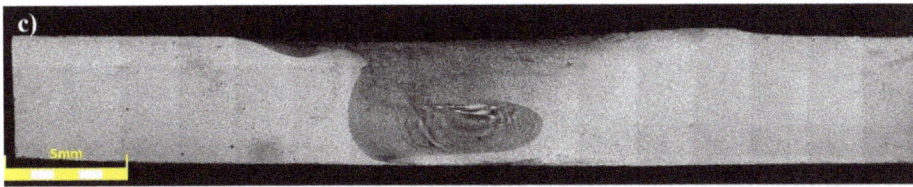

Figure 2. Light microscopy image of friction stir welded (FSW) joints macrostructure in the as-welded state: (**a**) 400-100-HT0, (**b**) 800-100-HT0, (**c**) 800-200-HT0.

Our results showed that the investigated joints are free of any imperfections, such as voids, cracks, kissing bound or tunneling effect. The shape of the stir zone for each sample differed due to different parameters of welding process. In case of joints welded with a tool rotation speed equal to 800 rpm it is possible to observe a much rounded shape on the advancing side of the stir zone compared to the angular shape of this area in 400-100-HT0 sample. The investigation on the macrostructure of joints 800-100-HT0 and 800-200-HT0 allowed us to conclude that size of the stir zone is closely related to the welding velocity. Sample 800-200-HT0 had a significant smaller stir zone than in the case of a joint which has been welded with a twofold lower welding velocity. The most noticeable differences concern the retreating side of the joint, as well as, the lower part of the joint. The light microscopy observations of the stir zone of 400-100-HT0 sample revealed the microstructure characterized by the presence of fine, dynamically recrystallized grains with their size about 5 μm (Figure 3a). The boundary between stir zone and thermo-mechanically affected zone in this sample was the subject of further investigation and its light microscopy image is presented in Figure 3b.

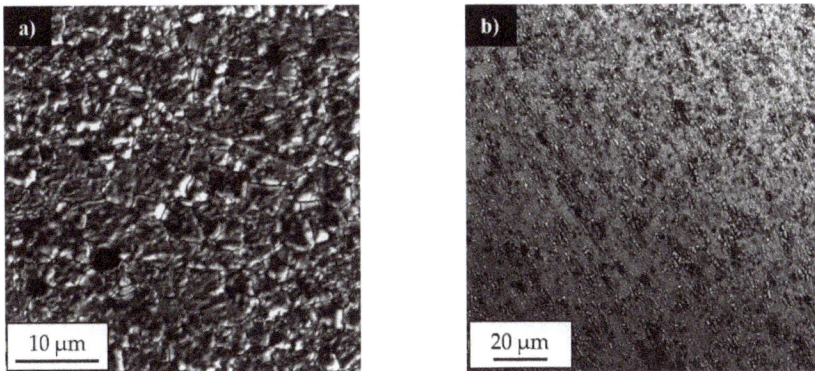

Figure 3. Light microscopy image of 400-100-HT0 sample microstructure in: (**a**) stir zone, (**b**) boundary between stir zone and thermo-mechanically affected zone. (Red marked areas in Figure 2a).

The light microscopy observations of this zone allowed us to observe a very specific boundary between deformed, elongated grains of thermo-mechanically affected zone and equiaxial, fine grains of the stir zone formed due to a dynamic recrystallization process. The stir zone microstructures of the joints obtained with tool rotation speed equal to 800 rpm are presented in Figure 4a,b.

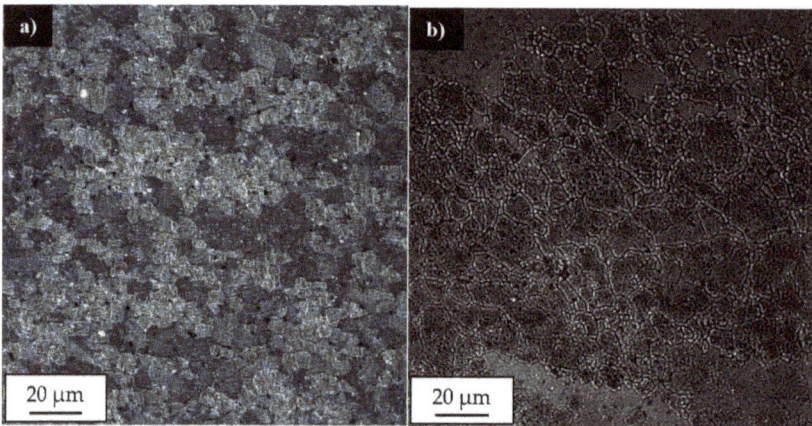

Figure 4. Light microscopy image of the stir zone microstructure in: (**a**) 800-100-HT0 sample, (**b**) 800-200-HT0 sample.

Although the stir zones were formed during friction stir welding due to dynamic recrystallization process it is possible to observe the differences between their grainy microstructures. In case of joint in 800-100-HT0 sample the microstructure of stir zone is more homogeneous than in 800-200-HT0 and its grains size about 5 μm is comparable to 400-100-HT0 sample. The stir zone microstructure of 800-200-HT0 is far more heterogeneous and it is possible to observe grains of 2–3 μm and the far larger ones with their size about 15–20 μm. Additionally, the scanning electron microscopy observations revealed the differences in concentration of the alloying elements between the thermo-mechanically affected zone and the stir zone. This phenomenon is most noticeable in case of 800-100-HT0 sample which has the highest ratio of tool rotation speed to tool traverse speed which results in the longest time affecting the workpiece material by the rotating tool (Figure 5a). The brighter image of the stir zone suggests a higher concentration of the elements heavier than aluminum in this area. It is also possible to observe lower participation of the large Al_2Cu precipitates in the stir zone compared to the thermo-mechanically affected zone. This observation allowed us to draw a conclusion that Al_2Cu precipitates dissolve in the stir zone during the friction stir-welding process. The results of the linear analysis of the chemical composition indicates decreasing participation of the aluminum in the stir zone (Figure 5b). At the same time, the fluctuations of aluminum and copper concentrations occurring due to the presence of Al_2Cu precipitates disappears.

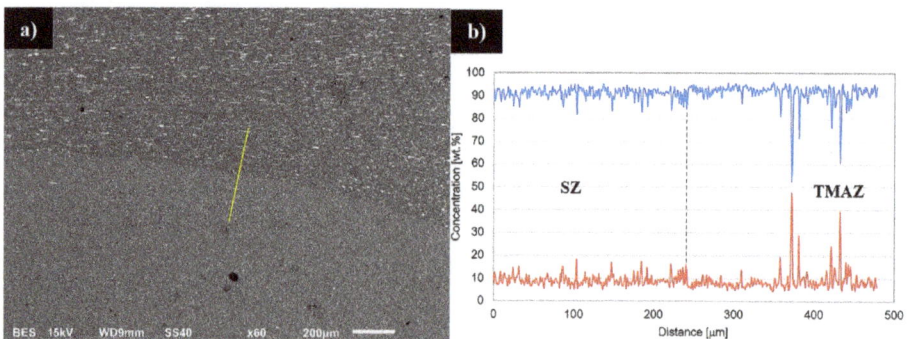

Figure 5. Scanning electron microscopy image of the boundary between stir zone and thermo-mechanically affected zone in 800-100-HT sample (**a**) with linear analysis of the chemical composition; (**b**) (yellow marker). Lines designation: Al (blue), Cu (red).

The scanning electron microscopy observations of the stir zone in the 800-100-HT0 sample allowed us to observe a fine dispersion of precipitates with its size about 1 μm, which have not been dissolved due to friction of the stir-welding process in this area (Figure 6a).

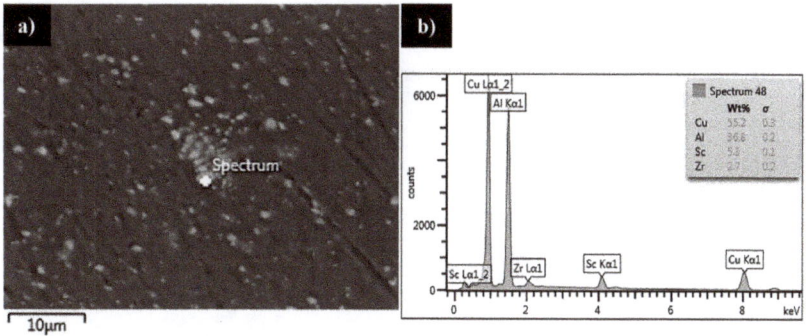

Figure 6. Scanning electron microscopy image of the precipitates in the stir zone (**a**), EDX area analysis of the chemical composition of the precipitate (**b**).

In order to establish the chemical composition of the precipitates occurring in the stir zone the EDX area analysis of the chemical composition was performed (Figure 6b). The results indicate a high concentration of zirconium and scandium in the precipitates which were not dissolved in the stir zone. It suggests the low solubility during severe plastic deformation in the elevated temperature of the precipitates rich in scandium and zirconium. The results of microhardness analysis of the samples in the as-welded state (HT0) are presented at Figures 7–9.

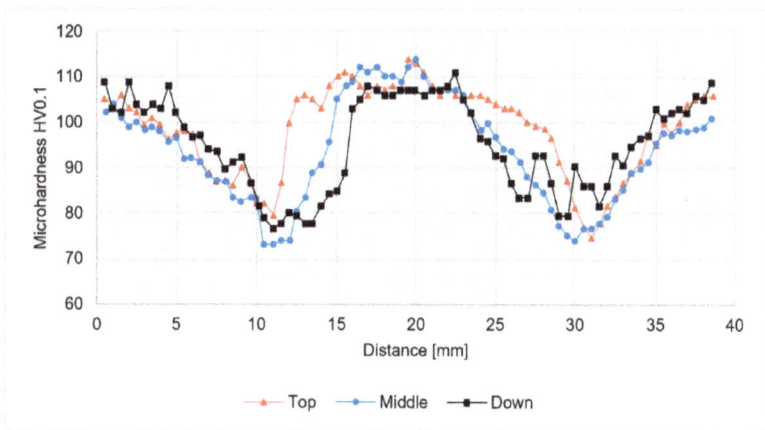

Figure 7. The results of 400-100-HT0 joint microhardness analysis.

The analysis of microhardness distribution at the cross-section of the joints allowed us to notice the increase of microhardness in the stir zone due to the occurence of fine, dynamically recrystallized grains in this area. The most significant increase was observed for the samples obtained using 800 rpm tool rotation speed (Figures 8 and 9). In case of 400-100-HT0 sample the microhardness of the stir zone is about 5-10 HV0.1 higher compared to the microhardness of the base material (Figure 7). At the same time, the increase of the microhardness in the stir zone for 800-100-HT0 and 800-200-HT0 samples was estimated to value 10-20 HV0.1 (Figures 8 and 9). The low hardness zone was localized at the boundary between thermo-mechanically affected zone and heat-affected zone and for each sample the lowest value in this area was noticed for the middle path of the measurements. The formation of this specific zone was related to the fact that in this area occurs the highest ratio of the heat input to the plastic deformation, which results in grain growth which is not compensated for by strain hardening. The lowest values of the hardness in the low hardness zone have been registered for 400-100-HT0 and 800-100-HT0 samples and are equal to 73 HV0.1 and 73.9 HV0.1 respectively (Figures 7 and 8). In the case of the 800-200-HT0 sample the lowest hardness is equal to 76 HV0.1 (Figure 9). The macrostructures of the joints subjected to the post-weld heat treatment (precipitation hardening) are presented at Figure 10.

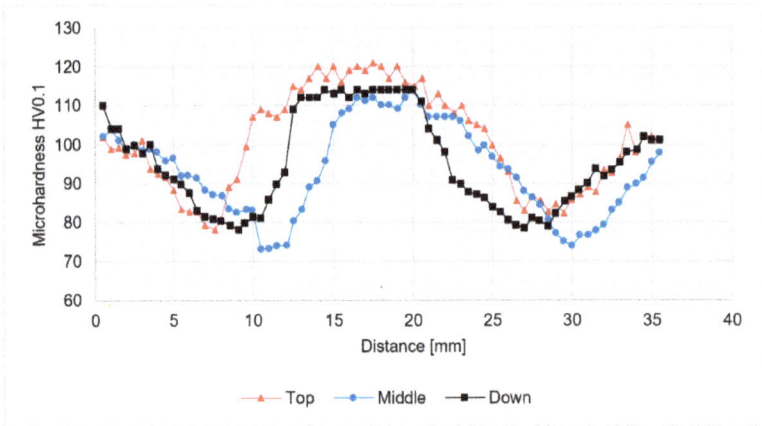

Figure 8. The results of 800-100-HT0 joint microhardness analysis.

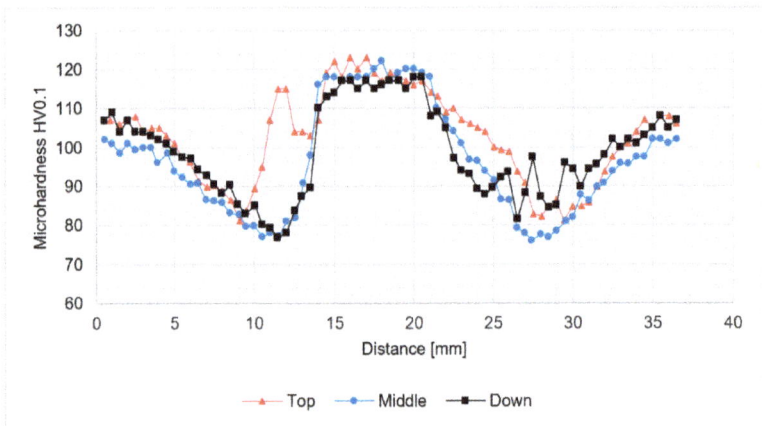

Figure 9. The results of 800-200-HT0 joint microhardness analysis.

The post-weld heat treatment changed the macrostructure of the joints significantly. Our results show that in the stir zone and the upper part of the thermo-mechanically affected zone the grainy microstructure suffered abnormal grain growth. This phenomenon occurred in each investigated sample, despite the welding parameters used. The grains in the stir zone undergo the greatest overgrowth and as a result their size has changed from the initial value in the as-welded state of 5 μm to the size even of 1–2 mm. The very specific structure of the grains in the upper part of the thermo-mechanically affected zone seemed to maintain texture which was formed due to the friction stir-welding process. The second important result revealed during light microscopy observations is the presence of the pores in the microstructure of joints subjected to the post-weld heat treatment (Figure 11).

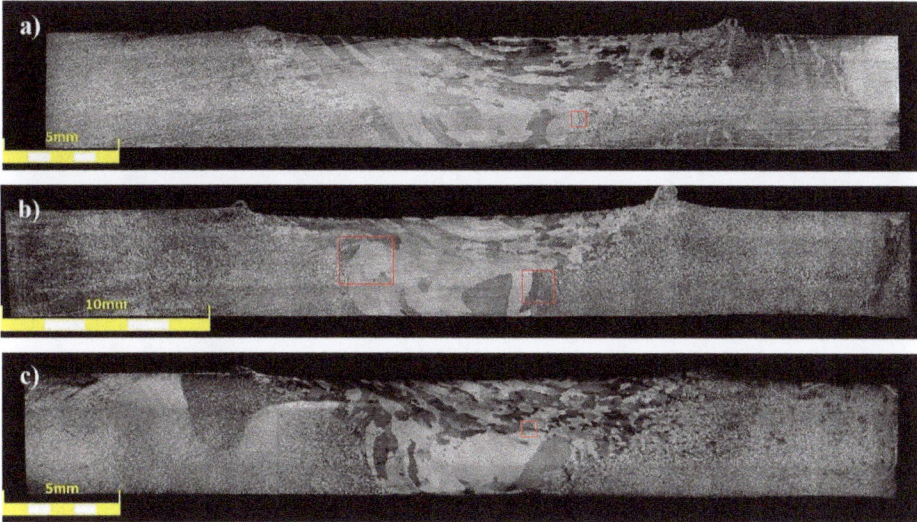

Figure 10. Light microscopy image of macrostructure of FSW joints subjected to the post-weld heat treatment: (**a**) 400-100-HT1, (**b**) 800-100-HT1, (**c**) 800-200-HT1.

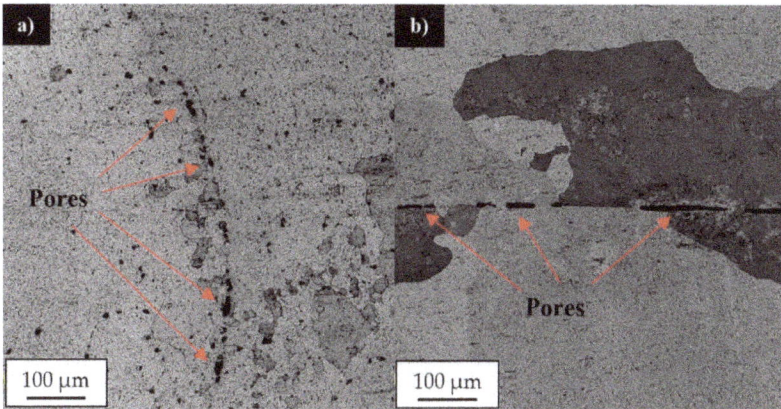

Figure 11. Light microscopy image of pores in the microstructure of: (**a**) 400-100-HT1 sample, (**b**) 800-200-HT1 sample. (Red marked areas in Figure 10a,c).

The pores were mainly localized at the boundary line between the stir zone and thermo-mechanically affected zone of the samples in the as-welded state. The pores have not been found in the base material after heat treatment. The sample 800-100-HT1 is characterized by occurrence of the smallest pores (Figure 12).

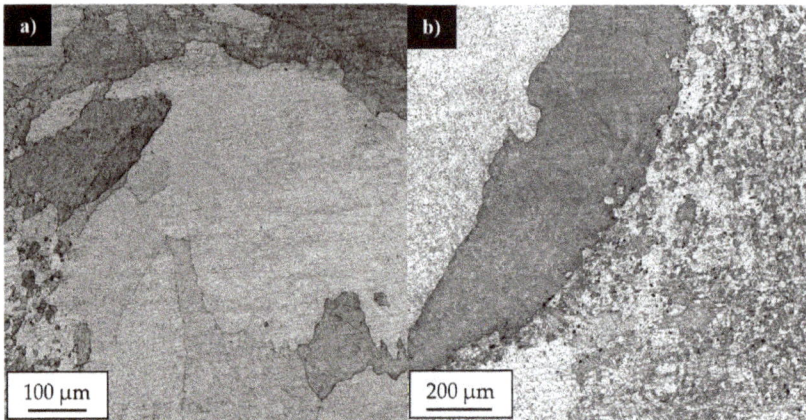

Figure 12. Light microscopy image of 800-100-HT1 sample microstructure in boundary between stir zone and thermo-mechanically affected zone at: (**a**) advancing side, (**b**) retreating side of the joint.

The light microscopy observations of 800-100-HT1 sample microstructure allowed us to observe a very specific boundary between zone consisting of fine grains with size about 5–10 μm and the large grain size of 2 mm (Figure 12b). Additionally, this area consists small pores. A similar boundary occurs in all investigated samples and it is possible to observe it at macrostructure images of the joints (Figure 10a–c). The results of microhardness analysis of the samples which have been subjected to the post-weld heat treatment (HT1) are presented in Figures 13–15.

The results of the microhardness distribution on the cross-section of the samples subjected to the post weld heat treatment indicate uniform hardening of AA2519 alloy. Our results show that the small fluctuations in microhardness occured in the stir zone which suffers the abnormal grain growth. The average microhardness of the material was about 130–140 HV0.1 which is a typical value for 2519 aluminum alloy after a precipitation-hardening process. Abnormal grain growth in the stir zone and the upper part of thermo-mechanically affected zone seems to have a very low influence on the distribution of the microhardness in the analyzed samples.

The results of the fatigue testing of joints in the as-welded state are presented at Figure 16.

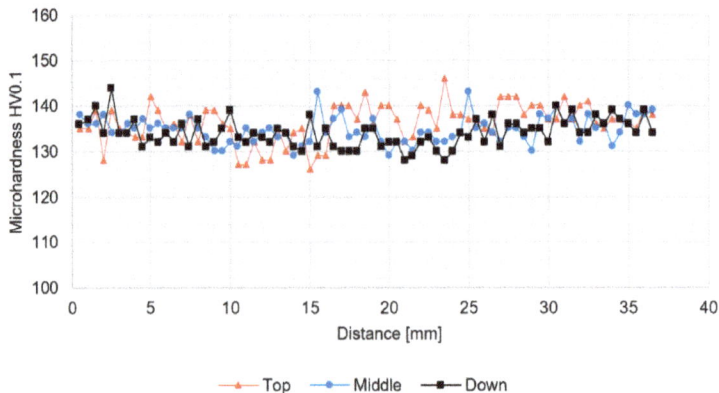

Figure 13. The results of 400-100-HT1 joint microhardness analysis.

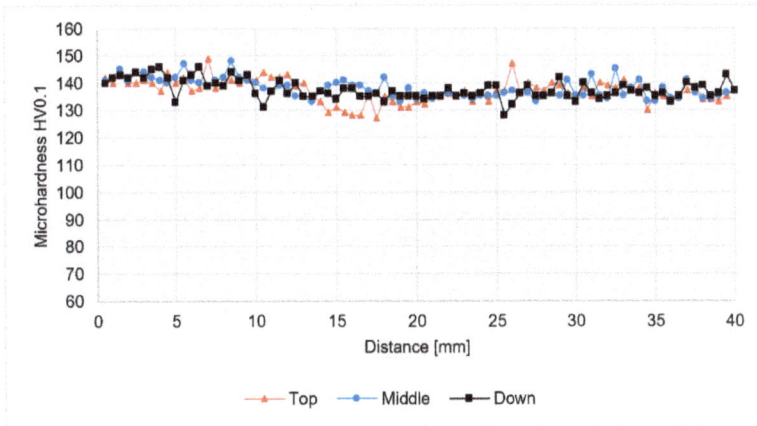

Figure 14. The results of 800-100-HT1 joint microhardness analysis.

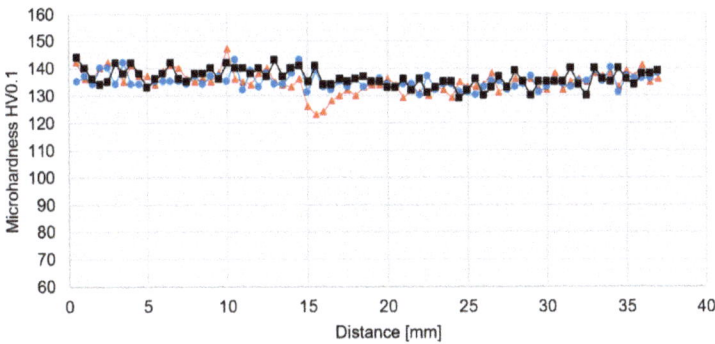

Figure 15. The results of 800-200-HT1 joint microhardness analysis.

The fatigue testing of the samples with constant true strain amplitude $\varepsilon = 0.25\%$ and the cycle asymmetry coefficient equal to R = 0.1 allowed us to investigate the welded joints response to cyclic loading. Our results show that the 400-100-HT0 sample has the lowest fatigue life compared to the joints obtained using the 800 rpm tool rotation speed. The analysis of maximum stress vs. number of cycles curve indicates that in the initial state of testing this joint is characterized by cyclic softening, then it is subjected to the cyclic hardening and the final stabilization (Figure 16a). Although 800-100-HT0 and 800-200-HT0 samples also underwent similar cyclic softening during the initial phase of fatigue testing, they gained their cyclic stability sooner (Figure 16c,e). The analysis of the hysteresis loops evolutions allowed us to state that initial loops have a high participation of the plastic strain (Figure 16b,d,f). The post weld heat treatment influenced the joints response to cyclic loading significantly and the results of the fatigue testing of joints subjected to the post-weld precipitation hardening are presented in Figure 17.

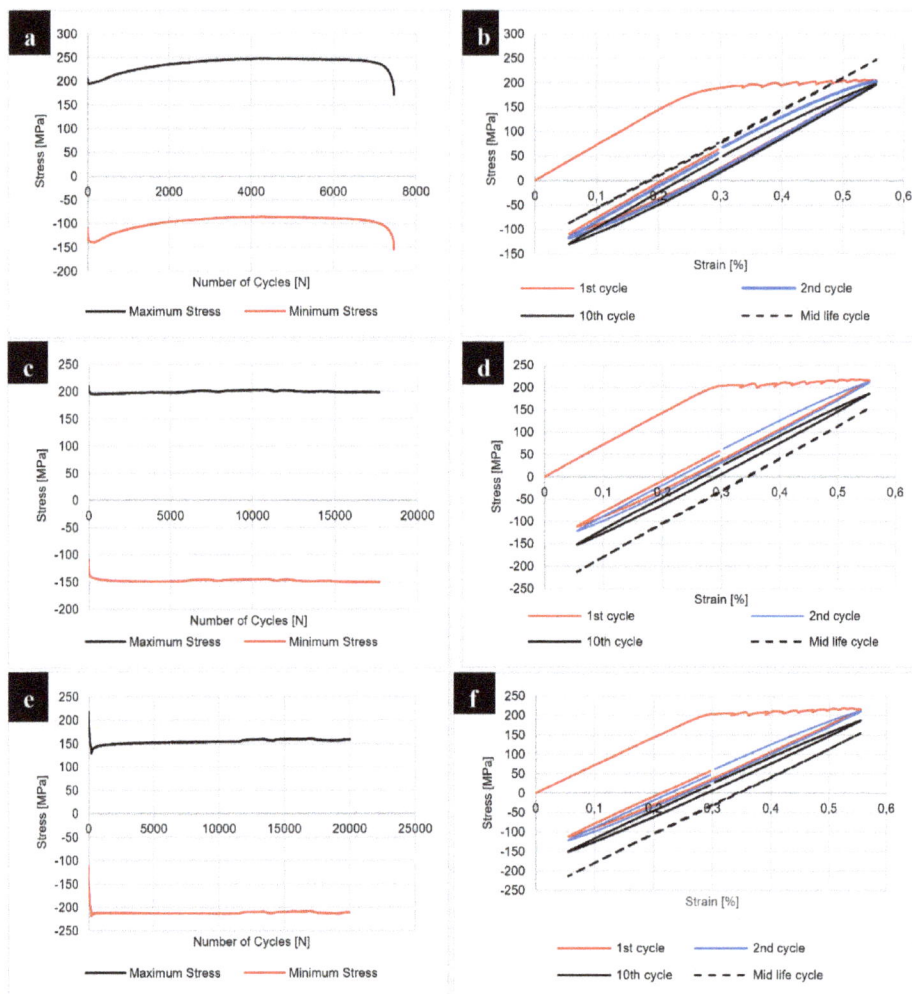

Figure 16. Results of the fatigue testing for 400-100-HT0 sample: (**a**) maximum and minimum stress vs number of cycles, (**b**) hysteresis loop evolution; for 800-100-HT0 sample: (**c**) maximum and minimum stress vs. number of cycles, (**d**) hysteresis loop evolution; for 800-200-HT0 sample: (**e**) maximum and minimum stress vs. number of cycles, (**f**) hysteresis loop evolution.

The post-weld heat treatment significantly reduced the fatigue life of all investigated joints. Our results show that samples 400-100-HT1 and 800-200-HT1 in the first phase of fatigue life undergo cyclic hardening (Figure 17a,e). The 800-100-HT1 sample had a cycle stability during fatigue testing but its fatigue life was extremely short (Figure 17c). Decrease in fatigue life of the joints is mostly related to the presence of pores in the structure of welds subjected to the post-weld precipitation hardening process. The analysis of hysteresis loops evolutions confirmed the cyclic hardening of the investigated joints (Figure 17b,d,f). Additionally, the hysteresis loops of the joint subjected to the post-weld heat treatment have a significantly lower participation of plastic strain compared to the joint in the as-welded state (Figure 16b,d,f). The fracture surfaces of the samples 800-200-HT0 and 800-200-HT1 have been subjected to scanning electron microscopy observations (Figures 18 and 19).

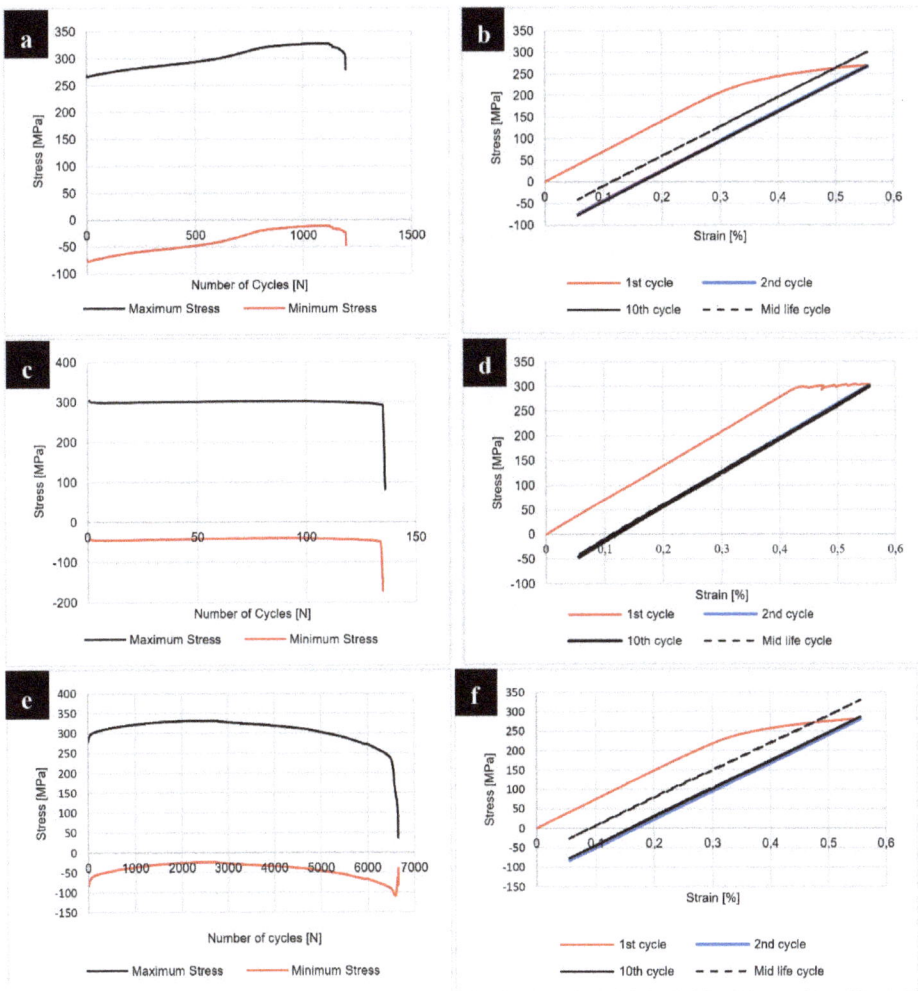

Figure 17. Results of the fatigue testing for 400-100-HT1 sample: (**a**) maximum and minimum stress vs number of cycles, (**b**) hysteresis loop evolution; for 800-100-HT1 sample: (**c**) maximum and minimum stress vs. number of cycles, (**d**) hysteresis loop evolution; for 800-200-HT1 sample: (**e**) maximum and minimum stress vs. number of cycles, (**f**) hysteresis loop evolution.

Our results show that the fracture surface of the 800-200-HT0 sample was characterized by mixed ductile and brittle fracture with the predominance of ductile fracture with dimpled texture, especially in the surrounding area of the Al$_2$Cu precipitates (Figure 18a,b). The reason of the increasing of plastic deformation participation is that non-coherent, large Al$_2$Cu particles caused local stress concentration (Figure 18c). The character of the fracture in the stir zone differs from the rest of the fatigue surface (Figure 18a,e). Additionally, a back-scattered electron (BSE) image of the stir zone fatigue surface allowed to observe that this zone consists very low amount of large precipitates, what confirmed previous scanning electron microscopy observations (Figures 18d and 5a). The magnification of the fatigue surface in the stir zone allowed us to observe that this area is characterized by ultrafine equiaxed dimples (Figure 18f).

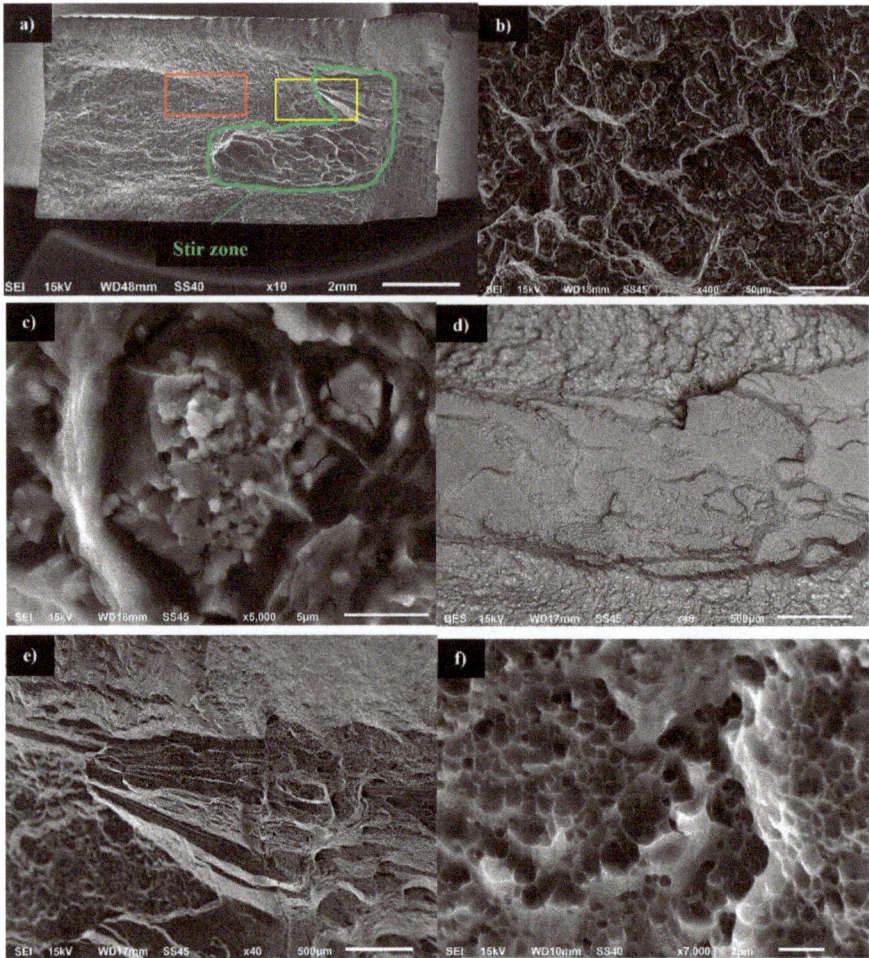

Figure 18. Scanning electron microscopy images of 800-200-HT0 fatigue surface: (**a**) overall view of the fracture surface, (**b**) image of the red marker area, (**c**) magnification of the precipitate, (**d**) back-scattered electron (BSE) image of the stir zone fatigue surface, (**e**) image of yellow area, (**f**) magnification of the fatigue surface in the stir zone.

Investigation of the fatigue surface of the 800-200-HT1 sample revealed the presence of characteristic bands localized in the center of the examined surface (Figure 19a,b). Compared to the fatigue surface of the joint in the as-welded state, the heat treatment changed the character of fracture to being more brittle and no presence of dimpled texture was observed. Additionally, the occurring of pores in the bands has been revealed (Figure 19c). Further investigation of the bands allowed us to observe a very high participation of intergranular brittle fracture in the center of this area (Figure 19d). The presence of both pores and fine grains in the crack propagating bands suggests that this area was a boundary between thermo-mechanically affected zone and the stir zone, in which after heat treatment the imperfections (pores) were found.

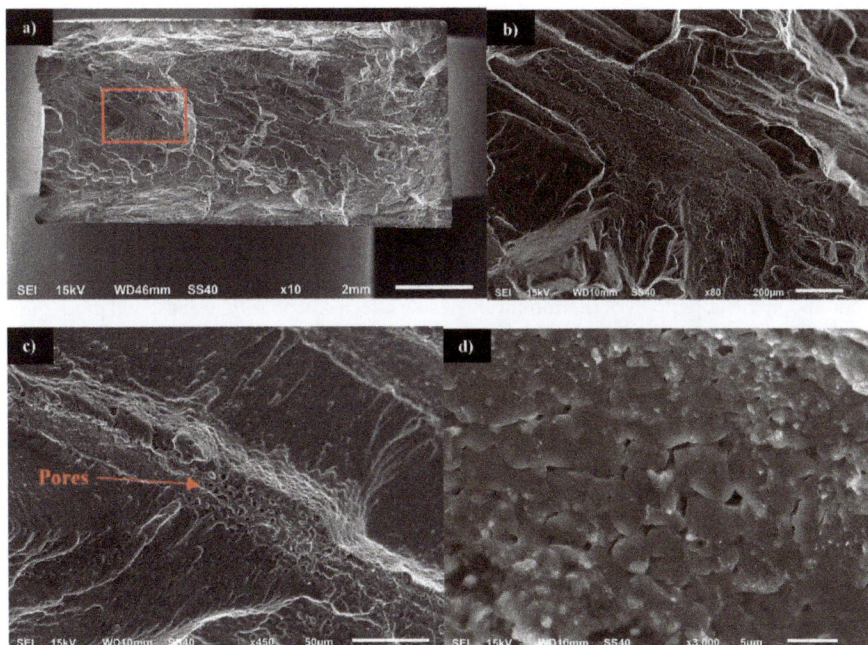

Figure 19. Scanning electron microscopy images of 800-200-HT1 fatigue surface: (**a**) overall view of the fracture surface, (**b**) image of the red marker area, (**c**) band presence in the fracture surface, (**d**) magnification of the band with visible fine grains.

4. Discussion

The friction stir-welding technology allowed us to produce Sc-modified AA2519-O joints with no presence of any imperfection in their macrostructure in the as-welded state. The microstructure analysis together with the results of the microhardness distribution on the cross-sections of the samples allowed us to identify typical zones: stir zone, thermo-mechanically affected zone and heat-affected zone. The size of each zone, as well as, their measured microhardness differ depending on the welding parameters used. Our results show that tool rotation speed has a most significant impact on the increase of microhardness in the stir zone, which is characterized by ultrafine grain microstructure with grains sized about 5 μm. Additionally, the results of the chemical composition analysis allowed us to state that in the stir zone Al_2Cu precipitates undergo the dissolution process due to intense stirring of the welded material in this area. The analysis of the solvus curve in the Al-Cu phase equilibrium diagram allows us to conclude that with the increase of the temperature above 300 °C the solubility of the Al_2Cu phase into Al alloy also increases significantly [3].

Although the dissolution of Al_2Cu precipitates into aluminum alloy during heat treatment takes time, in the case of the friction stir-welding process the intense mixing of material in the temperature about 400–450 °C promotes the formation of saturated solution in the stir zone. This phenomenon finds its consequences in the fatigue surface analysis. In the case of the zone occurring large Al_2Cu precipitates (such as base material, heat affected zone and thermo-mechanically affected zone) the fracture is characterized by dimpled texture with precipitates localized in the dimples. Stir zone, which has lack of large precipitates is characterized by ultrafine equiaxed dimples with no visible precipitate presence. This different size of dimples can be explained by the fact that non-coherent, large Al_2Cu particles cause local stress concentration and as a result the increase of plastic deformation participation in the surroundings of each precipitate. Scanning electron microscopy observations of the stir zone revealed the presence of ultrafine precipitates with about 1 μm size. The results of EDX area analysis of the chemical composition indicate an increased concentration of scandium and zirconium in the precipitates in the stir zone. The analysis of the Al-Sc-Zr phase equilibrium diagram indicates a low solubility of $Al_3(Sc,Zr)$ precipitates into Al alloy even at the temperature of 600 °C (which exceeded the friction stir-welding process temperature range) [26].

This phenomenon finds its confirmation in similar research concerned with friction stir welding of scandium-modified aluminum alloys [27–29]. The $Al_3(Sc,Zr)$ is the phase coherent with the aluminum matrix and as the results it not causes such stress concentration as non-coherent Al_2Cu precipitates. For this reason there is not predominant influence of $Al_3(Sc,Zr)$ precipitates on the dimpled texture of the stir zone in the fatigue surface of 800-200-HT0 sample. The post-weld heat treatment (precipitation hardening) influences the microstructure of joints in a significant way. The grainy structure in the stir zone and in the upper part of thermo-mechanically affected zone suffers abnormal grain growth. This effect of the post-weld heat treatment on the grainy microstructure of high-strength aluminum alloys has been the subject of research [25,30–33]. The reason of grain overgrowth is a significant thermal instability of grains size mostly in the stir zone. Grains in the stir zone have a large number of dislocations, as well as much distortion energy and on the other hand the energy of their boundary is low due to absence of second phase particles [33]. Abnormal grain growth results in the formation of a very specific boundary between fine, recrystallized grain microstructure and the abnormal grains with the size of millimeters. Additionally, the pores have been found at the boundary between the stir zone and the thermo-mechanically affected zone of the samples in the as-welded state. Despite the imperfections in the joints subjected to the post-weld heat treatment, the results of microhardness distribution analysis indicate a uniform hardening of AA2519 alloy due to precipitation-hardening process with small local fluctuations in microhardness in the abnormal grain growth zone. The changes in the microstructure of heat treated joints found their confirmation in the results of fatigue testing. The post-weld heat treatment significantly reduced the fatigue life of all analyzed samples despite the welding parameters. The observations of the fatigue surface of the samples subjected to heat treatment revealed the crack-propagating bands with the presence of pores and ultra-fine grains. These crack-propagating bands have been identified during microstructure examination as a pore-rich area between large grained and fine grained microstructure.

5. Conclusions

Investigation on Sc-modified AA2519 friction stir-welded joint in the as-welded state and after post-weld heat treatment (precipitation hardening) allowed the following conclusions to be drawn:

1. The friction stir-welding process of AA2519-O alloy causes the dissolution of Al_2Cu precipitates in the stir zone and formation of supersaturated solution in this area;
2. $Al_3(Sc,Zr)$ precipitates do not dissolve due to the FSW process and form dispersion of fine precipitate in the stir zone;
3. The post-weld heat treatment of obtained joints causes the abnormal grain growth in the stir zone and the upper part of the thermo-mechanically affected zone, as well as the formation of pores in this area;
4. The dissolution of Al_2Cu in the stir zone finds its consequence in the fracture of the joints. Non-coherent, large Al_2Cu particles' presence in the thermo-mechanically affected and heat-affected zone cause local stress concentration and as a result the increase of plastic deformation participation in the surroundings of each precipitate which results in the formation of characteristic dimples on the fatigue surface of the joint. At the same time the stir zone due to the lack of large Al_2Cu precipitates is characterized by ultrafine dimples;
5. The post-weld heat treatment results in decreasing fatigue life of all analyzed samples. The fatigue surface observations have confirmed that the main reason for this phenomenon is presence of pores and boundary between abnormal and fine grains in the structure of joints investigated.

Author Contributions: Conceptualization, R.K.; methodology, M.W. and J.T.; investigation, R.K., L.Ś., M.W. and J.T.; writing—original draft preparation, R.K.; writing—review and editing, L.Ś. and M.W.; visualization, M.W.; supervision, L.Ś. and J.T.

Funding: This research was funded by Polish Ministry of National Defence, grant number: PBG/13-998.

Conflicts of Interest: The authors declare no conflict of interest.

References

1. Crouch, I. *The Science of Armour Materials*; Woodhead Publishing: Duxford, England, 2016; ISBN 9780081010020.
2. Starke, E.A., Jr.; Staley, J.T. Application of modern aluminum alloys to aircraft. *Prog. Aerosp. Sci.* **1996**, *32*, 131–172. [CrossRef]
3. Hatch, J.E. *Aluminum: Properties and Physical Metallurgy*; ASM International: West Conshohocken, PA, USA, 1984; ISBN 0871701766.
4. Showalter, D.D.; Placzankis, B.E.; Burkins, M.S. *Ballistic Performance Testing of Aluminum Alloy 5059-H131 and 5059-H136 for Armor Applications*; US Army Research Laboratory: Aberdeen Proving Ground, MD, USA, 2008.
5. Fisher, J.; James, J. Aluminum alloy 2519 in military vehicles. *Mater. Sci. Forum* **2002**, *160*, 43–46.
6. Rozumek, D.; Marciniak, Z. Fatigue crack growth in AlCu4Mg1 under nonproportional bending-with-torsion loading. *Mat. Sci.* **2011**, *46*, 685–694. [CrossRef]
7. Wu, Y.P.; Ye, L.Y.; Jia, Y.; Liu, L.; Zhang, X.M. Precipitation kinetics of 2519A aluminum alloy based on aging curves and DSC analysis. *Trans. Nonferrous Metal. Soc. China* **2014**, *24*, 3076–3083. [CrossRef]
8. Mathe, J.W. *Precipitate Coarsening during Overaging of 2519 Al-Cu Alloy: Application to Superplastic Processing*; Institutional Archive of the Naval Postgraduate School: Monterey, CA, USA, 1992.
9. Zakharov, V.V. Combined alloying of aluminium alloys with scandium and zirconium. *Met. Sci. Heat Treat.* **2014**, *56*, 281–286. [CrossRef]
10. Davydov, V.G.; Elagin, V.I.; Zakharov, V.V.; Rostova, T.D. Alloying aluminium alloys with scandium and zirconium additives. *Met. Sci. Heat Treat.* **1996**, *38*, 347–352. [CrossRef]
11. Zakharov, V.V. Effect of Scandium on the structure and properties of aluminium alloys. *Met. Sci. Heat Treat.* **2003**, *45*, 246. [CrossRef]
12. Jia, Z.H.; Røyset, J.; Solberg, J.K.; Liu, Q. Formation of precipitates and recrystallization resistance in Al-Sc-Zr alloys. *Trans. Nonferrous Met. Soc. China* **2012**, *22*, 1866–1871. [CrossRef]
13. Sasabe, S. Welding of 2000 series aluminium alloy materials. *Weld. Int.* **2012**, *26*, 339–350. [CrossRef]
14. Kalita, W.; Hoffman, J.; Mucha, Z.; Czujko, T.; Jóźwiak, S.; Kusiński, J. Structural and mechanical properties of CO2-laser welded joints in difficult-to-weld metals. *Weld. Int.* **2009**, *10*, 257–261. [CrossRef]
15. Çam, G.; Mistikoglu, S. Recent Developments in Friction Stir Welding of Al-alloys. *J. Mater. Eng. Perform.* **2014**, *23*, 1936–1953. [CrossRef]
16. Mishra, R.S.; Mahoney, M.W. *Friction Stir Welding and Processing*; ASM International: Materials Park, OH, USA, 2007; ISBN 978-0-87170-840-3.
17. Xu, W.F.; Liu, J.H.; Chen, D.L.; Luan, G.H. Low-cycle fatigue of a friction stir welded 2219-T62 aluminium alloy at different welding parameters and cooling conditions. *Int. J. Adv. Manuf. Technol.* **2014**, *74*, 209–218. [CrossRef]
18. Radisavljevic, I.; Zikovic, A.; Radovic, N.; Grabulov, V. Influence of FSW parameters on formation quality and mechanical properties of Al 2042–T351 butt welded joints. *Trans. Nonferrous Met. Soc. China* **2013**, *23*, 3525–3539. [CrossRef]
19. Zhang, Z.; Xiao, B.L.; Ma, Z.Y. Effect of welding parameters on microstructure and mechanical properties of friction stir welded 2219Al-T6 joints. *J. Mater. Sci.* **2012**, *47*, 4075–4086. [CrossRef]
20. Kosturek, R.; Wachowski, M.; Ślęzak, T.; Śnieżek, L.; Mierzyński, J.; Sobczak, U. Research on the friction stir welding of Titanium Grade 1. In Proceedings of the International Conference on Advanced Functional Materials and Composites (ICAFMC2018), MATEC Web of Conferences 242, Barcelona, Spain, 5–6 September 2018. [CrossRef]
21. Liang, X.P.; Li, H.Z.; Li, Z.; Hong, T.; Ma, B.; Liu, S.D.; Liu, Y. Study on the microstructure in a friction stir welded 2519-T87 Al alloy. *Mater. Des.* **2012**, *35*, 603–608. [CrossRef]

22. Sabari, S.S.; Malarvizhi, S.; Balasubramanian, V. Characteristics of FSW and UWFSW joints of AA2519-T87 aluminium alloy: Effect of tool rotation speed. *J. Manuf. Process.* **2016**, *22*, 278–289. [CrossRef]

23. Sato, Y.S.; Park, S.H.C.; Kokowa, H. Microstructural Factors Governing Hardness in Friction-Stir Welds of Solid-Solution-Hardened Al Alloy. *Metall. Mater. Trans. A* **2001**, *32*, 3033–3042. [CrossRef]

24. Chu, G.; Sun, L.; Lin, C.; Lin, Y. Effect of Local Post Weld Heat Treatment on Tensile Properties in Friction Stir Welded 2219-O Al Alloy. *J. Mater. Eng. Perform.* **2017**, *26*, 5425–5431. [CrossRef]

25. Cerri, E. Effect of post-welding heat treatments on mechanical properties of double lap FSW joints in high strength aluminium alloys. *Metall. Sci. Technol.* **2011**, *29*, 32–40.

26. Røyset, J.; Ryum, N. Scandium in aluminium alloys. *Int. Mater. Rev.* **2005**, *50*, 19–44. [CrossRef]

27. Paglia, C.S.; Jata, K.V.; Buchheit, R.G. A cast 7050 friction stir weld with scandium: Microstructure, corrosion and environmental assisted cracking. *Mater. Sci. Eng. A-Struct. Mater.* **2006**, *424*, 196–204. [CrossRef]

28. He, Z.B.; Peng, Y.Y.; Yin, Z.M.; Lei, X.F. Comparison of FSW and TIG welded joints in Al-Mg-Mn-Sc-Zr alloy plates. *Trans. Nonferrous Met. Soc. China* **2011**, *21*, 1685–1691. [CrossRef]

29. Muñoz, C.A.; Rückert, G.; Huneau, B.; Sauvage, X.; Marya, S. Comparison of TIG welded and friction stir welded Al–4.5Mg–0.26Sc alloy. *J. Mater. Process. Technol.* **2008**, *197*, 337–343. [CrossRef]

30. Chen, P.S.; Russell, C.K. Controlling Abnormal Grain Growth In Friction Stir Welded Al-Li 2195 Spun Formed Domes. In Proceedings of the Aeromat 22 Conference and Exposition American Society for Metals, Long Beach, CA, USA, 23–26 May 2011.

31. Tayon, W.; Domack, M.; Hoffman, K.E.; Hales, S. Investigation of Abnormal Grain Growth in a Friction Stir Welded and Spin-Formed Al-Li Alloy 2195 Crew Module. In Proceedings of the 8th Pacific Rim International Congress on Advanced Materials and Processing, Waikoloa, HI, USA, 4–9 Auguest 2013. [CrossRef]

32. Mironov, S. About Abnormal Grain Growth in Joints Obtained by Friction Stir Welding. *Met. Sci. Heat Treat.* **2015**, *57*, 40–47. [CrossRef]

33. Luo, C.H.; Dong, F.B.; Guo, L.J.; Wei, X.F. Grain Thermal Instability and Heat Treatment Technology for Aluminum Alloy Welded by FSW. In Proceedings of the 2017 International Conference on Applied Mechanics and Mechanical Automation (AMMA 2017), Hong Kong, China, 23–24 June 2017. [CrossRef]

Article

Strain Rate during Creep in High-Pressure Die-Cast AZ91 Magnesium Alloys at Intermediate Temperatures

Mónica Preciado *, Pedro M. Bravo, José Calaf and Daniel Ballorca

Escuela Politécnica Superior, University of Burgos, 09006 Burgos, Spain; pmbravo@ubu.es (P.M.B.); josecc33@gmail.com (J.C.); danielballorca@gmail.com (D.B.)
* Correspondence: mpreciado@ubu.es

Received: 14 February 2019; Accepted: 12 March 2019; Published: 15 March 2019

Abstract: During creep, magnesium alloys undergo microstructural changes due to temperature and stress. These alterations are associated with the evolution of the present phases at a microstructural level, creating different strain rates during primary and tertiary creep, and with the stability of the inter-metallic phase $Mg_{17}Al_{12}$ formed at these temperatures. In this paper, the results of creep testing of high-pressure die-cast AZ91 magnesium alloys are reported. During creep, continuous and discontinuous precipitates grow, which influences creep resistance. The creep mechanism that acts at these intermediate temperatures up to 150 °C is termed dislocation climbing. Finally, the influence of the type of precipitates on the creep behavior of alloys is determined by promoting the formation of continuous precipitates by a short heat treatment prior to creep testing.

Keywords: AZ91; magnesium alloys; creep; high pressure die casting

---

1. Introduction

The use of magnesium in automotive industry components has increased significantly over recent years [1]. Although the cost of magnesium alloys is disadvantageous in comparison to steel, the weight reduction of the structural component, due to the lower density of magnesium, is highly advantageous. The magnesium alloy AZ91 is one of the most intensively employed magnesium alloys because of its very useful combination of properties such as castability, mechanical performance (at room temperature), corrosion resistance and competitive cost [2]. However, creep resistance begins to yield at temperatures over 127 °C, which limits the use of these alloys to components that are not in major areas [3]. Industrial production of this type of alloy, due to its high productivity and dimensional stability, has primarily been done through high-pressure die casting (HPDC), which introduces porosity in the microstructure [4].

One main reason that has been discussed to explain the low creep resistance of the AZ91 alloy is that the microstructure is mainly formed by solid solution α and inter-metallic β-phase ($Mg_{17}Al_{12}$) at the grain boundaries, the latter having a low melting point (437 °C). It therefore softens and thickens due to the temperature, resulting in a weakened grain boundary [5]. However, recent hardness tests in the β-phase at temperatures over 200 °C have shown the high deformation resistance of this inter-metallic phase [6], which contradicts the softening theory at the same temperatures.

Creep behavior can, in general, be explained in terms of microstructural stability [7] in connection with hardening mechanisms. Usually, two contradictory trends are observed in creep processes: one is represented by softening processes (cross slip, etc.) and the other by hardening processes (solid solution hardening, precipitation hardening, etc.). When the latter is dominant, creep resistance increases (primary creep), and when the softening processes are dominant, creep resistance decreases (tertiary creep). The minimum creep rate is reached when these two opposing mechanisms reach a balance (steady state or secondary creep) [8].

However, during creep tests, high temperatures are reached and this causes the amount of precipitates and their morphology to change. In the literature [9], information has been gathered on the way in which β precipitates, which can be either discontinuous and continuous, and their production is based on how the initial β-eutectic is modified during creep. Of these types of precipitates, it is the continuous precipitates (CP) that have a greater influence on the strength of magnesium alloys [10]. Continuous precipitation consists of alternating acicular-shaped precipitates and is much thinner than discontinuous precipitation [11]. Both types of precipitate coexist and compete in their growth, although there are temperature ranges that favor the growth of one type of precipitation over another. The coexistence of both types is observed at the aging temperature of 150 °C [12], which is the temperature of the creep tests developed in this paper.

The influence of aging treatment on tensile properties in as-cast samples was studied by some authors [8,9], although the time permanence was very high (several hours) and normally given after a solution treatment. In these cases, massive precipitation was formed. In this study, a group of samples was subjected to a very short pre-treatment prior to performing creep tests in order to slightly modify the initial state. The results enabled an evaluation of the role of the developing precipitate types on creep resistance. The studies with aging treatments prior to creep cannot be compared with the results obtained in this paper, as the pre-treatment given to the samples was only 1 h in duration. The purpose of this modification was to create small precipitate nuclei that could favor posterior precipitation in the form of continuous precipitates.

2. Materials and Methods

The composition of the AZ91D magnesium used in the present study was determined with an Arc Spark Analyzer (wt%): Al, 8.83; Be, 0.001; Cu, 0.007; Fe, 0.003; Mn, 0.32; Si, 0.028 and Zn, 0.6. Differential Scanning Calorimetry (DSC) tests were performed to obtain the temperature at which a phase change occurred, in this case associated with β-phase precipitation from the Al-supersaturated Mg solid solution. Differential Scanning Calorimetry (DSC) tests were performed on a Perkin Elmer DSC7 machine (Waltham, MA, USA) with an argon protective atmosphere. The samples were cut into small discs and heated to temperatures ranging from 30 °C to 230 °C with four different heating ramps: 20, 25, 30 and 40 °C/min. Pure aluminum discs were used as a reference material.

Verification of the precipitate growth in the β-phase was achieved through adiffractometry analysis of the two samples; one was obtained directly from injection and the other by pre-heating at 160 °C for 1 h. The equipment used was a Bruker D8 Advance (Davinci Design, Billerica, MA, USA). The creep tests were performed at constant loads at a temperature of 150 °C with initial stresses of 50, 60, 65 and 70 MPa. For the creep tests, the speed of load application was 1 mm/min and the strains were measured by an extensometer with a gage length of 25 mm. The temperature was maintained in a chamber around the sample and measured by a thermocouple. The specimens used in the creep tests were of circular geometry with a diameter of 6.5 mm. These samples (Figure 1) were obtained by a high-pressure injection process, so that the surface finish and the porosities were the same as the in-service components. These aspects are relevant for the creep behavior. The tests were performed on a Zwick/Roell Kappa 50DS machine (Ulm, Germany).

From the creep tests, the parameter n, stress exponent, can be calculated from the conventional power law in Equation (1):

$$\varepsilon_s = K\sigma^n \exp\left(-\frac{Q_c}{RT}\right) \tag{1}$$

where ε_s is the strain rate, K is a constant, Q_c is the activation energy, R is the gas constant and T is the temperature. By plotting the minimum ε_s versus σ logarithmically, the n exponent is calculated.

The existence of a threshold stress, σ_{th}, has been described for precipitate and particle-hardened alloys [13]. If this concept is taken into account, Equation (1) can be modified as:

$$\sigma_{eff} = \sigma - \sigma_{th} \qquad (2)$$

$$\varepsilon_s = k\sigma_{eff}^{n_t} \exp\left[-\frac{Q_c}{RT}\right] \qquad (3)$$

where n_t is the true exponent. The precipitates play an important role; however, the samples were not submitted to a solution treatment and most of the Al is in the form of the existing β-eutectic $Mg_{17}Al_{12}$ and does not participate in the precipitation produced during creep. As a result, the precipitation is not homogenous and is limited by the existing eutectic β-phase prior to creep. This is why, as estimated by Spigarelli et al. [14], $\sigma_{th} \approx 0$ for die-cast AZ91 alloys. When taking into account the relevant role of precipitates, the n_t exponent was calculated according to a method [15] that consists of an extrapolation of $\log\sigma - \log\varepsilon_s$ up to $\varepsilon_s = 10^{-10}$ s^{-1} (considered the lowest measurable creep rate) to obtain σ_{th}. It was estimated that the real stress exponent of the HPDC samples should be between conventional, n, and true, n_t, exponents.

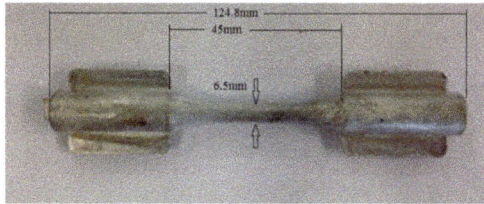

Figure 1. Creep test sample (high-pressure die casting process).

3. Results and Discussion

The microstructure of this alloy (Figure 2) consists of Mg-solid solution grains (α-phase) decorated with precipitates in the grain boundary that correspond to a divorced eutectic formed by β-phase ($Mg_{17}Al_{12}$) and an Mg-super saturated solid solution (eutectic α-phase). The porosity of the samples is inherent to the manufacturing process and the incompatibility between the hexagonal compact α-phase magnesium matrix and the cubic inter-metallic β-phase, $Mg_{17}Al_{12}$, is remarkable.

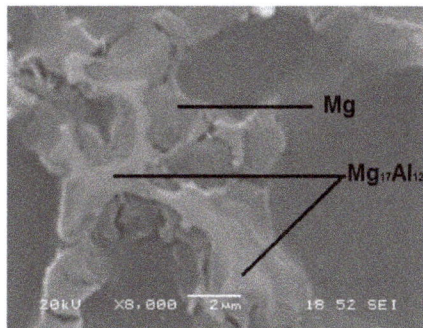

Figure 2. SEM image of divorced eutectic at the grain boundary.

The results of the DSC tests are shown in Figure 3. The values at which a small peak was observed were between 129 °C and 137 °C depending on the heating ramp.

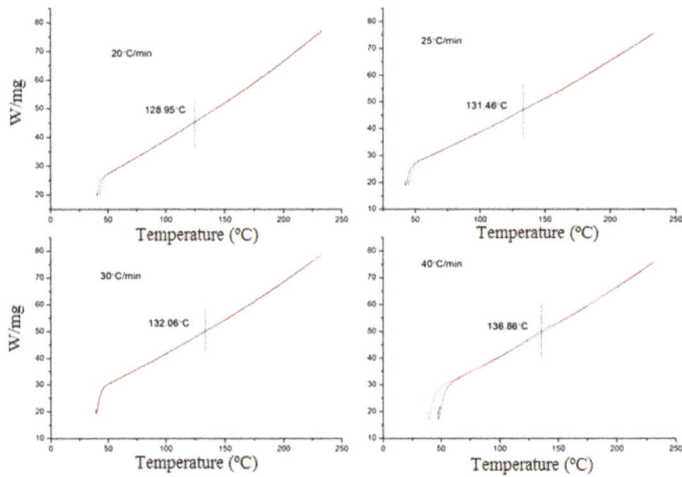

Figure 3. Differential Scanning Calorimetry (DSC) tests at 20 °C/min, 25 °C/min, 30 °C/min and 40 °C/min heating ramps (the vertical line corresponds to the temperature at which a phase change is detected).

The diffractometry of the samples in Figure 4 shows a higher amount of $Mg_{17}Al_{12}$ in the pre-treated sample (red line is over the black line).

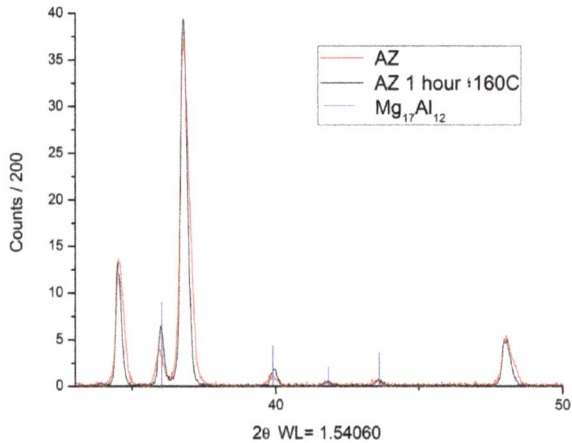

Figure 4. Diffractometry of the as-cast and pre-heated samples. The vertical lines represent $Mg_{17}Al_{12}$ location.

The results of creep are shown in Figure 5. For each condition three samples were tested. The intermediate curve was chosen as the average behavior.

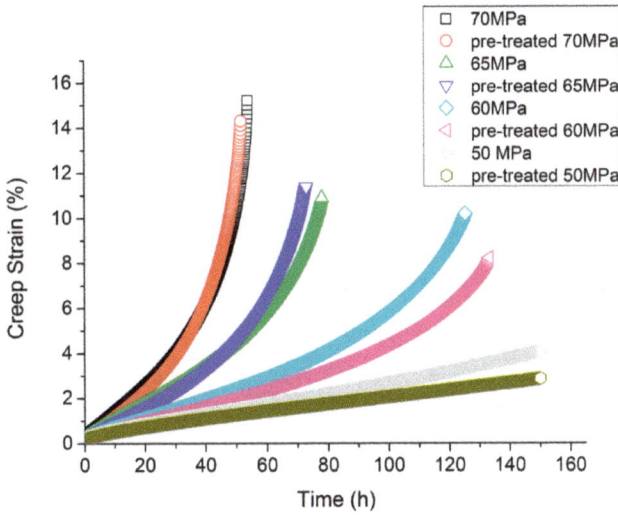

Figure 5. Creep tests in the different samples.

The creep resistance of the pre-treated samples was better for lower loads (50 and 60 MPa). This result is not in contradiction with the results of other authors that clearly showed that aging worsened the creep resistance, because the pre-treatment of 1 h at 160 °C was too short to be considered as aging.

To analyze the primary and secondary creep it is convenient to study the curves of strain rate versus strain (Figure 6).

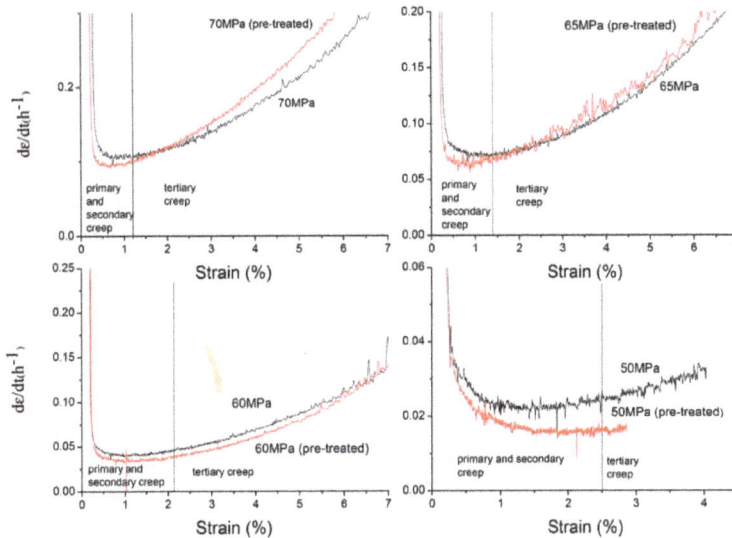

Figure 6. Strain rate variation in creep tests.

The strain rate during primary and secondary creep was lower for the pre-treated samples although at 65 and 70 MPa the strain rate increased at a higher rate than the non-treated samples once the minimum strain rate was reached. In Figure 6, it can be observed that the pre-treated samples curves were below the non-treated samples during the two first stages of creep at all the loads. At stresses of 65 and 70 MPa a stable secondary creep was not reached (this is more pronounced in the pre-treated samples). A plausible reason for this could be that the precipitates size was small and they are not able to arrest the dislocations movement. When the load decreased to 60 and 50 MPa three stages of creep was observed and the behavior was better for the pre-treated samples at all stages. The stress coefficient and the true stress coefficient was calculated and shown in Table 1. According to the literature (see references in Table 1), the deformation mechanism depended on the value of n. A value of n = 3 was related to the dislocation glide and n = 5 corresponded to dislocation climbing. It is not clear in the literature if the obtained values were based on the values of true n-exponent but in any case, the deformation mechanism seems to be controlled by the climbing of dislocations.

Table 1. Values of n coefficient and n_t true coefficient.

Samples	n	n_t
Non-treated samples	4.8	4
Treated samples	5.6	4.8
Spigarelli et al. [14]	5	-
Vagarali & Langdon [16], Mg-0.8%Al, T4	3.6	-
Ishikawa &Watanabe [17], AZ31 with T4	5–7	-
Kaveh et al. [18], as-cast AZ91	5.6	-

Microstructural Analysis

During creep there is precipitation of the intermetallic $Mg_{12}Al_{17}$ in the form of continuous and discontinuous precipitates, mostly from the former eutectic Al-supersaturated α-phase [19]. There was also loss of solution hardening with an increase in precipitation hardening in the evolution of the primary creep. It has been stated [20] that a decrease of the creep rate in the primary creep was due to an increase in dislocation density and to the precipitation of β-phase in different forms. At some point these precipitates coarsen with a loss of hardening effect and in some samples a short period of secondary creep was reached. This corresponded to the primary and secondary creep stage as seen in Figure 6. Once the precipitates coarsen and were not able to effectively stop the dislocation movement, the tertiary creep rapidly progressed until rupture. This last stage creep was assisted by the fact that the microstructure is porous. These pores are in the grain boundaries and the incoherence between intermetallic (also in the boundaries) and the matrix helped to make the final fracture.

SEM (FEI Quanta 600, Hillsboro, OR, USA) observation of the samples after the creep tests showed the microstructure with the precipitates that are formed. In Figure 7, it was observed that the precipitates from eutectic saturated solid solution α decorated both sides of the eutectic β. These samples were loaded at 70 MPa and it was visible that the precipitation was finer in the pre-treated sample. This was observed in greater detail in Figure 8 at a higher magnification.

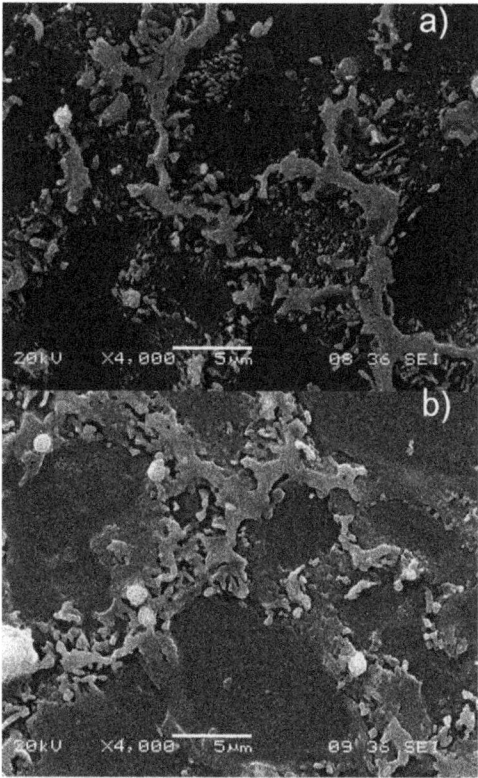

Figure 7. SEM: (**a**) Pre-treated sample after creep test at 70 MPa. (**b**) As-cast sample after creep test at 70 MPa.

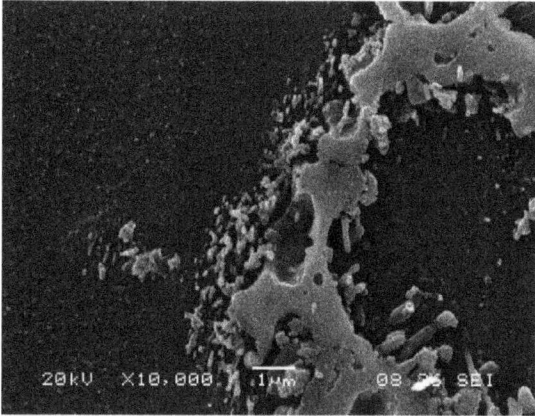

Figure 8. Detail of the precipitates in pre-treated sample at 70 MPa.

In a comparison of the microstructures at 50 MPa of a pre-treated sample with an as-cast sample (Figure 9), where the maximum difference in strain rate was encountered, the presence of coarsened continuous and discontinuous precipitates was observed in both. It has been noted that the continuous precipitation in the as-cast sample, when encountered, was smaller than that in the pre-treated sample.

A plausible reason for this could be the creation of small nucleus sites during the short pre-heating that in the subsequent creep testing lead to continuous precipitates. The absence of continuous precipitation without quenching (after solution annealing) was reported [21], due to the deficiency of nucleation sites. The same study also reported that continuous and discontinuous precipitation occurred when a supersaturated solution was heated. This needs to be taken into account in the present study, where visible SEM β-precipitates appeared in the rich aluminum solid solution zones that were eutectic α before creep. Furthermore, precipitation was assisted by stress, where strain fields resulted in multiple defects that, according to the same author, may act as heterogeneous nucleation sites in the precipitation of continuous β-precipitates.

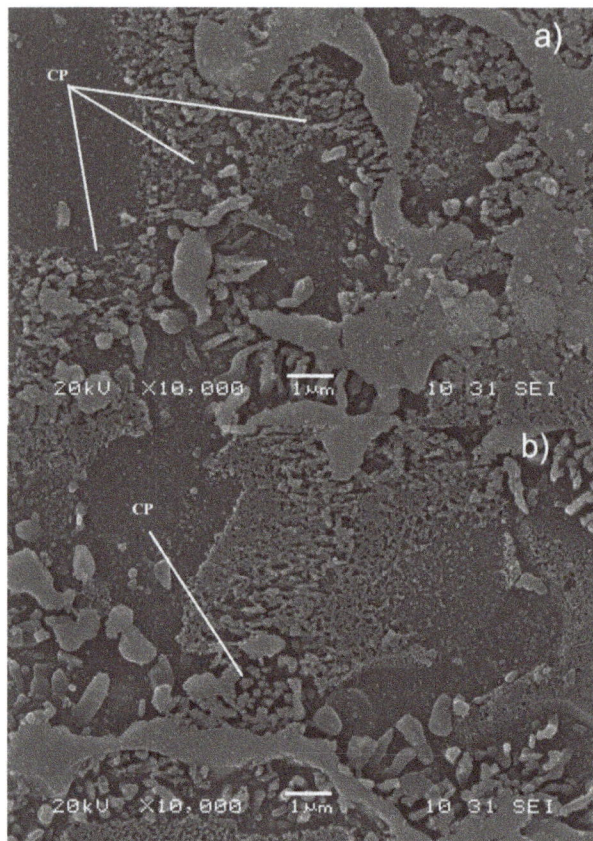

Figure 9. SEM: (**a**) pre-treated sample after creep test at 50 MPa; (**b**) sample after creep test at 50 MPa.

On the other hand, continuous precipitation was the most effective means of increasing mechanical resistance to this type of alloy [22]. During primary creep, the higher capacity of strain hardening for the pre-treated samples at all the loads was probably due to the continuous β-precipitates. However, 65 MPa and 70 MPa minimum creep strains were reached almost immediately, and then the softening processes occurred quickly. This suggests that at high stresses the continuous β-precipitates were not effective, probably because of the small size of the precipitates (smaller in the pre-heated samples), which provoked the dislocations to move more easily. At lower stresses the primary creep was longer (40 h for 50 MPa and 17 h for 60 MPa), allowing the precipitates to grow and be more effective in the hardening process. When the precipitates coarsened, the hardening effect decreased and finally the strain rate increased until fracture. The intragranular precipitation of β-phase was not considered in the discussion because the content of aluminum in the primary α was low, so the amount of precipitates would also be low.

The deformation mechanism of dislocation climbing was given by the stress exponent value, which was also higher in the case of the pre-treated samples. This related to how precipitates forced the dislocations to climb over particles and resulted in an increase in the creep resistance at least during the primary creep.

Finally, a fractographic analysis of the fracture surfaces after creep is shown in Figure 10, where microvoid coalescence can be observed. The existence of pores is also visible. This porosity accelerated the strain rate during tertiary creep until fracture.

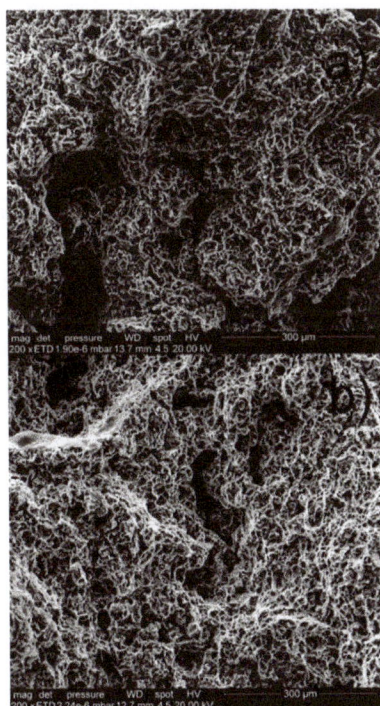

Figure 10. Fracture surfaces after creep. (**a**) Sample at 70 MPa; (**b**) pre-treated sample at 70 MPa.

4. Conclusions

The following conclusions were obtained:

(1) A short pre-treatment of 1 h at 160 °C stimulated the formation of continuous precipitation during creep in HPDC AZ samples. A plausible reason for this could be that small nuclei of β-phase were formed in the pre-treated samples and, when submitted to creep, the strains and temperatures enhanced the formation of continuous precipitates from these nuclei.

(2) The minimum creep rate that corresponded to the secondary creep region was lower for pre-treated samples than for the conventional samples. The higher stresses (70 and 65 MPa) did not permit a stable secondary creep region to be reached.

(3) The stress exponent, n, was higher for pre-treated samples and for both types of samples, treated and non-treated, is indicating that the mechanism of creep is dislocation climbing.

(4) The creep response of the pre-treated samples was better in the sense of having a lower strain rate at all stages of creep when the loads were less than 60 MPa.

Author Contributions: Investigation, M.P., J.C. and D.B.; Methodology, P.M.B.; Writing—original draft, M.P.

Funding: This research received no external funding.

Conflicts of Interest: The authors declare no conflict of interest.

References

1.   Fiedrich, H.; Shumann, S. Research for a "new age of magnesium" in the automotive industry. *J. Mater. Proc. Technol.* **2001**, *117*, 276–281. [CrossRef]
2.   Mordike, B.L.; Ebert, T. Magnesium: Properties—Applications—Potential. *Mater. Sci. Eng. A* **2001**, *302*, 37–45. [CrossRef]
3.   Luo, A.; Pekguleryuz, M.O.J. Cast magnesium alloys for elevated temperature applications. *Mater. Sci.* **1994**, *29*, 5259–5271. [CrossRef]
4.   Myshlyaev, M.M.; Mcqueen, H.J.; Mwembela, A.; Konopleva, E. Twinning, dynamic recovery and recrystallization in hot worked Mg-Al-Zn alloy. *Mater. Sci. Eng. A* **2002**, *337*, 121–133. [CrossRef]
5.   Raynor, G.V. *The Physical Metallurgy of Magnesium and Its Alloys*; Pergamon Press: Oxford, UK, 1959.
6.   Yoo, M.S.; Shin, K.S.; Kim, N.J. Effect of Mg_2Si particles on the elevated temperature tensile properties of squeeze-cast Mg-Al alloys. *Metall. Mater. Trans. A* **1982**, *35*, 1629–1632. [CrossRef]
7.   Srinivasan, A.; Ajithkumar, K.K.; Swaminathan, J.; Pillai, U.T.S.; Pai, B.C. Creep behaviour of AZ91 magnesium alloy. *Procedia Eng.* **2013**, *55*, 109–113. [CrossRef]
8.   Regev, M.; Rosen, A.; Bamberger, M. Qualitative model for creep of AZ91D magnesium alloy. *Metall. Mater. Trans. A* **2001**, *32A*, 1335–1345. [CrossRef]
9.   Blum, W.; Watzinger, B.; Zhang, P. Creep of die-cast light-weight Mg-Al-base alloy AZ91hp. *Adv. Eng. Mater.* **2000**, *2*, 349–355. [CrossRef]
10. Prakash, D.G.L.; Regener, D.; Vorster, W.J.J. Effect of long-term annealing on the microstructure of hpdc AZ91 Mg alloy: A quantitative analysis by image processing. *Comput. Mater. Sci.* **2008**, *3*, 759–766. [CrossRef]
11. Zhao, D.; Wang, Z.; Zuoand, M.; Geng, H. Effects of heat treatment on microstructure and mechanical properties of extruded AZ80 magnesium alloy. *Mater. Des.* **2014**, *56*, 589–593. [CrossRef]
12. Duly, D.; Simon, J.P.; Brechet, Y. On the competition between continuous and discontinuous precipitations in binary Mg–Al alloys. *Acta Metal. Mater.* **1995**, *43*, 101–106. [CrossRef]
13. Dieringa, H.; Huang, Y.; Wittke, P.; Klein, M.; Walther, F.; Dikovits, M.; Poletti, C. Compression-creep response of magnesium alloy DieMag 422 containing barium co MPared with the commercial creep-resistant alloys AE42 and MRI230D. *Mater. Sci. Eng. A* **2013**, *585*, 430–438. [CrossRef]
14. Spigarelli, S.; Regev, M.; Evangelista, E.; Rosen, A. Review of creep behaviour of AZ91 magnesium alloy produced by different technologies. *Mater. Sci. Technol.* **2001**, *17*, 627–638. [CrossRef]
15. Li, Y.; Langdon, T.G. A simple procedure for estimating threshold stresses in the creep of metal matrix composites. *Scr. Mater.* **1992**, *36*, 1457–1460. [CrossRef]
16. Vagarali, S.S.; Langdon, T.G. Deformation mechanisms in H.c.p. metals at elevated temperatures –II creep behavior of a Mg-0.8%Al solid solution alloy. *Acta Metall.* **1982**, *30*, 1157–1170. [CrossRef]
17. Ishikawa, K.; Watanabe, H.; Mukai, T. High temperature compressive properties over a wide range of strain rates in an AZ31 magnesium alloy. *Mater. Sci.* **2005**, *40*, 1577–1582. [CrossRef]
18. Kaveh, M.A.; Alireza, T.; Farzad, K. Effect of Deep cryogenic treatment on microstructure, creep and wear behaviors pf AZ91 magnesium alloy. *Mater. Sci. Eng. A* **2009**, *523*, 27–31. [CrossRef]
19. Blum, W.; Li, Y.J.; Zeng, X.H.; Zhang, P.; Von Großmann, B.; Haberling, C. Creep deformation mechanisms in high-pressure die-cast magnesium-aluminum-base alloys. *Metall. Mater. Trans. A* **2005**, *36*, 1721–1728. [CrossRef]
20. Zhang, P.; Watzinger, B.; Blum, W. Changes in the microstructure and deformation resistance during creep of die-cast Mg-Al-base alloy AZ91hp at intermediate temperatures up to 150 °C. *Phys. Stat. Sol.* **1999**, *175*, 481–489. [CrossRef]

21. Braszczynska-Malik, K.N. Discontinuous and continuous precipitation in magnesium-aluminum type Alloys. *J. Alloys Compd.* **2009**, *477*, 870–876. [CrossRef]
22. Han, G.M.; Han, Z.Q.; Luo, A.A.; Liu, B.C. Microstructure characteristic and effect of aging process on the mechanical properties of squeeze-cast AZ91 alloy. *J. Alloys Compd.* **2015**, *641*, 56–63. [CrossRef]

materials

MDPI

Article

Identification of Mechanical Properties for Titanium Alloy Ti-6Al-4V Produced Using LENS Technology

Aleksandra Szafrańska [1], Anna Antolak-Dudka [2], Paweł Baranowski [1], Paweł Bogusz [1], Dariusz Zasada [2], Jerzy Małachowski [1],* and Tomasz Czujko [2]

[1] Department of Mechanics and Applied Computer Science, Military University of Technology, Gen. W. Urbanowicza 2 St., 00-908 Warsaw, Poland; aleksandra.szafranska@wat.edu.pl (A.S.); pawel.baranowski@wat.edu.pl (P.B.); pawel.bogusz@wat.edu.pl (P.B.)
[2] Department of Advanced Materials and Technologies; Military University of Technology, Gen. W. Urbanowicza 2 St., 00-908 Warsaw, Poland; anna.dudka@wat.edu.pl (A.A.-D.); dariusz.zasada@wat.edu.pl (D.Z.); tomasz.czujko@wat.edu.pl (T.C.)
* Correspondence: jerzy.malachowski@wat.edu.pl; Tel.: +48-261-839-683

Received: 15 February 2019; Accepted: 13 March 2019; Published: 16 March 2019

Abstract: This paper presents a characterization study of specimens manufactured from Ti-6Al-4V powder with the use of laser engineered net shaping technology (LENS). Two different orientations of the specimens were considered to analyze the loading direction influence on the material mechanical properties. Moreover, two sets of specimens, as-built (without heat treatment) and after heat treatment, were used. An optical measurement system was also adopted for determining deformation of the specimen, areas of minimum and the maximum principal strain, and an effective plastic strain value at failure. The loading direction dependence on the material properties was observed with a significant influence of the orientation on the stress and strain level. Microstructure characterization was examined with the use of optical and scanning electron microscopes (SEM); in addition, the electron backscatter diffraction (EBSD) was also used. The fracture mechanism was discussed based on the fractography analysis. The presented comprehensive methodology proved to be effective and it could be implemented for different materials in additive technologies. The material data was used to obtain parameters for the selected constitutive model to simulate the energy absorbing structures manufactured with LENS technology. Therefore, a brief discussion related to numerical modelling of the LENS Ti-6Al-4V alloy was also included in the paper. The numerical modelling confirmed the correctness of the acquired material data resulting in a reasonable reproduction of the material behavior during the cellular structure deformation process.

Keywords: additive manufacturing; Ti-6Al-4V; LENS; mechanical characterization

1. Introduction

Additive manufacturing (AM) is currently the fastest growing manufacturing method. It allows for the obtainment of products with a complex geometry and a high dimensional accuracy, as well as with high strength parameters [1]. The applied technology depends mainly on the batch material used. Products made of polymeric materials are obtained with such techniques as fused deposition modeling (FDM), where the material is a thermoplast in a circular wire; stereolitography (SLA), where the material is a light-cured resin; and selective laser sintering (SLS), where the batch material is a powdered polymer [2,3]. For metal alloys, techniques including powder bed fusion (PBF), selective laser melting (SLM), electron beam melting (EBM) and direct energy deposition (DED), for instance, laser engineered net shaping (LENS) [4–7] are used.

The tested specimens were prepared from the powder of Ti-6Al-4V alloy using LENS technology, where the metal powder is melted with a high-power laser beam, layer by layer. The technology enables repairing the damaged parts and manufacturing new fully functional elements, with properties generally comparable to those made by conventional methods [8–10]. As a result of high cooling rates during the process, the products have a fine-grained structure, which increases their mechanical properties. The disadvantage of these products are high surface roughness and reduced plasticity associated to the residual stress generated during the rapid material solidification. Additionally, fatigue strength is limited but their higher strength is significant in high cycle fatigue [11–13]. It is possible to use a wide range of materials such as stainless and tool steel, titanium alloys, aluminum, copper, nickel, or engineering ceramics [8,14].

The tested alloy is Ti-6Al-4V, which is a lightweight alloy with high strength properties. It is quite a good material for manufacturing thin wall structures [15,16]. The mechanical properties of the alloy depend on the phase arrangement as well as on the grain morphology. Shortly after the production process, as a result of the LENS process, the tested alloy is characterized by a fine-grain martensitic structure α', as in the case of the SLM process [7]. The samples have an increased yield point and a tensile strength with reduced plasticity. In order to improve the LENS Ti-6Al-4V plasticity and anisotropic response, a heat treatment [7,10,13,17–20] is applied.

AM development contributed directly to the origin of a new group of functional materials, such as regular cellular structures. The mechanical properties of the structures shall be designed depending on their future application. A characteristic feature of the products is a low relative density and high strength [3,21–24]. Currently, there are many studies on energy-consuming structures produced using AM in various technologies, using a whole range of materials [2,21,23,25]. These structures are often inspired by nature, produced in the form of micro-trusses [21,26–28] or honeycomb structures [2,3,22,23,25], and their complex geometry would not be possible to obtain with conventional manufacturing methods. Energy-absorbing properties are closely related to the adopted topology of structures as well as processing parameters of the manufacturing process.

In the process of designing structures produced by additive techniques, numerical methods have been increasingly used to optimize the process parameters and the structure morphology in terms of expected utility features [3,8,23–25]. To develop an accurate numerical model, it is necessary to conduct a series of material tests to identify the basic material features. Through research of the microstructure, it is possible to take into account local effects (pores, inclusions, phase share) that have an impact on the global behavior of the whole structure [29–35].

The paper presents a comprehensive methodology for testing, modelling, and analyzing the Ti-6Al-4V alloy obtained by LENS technology, belonging to DED techniques. This technique is characterized by high density of a laser beam (thousands of $J \cdot mm^{-1}$) and a heat flow from a molten metal pool, which is controlled by conduction through a manufactured detail and the applied substrate, as well as by conduction through shielding gas and a nozzle supplying metal powder. It is possible to acquire a layer thickness of the level of 0.3–1 mm [11]. A proper selection of technological parameters and their impact on the obtained microstructure and mechanical properties [13,19,36,37] is also discussed. During the investigations, the relationship between the assumed wall thickness and the obtained microstructure [37,38], defect analysis, and fatigue strength of the samples [4,12,13,39] was analyzed.

The tests carried out using light microscopy aimed to determine the homogeneity of the microstructure of the analyzed alloy. Detailed structural studies were carried out on a scanning microscope equipped, among others, with an electron backscattered diffraction (EBSD), which allowed for a precise determination of the alloy phase composition, shape, and grain size. In addition, the fractures of the stretched samples were examined, which allowed linking of the obtained fracture type (brittle, plastic, or mixed) to the microstructure type (α' or $\alpha + \beta$) [12,13,39,40].

Measurements performed during the tensile test were supported by the Aramis system based on the digital image correlation (DIC) method. Such measurement equipment was successively used for measuring strains for specimens made of steel [41,42], wood [43], rubber [44], and foam [45]. It can be also adopted to observe the strains level within the tested structures, such as a seat belt, lattice structures, or composite joints [45,46]. This allowed for the determination of the plastic strain at the samples failure, which is impossible to be captured by means of traditional measurement methods (resistance tensometry and extensometry). The tensile test allows for the determination of the basic mechanical properties, such as Young's modulus, yield strength, tensile strength, maximum elongation, and Poisson's ratio. The obtained results were implemented into the constitutive material model to perform finite element analyses (FEA) of a uniaxial compression test of the selected energy-absorbing structures.

On the basis of the conducted tests, the obtained mechanical properties were confirmed and only slight differences were noted, due to the assumed technological parameters of the processes as well as the technology itself (SLM, EBM, LENS). These tests focused on identifying basic material features without indicating specific applications.

The paper is organized as follows: In Section 2 the specimens manufacturing process as well as the microstructural and tensile tests are described in detail. In Section 3, the results of uniaxial tensile tests and the microstructure photos are presented and discussed. The conclusions are included in the final section, Section 4.

2. Materials and Methods

2.1. Specimens Manufacturing Technology

The commercial Ti-6Al-4V powder, produced with an argon atomization method, was used for manufacturing the above mentioned square box structures. It was delivered by TLS Technik GmBH and Co (Bitterfeld-Wolfen, Germany) and the size of the particles was in the range of 44 to 105 µm. Optomec LENS MR 7 device (Albuquerque, NM, USA) with a 500 W laser (Figure 1) was used for the manufacturing of the thin-walled square boxes, with dimensions 44 mm × 44 mm × 40 mm (width × length × height) and walls' thickness equal to 1.5 mm (Figure 2a). The Ti-6Al-4V substrate plate was sandblasted and degreased with acetone before manufacturing. The thin-walled components were manufactured with the process parameters determined experimentally and they are listed in Table 1.

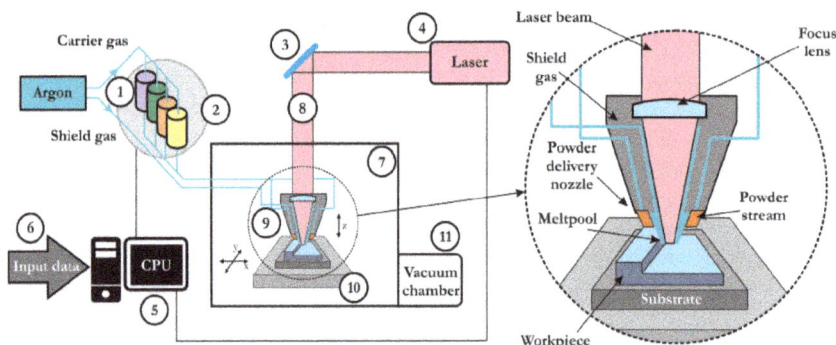

Figure 1. Laser engineered net shaping (LENS) system scheme [8].

Table 1. Process parameters.

Process Parameters	Value
Laser power (W)	400
Powder flow rate (rpm)	11.5
Layer thickness (µm)	30
Feed rate (mm/s)	20
Oxygen concentration (ppm)	>5

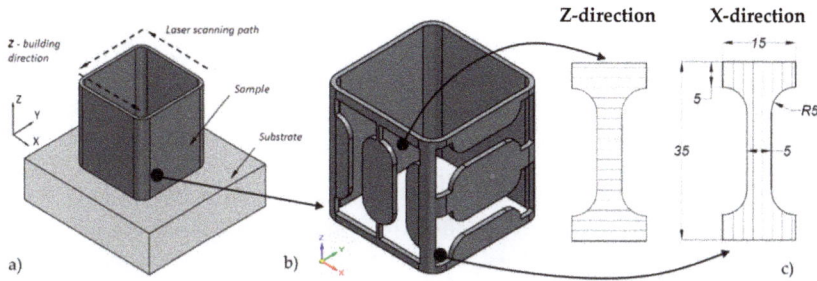

Figure 2. Strategy for manufacturing samples with the LENS technology: (**a**) Thin-walled square box; (**b**) method for cutting out specimens in two directions; and (**c**) geometry of the specimens with given dimensions.

One of the manufactured thin-walled boxes was subjected to heat treatment to obtain a two-phase structure and to improve the material ductility. The heat treatment process was conducted in a tubular furnace at low vacuum ($\times 10^{-2}$ mbar). Before heating, the chamber was purged by flow of argon. The square box samples were annealed in a 1050 °C for 2 h and cooled down with the furnace. Heat treatment conditions were selected based on the previous studies [19,47]. The square box samples were annealed in a 1050 °C for 2 h and cooled down with the furnace.

After heating, the components were sandblasted to remove the surface impurities. The sandblasting process was performed using a sandblast machine P-05 (ZAP-BP, Kutno, Poland). The Al_2O_3 powder, with granulation of 90–150 µm, was applied as the abrasive agent and the surface of the details was sandblasted until it was matted.

The square-box geometry was developed in such a manner that the dumbbell specimens can be easily cut out in two directions: X-direction according to the laser beam scanning path and Z-direction consistent with the direction of the structure manufacturing (Figure 2a). The BP-97d electro discharge machining device (SEW MET, Ligota, Poland) was used for the cutting process. It is worth noticing that a smaller thickness was impossible to be obtained due to technological constraints. It was also required that the corner radius was large enough to not initiate cracking during tensile tests. Specimen geometry was designed to satisfy a 1:5 ratio of width to length measurement condition included in PN-EN ISO 6892-1. The specimen geometry was the smallest possible, in which the Aramis system could correctly measure the deformation. The dog bone samples were cut out in two directions to analyze the anisotropy of the material and its impact on the stiffness parameters of the samples [12,35,36]. The samples were tested for the material before and after heat treatment [16,34]. In addition, the roughness was measured, which allowed for the determination of the quality of the surface after production.

Before testing, each specimen was subjected to manual grinding with 120, 240, 600, and 1200 SiC papers, in order to eliminate crack initiator, due to a high roughness of the surface, which was measured using the Keyence VHX-6000 microscope (Keyence, Mechelen, Belgium) with a 1000 times magnification and with a special software (using depth from defocus algorithm, which enabled the obtainment of 3D information based on 2D image sharpness). Table 2 shows the results of measurements.

Table 2. Surface roughness of the LENS samples.

Surface Roughness Ra (µm)	Value
As-built X-direction	50 ± 10
As-built Z-direction	48 ± 8
After heat-treatment X-direction	40 ± 5
After heat-treatment Z-direction	43 ± 3
After polishing	8 ± 4

2.2. Microstructure Analysis

Before conducting the microstructural examination, the samples were subjected to grinding with 600–4000 SiC papers, polishing with 3 and 0.25 μm diamond suspensions, and final polishing with colloidal silica suspension. After performing the EBSD investigation, the samples were etched with the Kroll's reagent to examine the microstructure with an optical microscope Nikon MA 200 (Nikon Metrology, CA, USA). To study the crystallographic orientation, imaging and texture were performed with EBSD FEI quanta 3D dual beam field emission gun scanning electron (FEI, OR, USA).

The fracture surface analyses were performed using secondary electron imaging on Quanta 3D FEI dual beam. Such a procedure allowed for qualitative evaluation, a fracture mode, and fracture surface details (e.g., defects, porosity).

2.3. Uniaxial Tensile Testing for Material Data Aquisition

In order to identify the material parameters of the LENS Ti-6Al-4V alloy that had an impact on the behavior of energy absorbing structures under load conditions, a uniaxial tensile test was carried out according to PN-EN ISO 6892-1 standard, using a universal strength machine INSTRON 8862 at a room temperature of 23 °C, at a cross-head displacement rate of 1 mm/s. Twelve specimens (6 from each direction and 6 before and after heat treatment) were tested. Yield stress, ultimate tensile stress, elongation at failure, and Young's modulus were determined using Aramis optical deformation measurement system and the measured cross-head displacement (Figure 3a). In Figure 3b, the surface condition of the specimen after cutting out from a thin square box, after mechanical grinding, and followed by stochastic pattern application on the measuring section (gauge length) is presented on the left, in the center, and on the right, respectively.

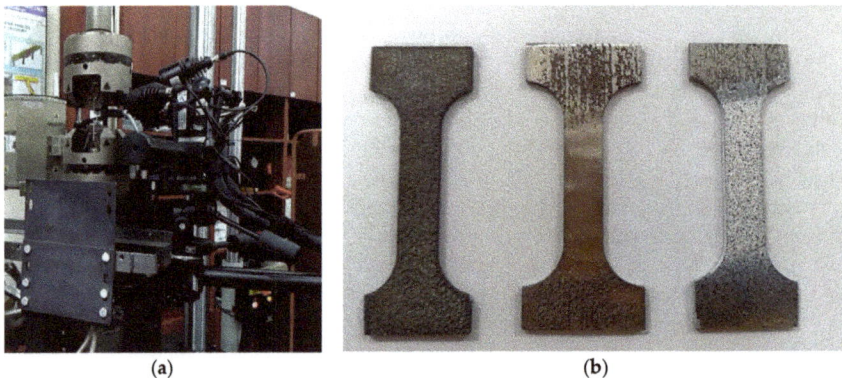

(a)   (b)

Figure 3. (a) Measuring position; and (b) preparation of the sample for test with digital image correlation (DIC) Aramis.

The advantages of using optical methods that provide a partial image of deformation distribution using DIC methods are associated with the ability to identify changes in the material structure at the micro-scale level. Based on the contrasting random surface texture of the side lock elements (pattern random point's paint to the test object), the system divides the image into the working areas called facets. These facets can be correlated with the corresponding areas on the successive captured images. Subdivisions are of several pixels in size. The optical deformation measurement system compares the successive images to the photo taken before the initial application of the load. Subsequently, three-dimensional maps of displacements and deformations for all the facets were computed.

Owing to Aramis system, deformation of the tensile bars, the areas of minimum, and the maximum principal strain were determined. The Aramis optical-measurement system is designed for non-contact displacement measurements in both flat and spatial elements subjected to loading. It consists of a set of cameras registering the shape changes of the tested object and a suitably adapted and programmed computer that stores and processes the recorded images. Depending on the configuration (i.e., the number and speed of cameras), the system can be used to analyze the displacement and deformation fields of both flat or spatial elements subjected to static or dynamic load [42].

The optical system was calibrated for measurements using a calibration plate with dimensions of 30 × 24 mm before the test. This procedure enabled obtainment of the measurement area with dimensions of 35 × 26 mm, where the tested elements of the side lock were located. The lenses with a focal length of 50 mm were used. Cameras, distanced from the specimen front surface by 225 mm, registered the images with a frequency of 5 frames per second. In the analysis, the facets of 0.5 × 0.5 pixel size (about 0.05 × 0.05 mm) were employed. The surface of samples prepared to obtain a stochastic pattern is shown in Figure 3b. It was required to color the surface white, which is the background, and, in the next step, the specimen was sprayed with black dots, which created a random pattern. This allowed the system to identify each fragment of the specimen and to split the images into rectangular areas called facets, which were correlated with the corresponding areas on the other frames. The specimen gauge length was determined in the post-processor and amounted to approximately 10 mm (Figure 4). The results of strain measurements in Aramis system were synchronized in time with displacement and force signals from the testing machine (SATEC).

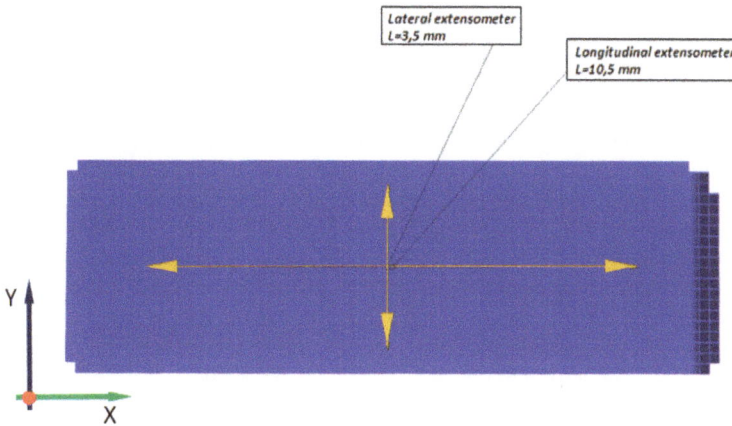

Figure 4. The specimen gauge length.

Logarithmic strains are computed locally in X- and Y-directions of the selected coordinate system in each photo as follows [42,46]:

$$\varepsilon_{log} = \log \lambda \tag{1}$$

The value of λ is calculated based on the following formula:

$$\lambda = \lim_{l \to 0} \frac{l + \Delta l}{l} \tag{2}$$

Symbol l_0 denotes the initial distance between two neighboring facets and Δl is the distance increased during the test.

Engineering strains are computed locally in X- and Y-directions of the selected coordinate system in each photo as follows:

$$\varepsilon_{eng} = \lambda - 1 \tag{3}$$

A coordinate system, in which X-axis extends along the load axis of the sample, was assumed. The measurement field was selected in the middle part of the sample.

The local value of the longitudinal and transverse strains was averaged within the area of measurement and used for further calculations, as well as for drawing the graphs of the tensile curves. Actual engineering stress values were calculated according to the following equation:

$$\sigma_{eng} = \frac{P}{A_0} \tag{4}$$

where *P*—current value of the tensile force, A_0—initial cross-section.

Tensile strength was determined according to the following equation:

$$R_m = \frac{P_{max}}{A_0} \tag{5}$$

where P_{max} is the maximum force registered during the test.

3. Results and Discussion

3.1. Microstructure Analysis of As-Built and after Heat-Treatment LENS Ti-6Al-4V

In the LENS process, application of the successive powder layers re-melts the grains, which serve as seeds, causing the epitaxial growth of subsequent columnar grains. Figure 5 shows the results of optical microscopy exam microstructure of the alloy before heat treatment in the longitudinal section (Figure 5a) and in the cross-section (Figure 5b). The analyzed samples were characterized by a coniferous microstructure, which was formed, first of all, on the grain boundaries of β phase and gradually filled up the entire grain space. The obtained microstructure of LENS Ti-6Al-4V is similar to a microstructure reported during research [10,11,13].

This microstructure is a characteristic feature of α′ phase, which is defined in the literature as a consequence of α phase hardening [9]. Obtainment of the microstructure is possible during fast cooling above the critical rate which for this type of material is at the level of about 1000 K/s [48]. The cooling rate above 1000 K/s is characteristic for processes using laser radiation, including LENS methods. Then, β phase is transformed directly into α′ phase, instead of α + β phase, as it is the case of the furnace cooling.

The analysis of the microstructure at low magnification allowed for the observation of the original β phase boundaries. The prior β grains had long axes of 0.25 to 0.75 mm, aligned with the build direction, and short axes from 0.2 to 0.3 mm in width. The conducted comprehensive literature study on the primary β-grain structure and texture demonstrated a strong $\langle 001 \rangle \beta$//Nz fiber texture of prior β grains, with the longitudinal axis parallel to the building direction in Ti-6Al-4V alloy fabricated by additive manufacturing [11,49–51]. Moreover, the α′ phase of the as-built components had a weak texture because of the relatively high number of variants that precipitate within each columnar β grain [50].

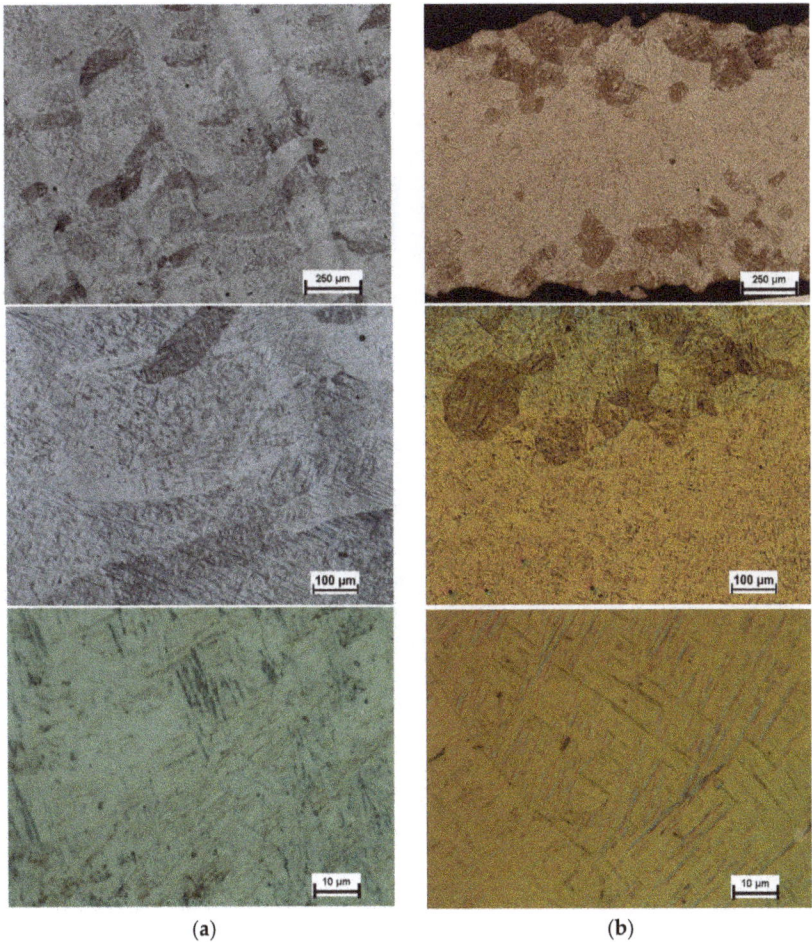

(a)   (b)

Figure 5. Optical micrographs of Ti-6Al-4V microstructure as-built: (**a**) longitudinal section; and (**b**) cross-section.

Figure 6 shows the microstructure of the alloy after heat treatment examined by optical microscopy. In this case, the presence of a two-phase structure is clearly visible and results are corresponding with those reported in [10,52]. The microstructure contains α-phase lamellas (white areas) and β-phase lamellas (black areas). Visible primary grains of the β-phase indicate a clear growth of these grains during the applied heat treatment. The barely visible black spots are few small pores.

Strongly-textured structures can lead to significant anisotropic mechanical properties causing different mechanical responses to the external loading along different sample orientations [50].

The EBSD data produced the inverse pole figure maps, pole figures, and orientation distribution functions (represented by generalized spherical harmonics). Figure 7 illustrates an EBSD orientation map of the LENS Ti-6Al-4V as-built. EBSD analysis proved the absence of β phase (phase fraction below 1%) in the Ti-6Al-4V alloy structure immediately after production. The martensite needles with different crystallographic orientations are clearly visible. The presence of wide-angle boundaries (above 15°) was detected. It means that a single grain was martensite needle, whose typical structure and morphology are shown in Figure 7b.

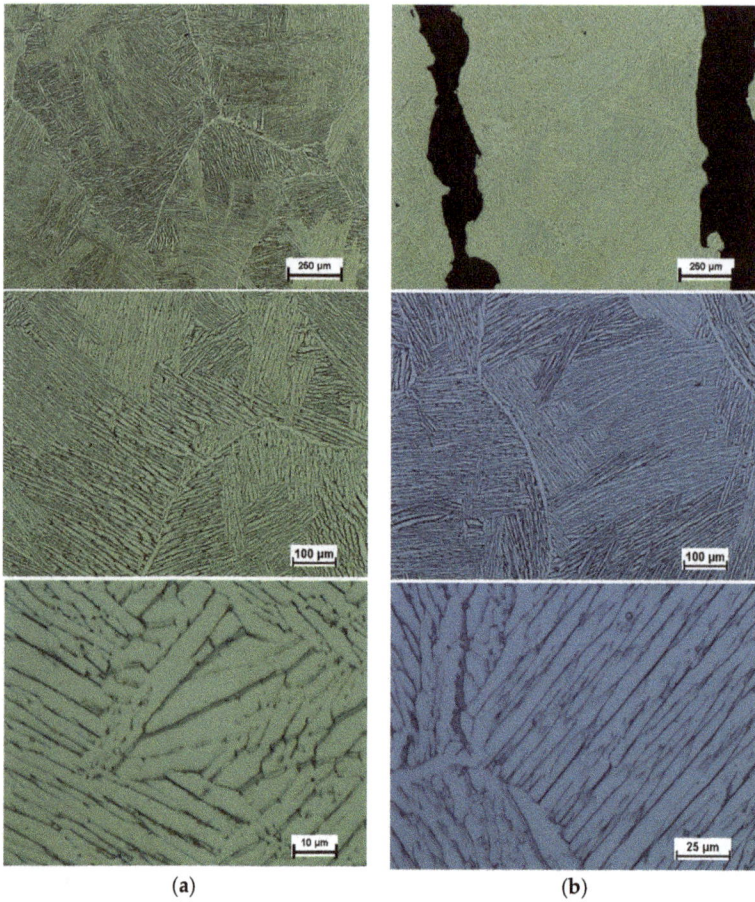

Figure 6. Optical micrographs of Ti-6Al-4V microstructure after heat-treatment: (**a**) longitudinal section; and (**b**) cross-section.

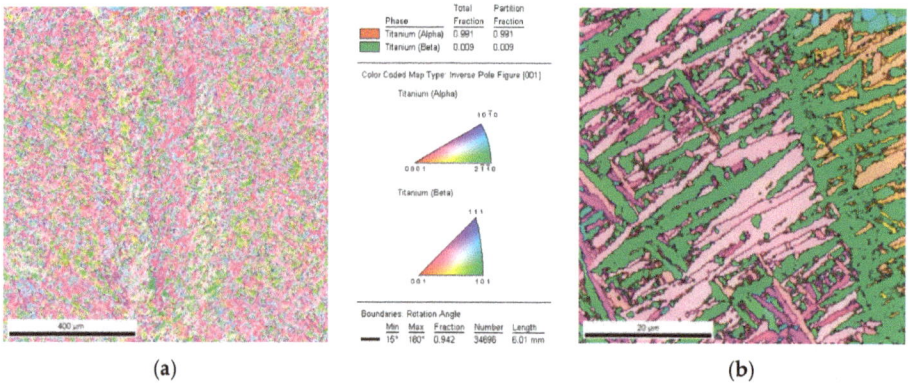

Figure 7. Electron Backscatter Diffraction (EBSD) orientation map of LENS Ti-6Al-4V (**a**) as-built; and (**b**) grain boundary reconstruction.

Figure 8 presents the EBSD orientation map of the LENS Ti-6Al-4V after heat treatment. The data obtained from EBSD analysis proved the presence of both α and β phase in the samples after heat treatment. The map of the phase distribution presented in Figure 8b shows that the amount of β phase was clearly smaller than α phase, which was about 8%.

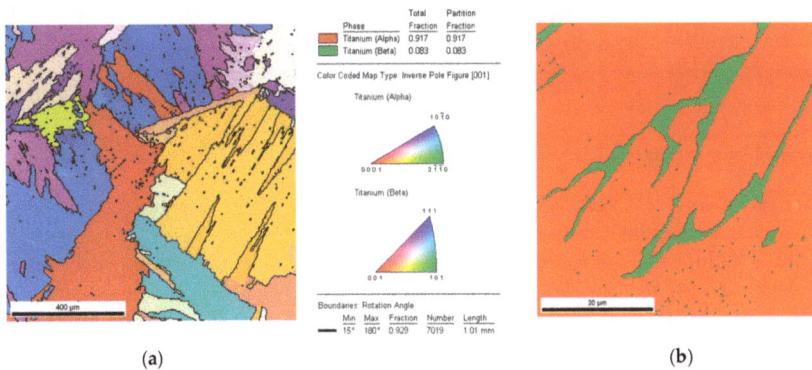

(a)   (b)

Figure 8. EBSD orientation map of LENS Ti-6Al-4V (**a**) after heat treatment; and (**b**) phase distribution map.

3.2. Uniaxial Tensile Testing of As-Built and after Heat-Treatment LENS Ti-6Al-4V Material

Based on Aramis measurements, local axial strains and local shear strains, which occurred in the plasticized area of the neck, were obtained. Figure 9 presents the selected stages of the tensile test. Figure 9a presents an unloaded specimen at the beginning of the test, Figure 9b presents the stage before the failure and necking is also observed. The final stage is illustrated in Figure 9c and presents specimens after failure. Within the analyzed area, marked with a green color, local strain in facets were measured.

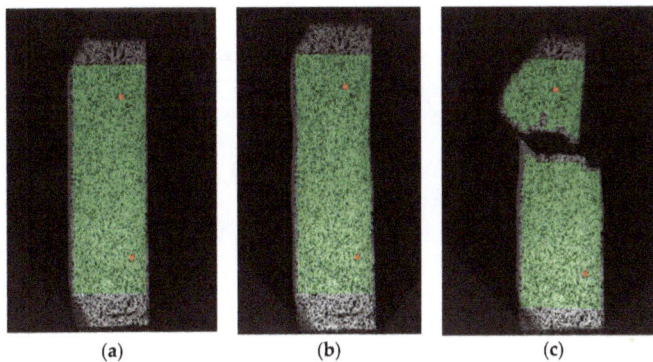

(a)   (b)   (c)

Figure 9. Picture of the specimen as-built with the analyzed area, marked with a green color: (**a**) Stage "0" at the beginning of the test; (**b**) stage in the middle of the test; and (**c**) stage registered in the first frame after failure.

Distribution of principal logarithmic strains are presented in Figure 10a,b, which correspond to the photos taken on frame before failure (Figure 9b). Local axes are also included in the results. Figure 10a presents specimen as-built stretched at X-direction and Figure 10b presents the same results for the material manufactured in Z-direction. Plastic strain for X-direction was approximately 21.7% before failure and for Z-direction about 17%. The Z-direction specimen failed near the grip.

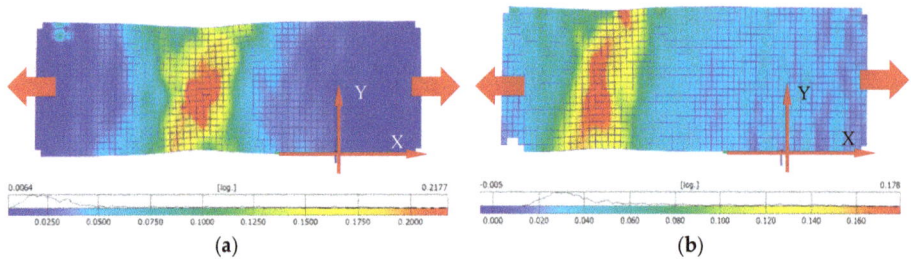

Figure 10. Map of principal strain before failure in specimen as-built: (**a**) X-direction; and (**b**) Z-direction.

Figure 11a presents the specimen after heat treatment stretched in the X-direction and Figure 11b presents stretching it in the Z-direction. Plastic strain for the X-direction was approximately 20.6% before failure and for the Z-direction it was about 24%. The strain at fracture for the Z-direction increased compared to the sample before heat treatment. Strain concentration the specimens after heat treatment occurred outside their central part, compared to as-built specimens with central strain concentration.

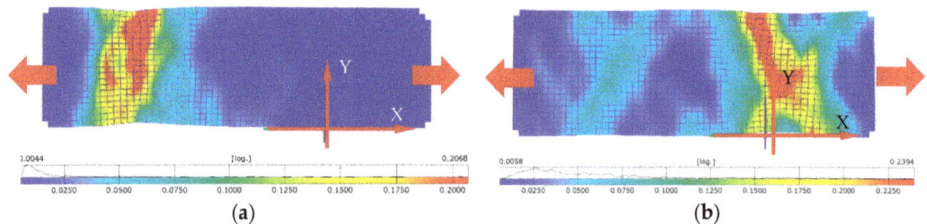

Figure 11. Map of principal strain before failure in the specimen after heat treatment: (**a**) X-direction; and (**b**) Z-direction.

In Figure 12, engineering stress vs. strain characteristics, calculated from the uniaxial tests, are presented for X and Z specimens. The results before and after heat treatment are shown in Figure 12a,b, respectively. For the as-built specimens, X-direction of cutting resulted in slightly higher ductility, whereas in the Z-direction an opposite correlation was observed. A visible influence of the heat process was observed in the decreasing of strength while increasing the ductility.

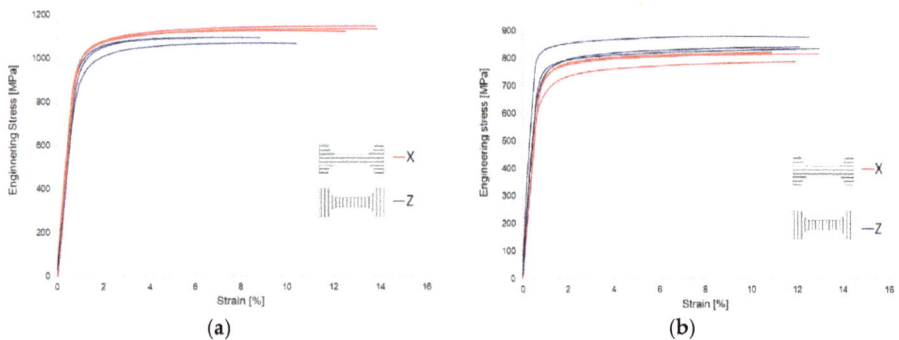

Figure 12. Engineering tensile stress–strain curves of the LENS Ti-6Al-4V: (**a**) As-built; and (**b**) after heat treatment.

The results obtained from the tensile test of LENS Ti-6Al-4V as-built are presented in Table 3. The sample direction (X vs. Z) significantly influenced the mechanical properties. The elongation of the Z-direction sample was on average 30% lower compared to the X-direction sample. Stiffness of the X-direction sample was about 8% higher than in the case of the Z-direction sample. The value of effective plastic strain was read from Figure 10a,b.

Table 3. Tensile properties of LENS Ti-6Al-4V as-built.

Direction	Yield Stress σ_y (MPa)	Ultimate Tensile Stress UTS (MPa)	Elongation ε (%)	Effective Plastic Strain EPFS (%)	Young's Modulus E (GPa)
X	1080 ± 4	1139 ± 12	13.1 ± 0.7	21.7	118 ± 2
Z	1038 ± 28	1080 ± 13	8.7 ± 1.8	17.8	109 ± 5

The results obtained from the tensile test of LENS Ti-6Al-4V after heat treatment are presented in Table 4. The evaluated heat treatment clearly influences all the measured properties and; therefore, it can be concluded that the anisotropic features were minimized. After heat treatment, a significant increase in ductility was observed as a result of α' decomposition into $\alpha + \beta$ structure. The value of effective plastic strain was read from Figure 11a,b.

Table 4. Tensile properties of LENS Ti-6Al-4V after heat treatment.

Direction	Yield Stress σ_y (MPa)	Ultimate Tensile Stress UTS (MPa)	Elongation ε (%)	Effective Plastic Strain EPFS (%)	Young's Modulus E (GPa)
X	762 ± 20	813 ± 16	13.8 ± 2.3	20.6	100 ± 4
Z	808 ± 38	858 ± 27	14.3 ± 3.5	23.9	108 ± 10

Table 5 compares the results of tests available in literature. There were selected positions, in which the specimens of similar geometry and dimensions were uniaxially stretched. Additionally, the loading direction was defined and denoted according to terminology adopted in the article. The data included in the Table 5 show that there was anisotropy of mechanical properties in the specimens and the greatest differences were observed on their elongation. There was also observed an influence of heat treatment owing to which an anisotropy effect in the structure was reduced and larger elongation of the specimens was achieved.

Table 5. Tensile properties of Ti-6Al-4V obtained with different AM techniques.

Method		Geometry of Sample	Direction	Yield Stress σ_y (MPa)	Ultimate Tensile Stress UTS (MPa)	Elongation ε (%)
Direct Energy Deposition (DED) [11]	As-built	Tall wall	X	960 ± 26	1063 ± 20	13.3 ± 1.8
			Z	945 ± 13	1041 ± 12	18.7 ± 1.7
Powder Bed Fusion—Selective Laser Melting (SLM) [50]	As-built	Dog bone	X	978 ± 5	1143 ± 6	11.8 ± 0.5
	Heat treated		Z	967 ± 10	1117 ± 49	8.9 ± 0.4
			X	958 ± 6	1057 ± 8	12.4 ± 0.7
			Z	937 ± 9	1052 ± 11	9.6 ± 0.9
Electron Beam Melting (EBM) [10]	As-built	Dog bone	X	1006	1066	15
			Z	1051	1116	11
	Heat treated		-	1039	1294	10

On the bases of summarization in Table 5, it can be concluded that the obtained results are comparable with data reported in literature.

3.3. Fracture Mechanism

Fracture surface analyses were conducted using secondary electron imaging on a scanning microscope. In the case of as-built samples, a brittle-ductile fracture, with a predominantly brittle fracture, was obtained (Figure 13a). Although the dimple rupture dominated, quasi-cleavage facets features of a brittle type were observed. After the tensile tests, the fracture surfaces in the as-built and stress-relieved samples resembled a "cup and cone", which is a characteristic for ductile types of the fractures (Figures 13b and 14b). The presented fractographs are similar to the previous studies [10,11].

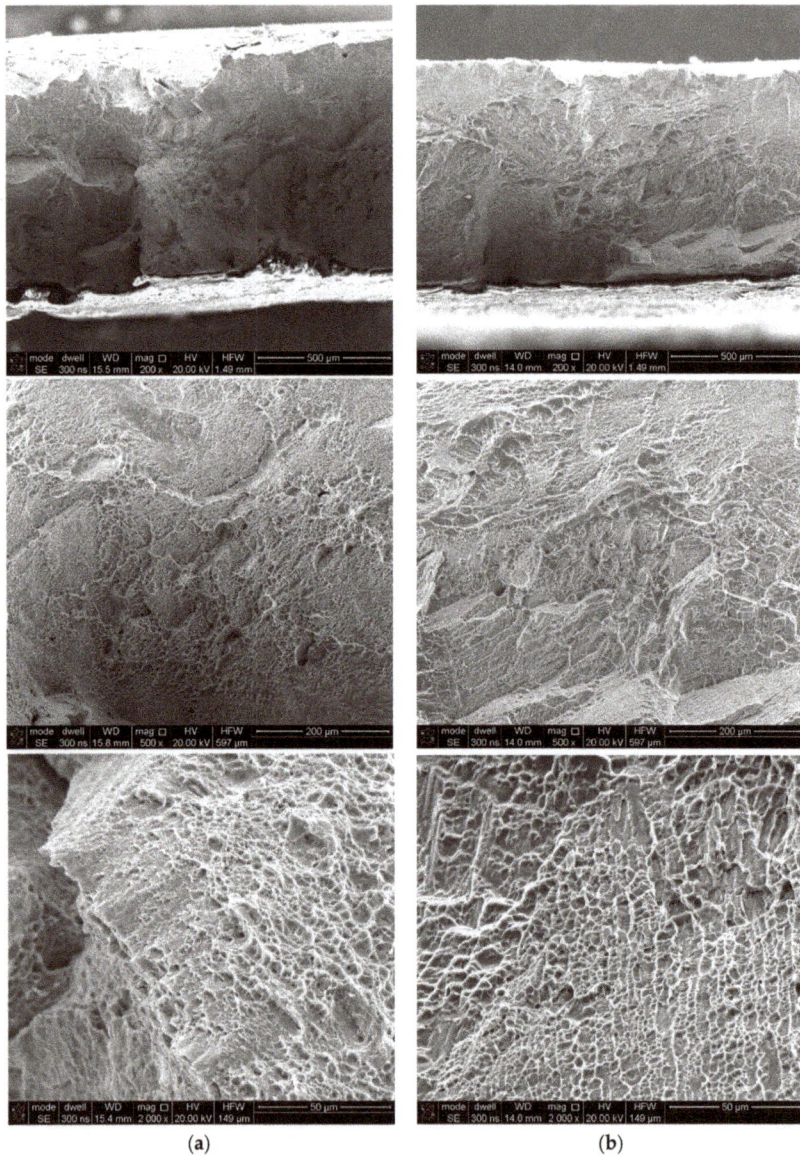

(a)   (b)

Figure 13. Fracture surfaces of the LENS Ti-6Al-4V longitudinal section: (a) As-built; and (b) after heat treatment.

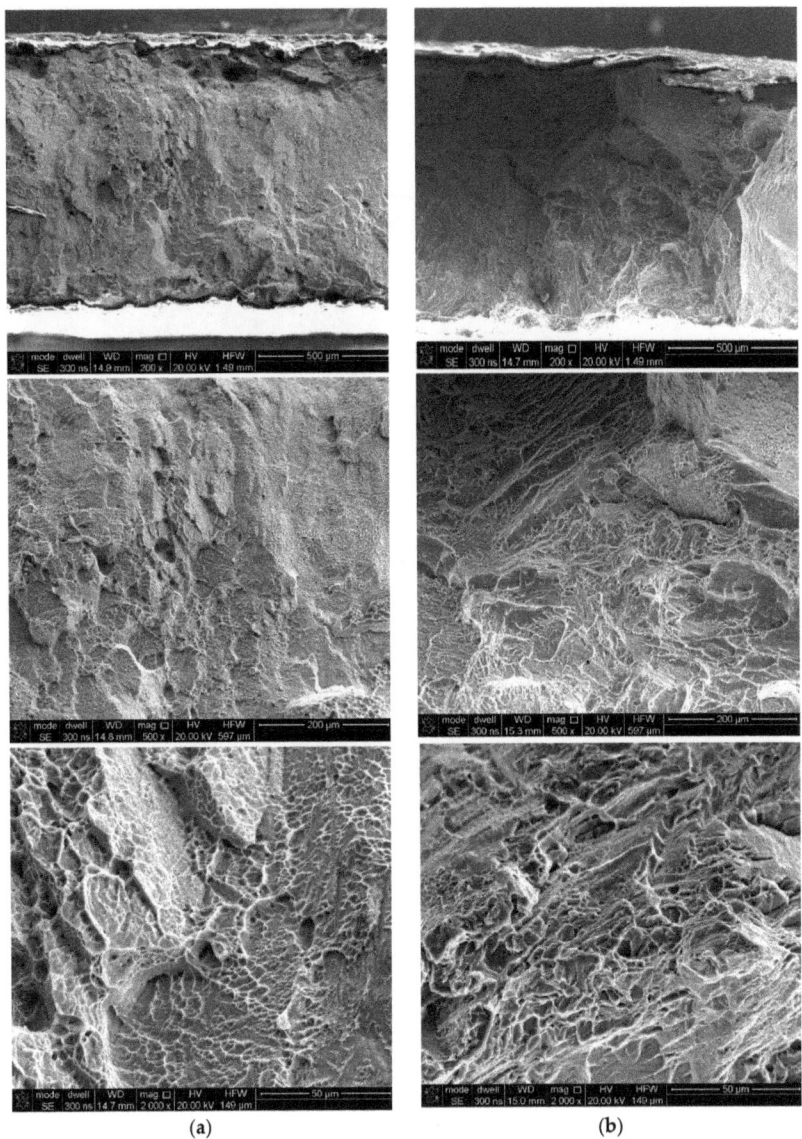

Figure 14. Fracture surfaces of the LENS Ti-6Al-4V cross section (**a**) as-built (**b**) after heat treatment.

3.4. Constitutive Modelling with FEA

The discussed material data was used for determining the parameters for the selected constitutive model. An elasto-visco-plastic material model was adopted for predicting the Ti-6Al-4V behavior. In the first step of FEA, the material model was correlated based on the uniaxial tensile tests. The elasticity modulus, Poisson's ratio, yield stress, and EPFS were determined from an experimental uniaxial tensile test as the average values from the two directions of sample (see Table 4). The adopted constitutive model uses an effective stress (ES) vs. effective plastic strain (EPS) curve, in which the last point corresponds to experimental values of ultimate stress and EPFS. The results obtained from the numerical simulation of the uniaxial tensile tests were compared with the Aramis measurements

(Figure 15). The elastic and plastic ranges were in excellent agreement with the experimental curve. Moreover, strain distributions within the neck of the dumbbell specimen obtained from FEA and Aramis measurement were also very similar. This indicates that the constitutive model was correct and could be applied for simulating the cellular structure.

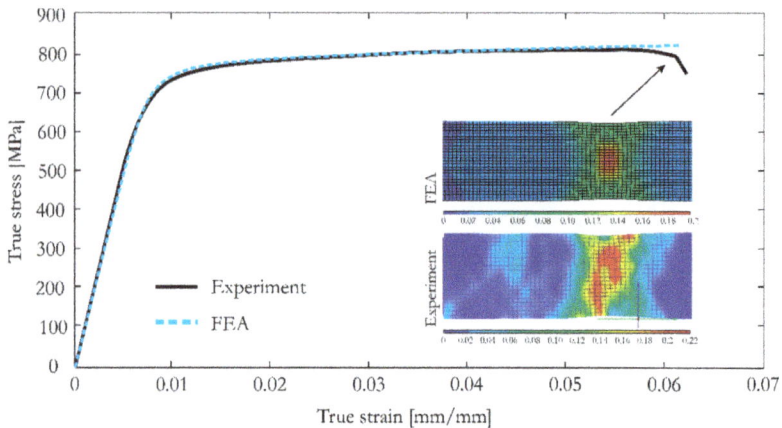

Figure 15. Correlation of the constitutive model based on true stress vs. true strain curves and strain distribution within the specimen neck: Finite Element Analysis (FEA) and experiment.

Obtainment of the correlated material model was necessary to carry out numerical simulation of the compression tests for a selected honeycomb cellular structure, with a geometry similar to a cube with dimensions as follows: width: 40 mm, height: 40 mm, and thickness: 10 mm. The wall thickness was approximately 0.7 mm and the elementary cell diameter was 5 mm. The results of the FEA were compared with the outcomes from the actual tests, which were conducted on a universal MTS Criterion C45 strength machine (MTS, MN, USA). During the experimental test the loading velocity was 1 mm/s. The honeycomb structure was compressed until reaching 50% of its height. For recording the structure deformation during experimental testing, the fast Phantom V12 camera (Vision Research, NJ, USA) was used.

The results of experimental tests and numerical simulations were compared. In Figure 16, force characteristics are presented, and the experiments with the honeycomb structure after (depicted as HT) and without the heat treatment (depicted as NHT) are also included. A satisfactory reproduction of the actual results can be observed, and the influence of heat-treatment is visible, resulting in a more brittle fracture of the structure with no heat-treatment effect. Thus, the obtained results confirmed the observations during the tensile test (see Section 3.2). In Figure 17, the comparison between the deformations of the structure observed during experimental and numerical tests are presented. Four selected time frames of the compression process were considered for comparison purposes. An excellent reproduction of the structure behavior can be observed with a similar fracture characteristic in all stages. The FEA results proved that the discrete representation and the constitutive model, with the properties obtained from the discussed uniaxial tensile tests, were correctly developed.

A detailed description of the FE model, including adopted modeling techniques, methodology for determining properties for the constitutive model, and a validation procedure can be found in [25].

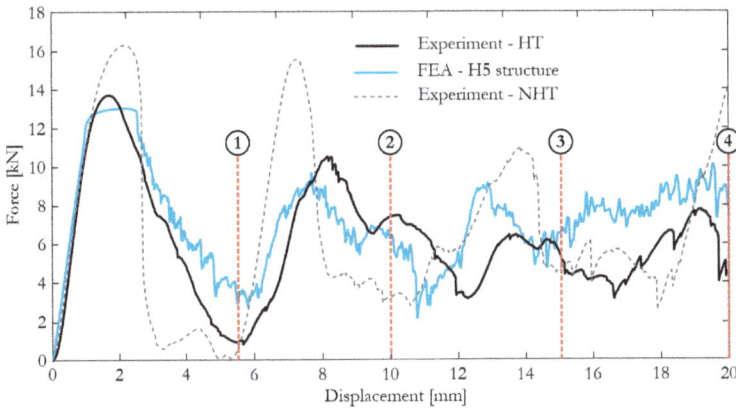

Figure 16. Force characteristics obtained from FEA and experiments for the H5 structure [25] (Elsevier, 2018, Additive Manufacturing).

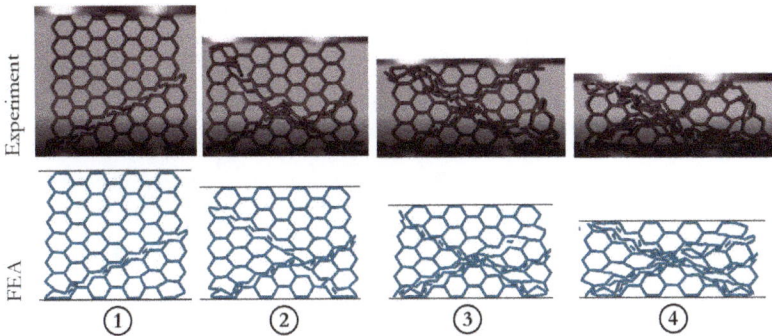

Figure 17. Selected time frames of the H5 structure compression process: Comparison between the FEA and the experiment [25] (Elsevier, 2018, Additive Manufacturing).

4. Conclusions

In this study, specimens for tensile test, cut in two build orientations, were evaluated in two heat treatment conditions: the as-built condition and after heat treatment. Their microstructure as well as mechanical properties were evaluated. For this purpose, a tensile test was carried out, which showed that both the direction of the load and the heat treatment conditions affect the mechanical properties of the alloy.

- In the case of a microstructure, there was a clear difference between as-built and after heat treatment specimens. In both conditions, prior-β grains were found to be aligned with the manufacturing direction. Ti-6Al-4V fabricated by LENS consists of columnar prior-β grains filled with acicular α martensite, and showed high yield strength with reduced plasticity compared with alloy manufactured conventionally.
- The analysis of the results obtained from the static tensile test indicated a significant influence of the as-built sample loading direction, since specimen elongation in the Z-direction was 30% smaller.
- The results show a significant influence of heat treatment, both on the alloy microstructure and its mechanical properties. Obtainment of a dual-phase structure increased specimen plasticity and

reduced anisotropy of mechanical properties along, with a decrease of yield stress and ultimate tensile stress.

- It was noticed that in the Ti-6Al-4V alloy obtained by traditional methods, ductile fractures were obtained in which the characteristic elements of the brittle fracture are not visible. This difference is due to the existence of α' martensite, which is characterized by higher durability but worse plasticity. The partial re-melting of the previous layer of applied powder caused the grain to lengthen according to the direction of the model's production.
- The adopted numerical methodology proved to be suitable for simulating energy-absorbing cellular structures. Furthermore, the implemented constitutive model described, based on the experimental material tests, resulted in a reasonable reproduction of the material behavior during the deformation process. Application of numerical methods enabled the reduction of time consuming and expensive experimental methods.

In future studies, obtained results will be used to create a large-scale cell structure model from the LENS Ti-6Al-4V alloy subjected to a compressive load. Additionally, knowing the technological limitations, the geometry of the structure will be optimized in terms of maximizing its energy consumption while maintaining the minimum weight.

Author Contributions: Conceptualization, A.S., P.B. (Paweł Baranowski), J.M. and T.C.; Methodology, A.S., P.B. (Paweł Baranowski), A.A.-D., D.Z. and P.B. (Paweł Bogusz); Software, P.B. (Paweł Baranowski), P.B. (Paweł Bogusz), D.Z. and A.S.; Validation, P.B. (Paweł Baranowski) and T.C.; J.M. and T.C.; Investigation, A.S., P.B. (Paweł Baranowski), D.Z. and P.B. (Paweł Bogusz); Resources, A.S. and A.A.-D.; Data Curation, A.S. and D.Z.; Writing-Original Draft Preparation, A.S. and P.B. (Paweł Baranowski); Writing-Review & Editing, A.S., P.B. (Paweł Baranowski), T.C.; Visualization, P.B. (Paweł Baranowski); Supervision, J.M. and T.C.; Project Administration, J.M. and T.C.; Funding Acquisition, J.M. and T.C.

Funding: The research was funded by the National Science Centre under research grant no. 2015/17/B/ST8/00825.

Conflicts of Interest: The authors declare no conflict of interest.

References

1. Zhai, Y.; Lados, D.A.; LaGoy, J.L. Additive Manufacturing: Making Imagination the Major Limitation. *JOM* **2014**, *66*, 808–816. [CrossRef]
2. Tabacu, S.; Ducu, C. Experimental testing and numerical analysis of FDM multi-cell inserts and hybrid structures. *Thin-Walled Struct.* **2018**, *129*, 197–212. [CrossRef]
3. Brenne, F.; Niendorf, T.; Maier, H.J. Additively manufactured cellular structures: Impact of microstructure and local strains on the monotonic and cyclic behavior under uniaxial and bending load. *J. Mater. Process. Technol.* **2013**, *213*, 1558–1564. [CrossRef]
4. Mirone, G.; Barbagallo, R.; Corallo, D.; Di Bella, S. Static and dynamic response of titanium alloy produced by electron beam melting. *Procedia Struct. Integr.* **2016**, *2*, 2355–2366. [CrossRef]
5. Thijs, L.; Verhaeghe, F.; Craeghs, T.; Van Humbeeck, J.; Kruth, J.P. A study of the microstructural evolution during selective laser melting of Ti-6Al-4V. *Acta Mater.* **2010**, *58*, 3303–3312. [CrossRef]
6. Durejko, T.; Zietala, M.; Łazińska, M.; Lipiński, S.; Polkowski, W.; Czujko, T.; Varin, R.A. Structure and properties of the Fe3Al-type intermetallic alloy fabricated by laser engineered net shaping (LENS). *Mater. Sci. Eng. A* **2016**, *650*, 374–381. [CrossRef]
7. Mertens, A.; Reginster, S.; Paydas, H.; Contrepois, Q.; Dormal, T.; Lemaire, O.; Lecomte-Beckers, J. Mechanical properties of alloy Ti–6Al–4V and of stainless steel 316L processed by selective laser melting: Influence of out-of-equilibrium microstructures. *Powder Metall.* **2014**, *57*, 184–189. [CrossRef]
8. Ziętala, M.; Durejko, T.; Polański, M.; Kunce, I.; Płociński, T.; Zieliński, W.; Łazińska, M.; Stępniowski, W.; Czujko, T.; Kurzydłowski, K.J.; et al. The microstructure, mechanical properties and corrosion resistance of 316L stainless steel fabricated using laser engineered net shaping. *Mater. Sci. Eng. A* **2016**, *677*, 1–10. [CrossRef]
9. Welsch, G.; Boyer, R.; Collings, E.W. *Materials Properties Handbook: Titanium Alloys*; Materials properties handbook; ASM International: Russell Township, OH, USA, 1993; ISBN 9780871704818.

10. Zhai, Y.; Galarraga, H.; Lados, D.A. Microstructure, static properties, and fatigue crack growth mechanisms in Ti-6Al-4V fabricated by additive manufacturing: LENS and EBM. *Eng. Fail. Anal.* **2016**, *69*, 3–14. [CrossRef]

11. Carroll, B.E.; Palmer, T.A.; Beese, A.M. Anisotropic tensile behavior of Ti-6Al-4V components fabricated with directed energy deposition additive manufacturing. *Acta Mater.* **2015**, *87*, 309–320. [CrossRef]

12. Sterling, A.; Shamsaei, N.; Torries, B.; Thompson, S.M. Fatigue Behaviour of Additively Manufactured Ti-6Al-4 v. *Procedia Eng.* **2015**, *133*, 576–589. [CrossRef]

13. Zhai, Y.; Galarraga, H.; Lados, D. Microstructure Evolution, Tensile Properties, and Fatigue Damage Mechanisms in Ti-6Al-4V Alloys Fabricated by Two Additive Manufacturing Techniques. *Procedia Eng.* **2015**, *114*, 658–666. [CrossRef]

14. Palčič, I.; Balažic, M.; Milfelner, M.; Buchmeister, B. Potential of Laser Engineered Net Shaping (LENS) Technology. *Mater. Manuf. Process.* **2009**, *24*, 750–753. [CrossRef]

15. Ziętala, M.; Durejko, T.; Polański, M.; Kunce, I.; Płociński, T.; Zieliński, W.; Łazińska, M.; Stępniowski, W.; Czujko, T.; Kurzydłowski, K.J.; et al. Fatigue behavior of thin-walled grade 2 titanium samples processed by selective laser melting. Application to life prediction of porous titanium implants. *J. Mech. Behav. Biomed. Mater.* **2009**, *677*, 750–753.

16. Niesłony, P.; Grzesik, W.; Laskowski, P.; Sienawski, J. Numerical and Experimental Analysis of Residual Stresses Generated in the Machining of Ti6Al4V Titanium Alloy. *Procedia CIRP* **2014**, *13*, 78–83. [CrossRef]

17. Vilaro, T.; Colin, C.; Bartout, J.D. As-Fabricated and Heat-Treated Microstructures of the Ti-6Al-4V Alloy Processed by Selective Laser Melting. *Metall. Mater. Trans. A* **2011**, *42*, 3190–3199. [CrossRef]

18. Maamoun, A.H.; Elbestawi, M.; Dosbaeva, G.K.; Veldhuis, S.C. Thermal post-processing of AlSi10Mg parts produced by Selective Laser Melting using recycled powder. *Addit. Manuf.* **2018**, *21*, 234–247. [CrossRef]

19. Vrancken, B.; Thijs, L.; Kruth, J.P.; Van Humbeeck, J. Heat treatment of Ti6Al4V produced by Selective Laser Melting: Microstructure and mechanical properties. *J. Alloy Compd.* **2012**, *541*, 177–185. [CrossRef]

20. Galarraga, H.; Warren, R.J.; Lados, D.A.; Dehoff, R.R.; Kirka, M.M.; Nandwana, P. Effects of heat treatments on microstructure and properties of Ti-6Al-4V ELI alloy fabricated by electron beam melting (EBM). *Mater. Sci. Eng. A* **2017**, *685*, 417–428. [CrossRef]

21. Hussein, A.; Hao, L.; Yan, C.; Everson, R.; Young, P. Advanced lattice support structures for metal additive manufacturing. *J. Mater. Process. Technol.* **2013**, *213*, 1019–1026. [CrossRef]

22. Bauer, J.; Hengsbach, S.; Tesari, I.; Schwaiger, R.; Kraft, O. High-strength cellular ceramic composites with 3D microarchitecture. *Proc. Natl. Acad. Sci. USA* **2014**, *111*, 2453–2458. [CrossRef]

23. Kucewicz, M.; Baranowski, P.; Małachowski, J.; Popławski, A.; Płatek, P. Modelling, and characterization of 3D printed cellular structures. *Mater. Des.* **2018**, *142*, 177–189. [CrossRef]

24. Yu, X.; Zhou, J.; Liang, H.; Jiang, Z.; Wu, L. Mechanical metamaterials associated with stiffness, rigidity and compressibility: A brief review. *Prog. Mater. Sci.* **2017**, *94*, 114–173. [CrossRef]

25. Baranowski, P.; Płatek, P.; Antolak-Dudka, A.; Sarzyński, M.; Kucewicz, M.; Durejko, T.; Małachowski, J.; Janiszewski, J.; Czujko, T. Deformation of honeycomb cellular structures manufactured with Laser Engineered Net Shaping (LENS) technology under quasi-static loading: Experimental testing and simulation. *Addit. Manuf.* **2019**, *25*, 307–316. [CrossRef]

26. Ullah, I.; Elambasseril, J.; Brandt, M.; Feih, S. Performance of bio-inspired Kagome truss core structures under compression and shear loading. *Compos. Struct.* **2014**, *118*, 294–302. [CrossRef]

27. Xiao, L.; Song, W. Additively-manufactured functionally graded Ti-6Al-4V lattice structures with high strength under static and dynamic loading: Experiments. *Int. J. Impact Eng.* **2018**, *111*, 255–272. [CrossRef]

28. Yuan, C.; Mu, X.; Dunn, C.K.; Haidar, J.; Wang, T.; Jerry Qi, H. Thermomechanically Triggered Two-Stage Pattern Switching of 2D Lattices for Adaptive Structures. *Adv. Funct. Mater.* **2018**, *28*, 1705727. [CrossRef]

29. Babu, B. *Dislocation Density Based Constitutive Model for TI-6AL-4V: Including Recovery and Recrystallisation*; International Center for Numerical Methods in Engineering: Barcelona, Spain, 2007; pp. 631–634.

30. Madej, Ł.; Cybulka, P.; Perzynski, K.; Rauch, Ł. Numerical analysis of strain inhomogeneities during deformation on the basis of the three dimensional Digital Material Representation. *Comput. Methods Mater. Sci.* **2011**, *11*, 375–380.

31. Thomas, J.; Groeber, M.; Ghosh, S. Image-based crystal plasticity FE framework for microstructure dependent properties of Ti-6Al-4V alloys. *Mater. Sci. Eng. A* **2012**, *553*, 164–175. [CrossRef]

32. Piazolo, S.; Jessell, M.W.; Prior, D.J.; Bons, P.D. The integration of experimental in-situ EBSD observations and numerical simulations: A novel technique of microstructural process analysis. *J. Microsc.* **2004**, *213*, 273–284. [CrossRef]

33. Simonelli, M.; Tse, Y.Y.; Tuck, C.; Asgari, A.; Ghadbeigi, H.; Pinna, C.; Hodgson, P.D.; Xu, W.; Brandt, M.; Sun, S.; et al. Analysis of ballistic resistance of composites based on EN AC-44200 aluminum alloy reinforced with Al_2O_3 particles. *Acta Mater.* **2014**, *85*, 24–30.

34. Pach, J.; Pyka, D.; Jamroziak, K.; Mayer, P. The experimental and numerical analysis of the ballistic resistance of polymer composites. *Compos. Part B Eng.* **2017**, *113*, 24–30. [CrossRef]

35. Kurzawa, A.; Pyka, D.; Jamroziak, K.; Bocian, M.; Kotowski, P.; Widomski, P. Analysis of ballistic resistance of composites based on EN AC-44200 aluminum alloy reinforced with Al_2O_3 particles. *Compos. Struct.* **2018**, *201*, 834–844. [CrossRef]

36. Moletsane, M.G.; Krakhmalev, P.; Kazantseva, N.; du Plessis, A.; Yadroitsava, I.; Yadroitsev, I.; Yadroitsev, I. Tensile properties and microstructure of direct metal laser-sintered ti6al4v (eli) alloy. *S. Afr. J. Ind. Eng.* **2016**, *27*, 110–121. [CrossRef]

37. Antonysamy, A.A.; Prangnell, P.; Meyer, J. Effect of Wall Thickness Transitions on Texture and Grain Structure in Additive Layer Manufacture (ALM) Ti-6Al-4V. *Mater. Sci. Forum* **2012**, *706–709*, 205–210. [CrossRef]

38. Rafi, H.K.; Karthik, N.V.; Gong, H.; Starr, T.L.; Stucker, B.E. Microstructures and Mechanical Properties of Ti6Al4V Parts Fabricated by Selective Laser Melting and Electron Beam Melting. *J. Mater. Eng. Perform.* **2013**, *22*, 3872–3883. [CrossRef]

39. Chastand, V.; Tezenas, A.; Cadoret, Y.; Quaegebeur, P.; Maia, W.; Charkaluk, E. Fatigue characterization of Titanium Ti-6Al-4V samples produced by Additive Manufacturing. *Procedia Struct. Integr.* **2016**, *2*, 3168–3176. [CrossRef]

40. Seifi, M.; Salem, A.; Satko, D.; Shaffer, J.; Lewandowski, J.J. Defect distribution and microstructure heterogeneity effects on fracture resistance and fatigue behavior of EBM Ti–6Al–4V. *Int. J. Fatigue* **2017**, *94*, 263–287. [CrossRef]

41. Barnat, W.; Krasoń, W.; Bogusz, P.; Stankiewicz, M. Experimental and numerical tests of separated side lock of intermodal wagon. *J. KONES Powertrain Transp.* **2014**, *21*, 15–22. [CrossRef]

42. Bogusz, P.; Popławski, A.; Morka, A.; Niezgoda, T. Evaluation of true stress in engineering materials using optical deformation measurement methods. *J. KONES Powertrain Transp.* **2015**, *19*, 53–64. [CrossRef]

43. den Bulcke, J.; Biziks, V.; Andersons, B.; Mahnert, K.-C.; Militz, H.; Denis, V.; Dierick, M.; Masschaele, B.; Boone, M.; Brabant, L.; et al. Potential of X-ray computed tomography for 3D anatomical analysis and microdensitometrical assessment in wood research with focus on wood modification. *Int. Wood Prod. J.* **2013**, *4*, 183–190. [CrossRef]

44. Baranowski, P.; Bogusz, P.; Gotowicki, P.; Malachowski, J. Assessment of mechanical properties of offroad vehicle tire: Coupons testing and FE model development. *Acta Mech. Autom.* **2012**, *6*, 17–22.

45. Hensley, S.; Christensen, M.; Small, S.; Archer, D.; Lakes, E.; Rogge, R. Digital image correlation techniques for strain measurement in a variety of biomechanical test models. *Acta Bioeng. Biomech.* **2017**, *19*, 187–195.

46. Derewońko, A.; Niezgoda, T.; Bogusz, P. Uniaxial Tests of Limp Elastic Multi-Layer Materials. 2013. Available online: http://dpi-proceedings.com/index.php/ICCST9/article/view/212 (accessed on January 2013).

47. Donachie, M.J. *Titanium: A Technical Guide 2000*; ASM International: Russell Township, OH, USA, 2000.

48. Yin, H.; Felicelli, S.D. Dendrite growth simulation during solidification in the LENS process. *Acta Mater.* **2010**, *58*, 1455–1465. [CrossRef]

49. Liu, S.; Shin, Y.C. Additive manufacturing of Ti6Al4V alloy: A review. *Mater. Des.* **2019**, *164*, 107552. [CrossRef]

50. Simonelli, M.; Tse, Y.Y.; Tuck, C. Effect of the build orientation on the mechanical properties and fracture modes of SLM Ti–6Al–4V. *Mater. Sci. Eng. A* **2014**, *616*, 1–11. [CrossRef]

51. Waryoba, D.R.; Keist, J.S.; Ranger, C.; Palmer, T.A. Microtexture in additively manufactured Ti-6Al-4V fabricated using directed energy deposition. *Mater. Sci. Eng. A* **2018**, *734*, 149–163. [CrossRef]

52. Kelly, S.M.; Kampe, S.L. Microstructural evolution in laser-deposited multilayer Ti-6Al-4V builds: Part I. Microstructural characterization. *Metall. Mater. Trans. A* **2004**, *35*, 1861–1867. [CrossRef]

materials

MDPI

Article

Microstructure and Hot Deformation Behavior of Twin Roll Cast Mg-2Zn-1Al-0.3Ca Alloy

Kristina Kittner *, Madlen Ullmann, Thorsten Henseler, Rudolf Kawalla and Ulrich Prahl

Institute of Metal Forming, Technische Universität Bergakademie Freiberg, Bernhard-von-Cotta-Straße 4, 09599 Freiberg, Germany; madlen.ullmann@imf.tu-freiberg.de (M.U.); thorsten.henseler@imf.tu-freiberg.de (T.H.); rudolf.kawalla@imf.tu-freiberg.de (R.K.); ulrich.prahl@imf.tu-freiberg.de (U.P.)
* Correspondence: kristina.kittner@imf.tu-freiberg.de; Tel.: +49-3731-39-3308

Received: 27 February 2019; Accepted: 26 March 2019; Published: 28 March 2019

Abstract: In the present work, the microstructure, texture, mechanical properties as well as hot deformation behavior of a Mg-2Zn-1Al-0.3Ca sheet manufactured by twin roll casting were investigated. The twin roll cast state reveals a dendritic microstructure with intermetallic compounds predominantly located in the interdendritic areas. The twin roll cast samples were annealed at 420 °C for 2 h followed by plane strain compression tests in order to study the hardening and softening behavior. Annealing treatment leads to the formation of a grain structure, consisting of equiaxed grains with an average diameter of approximately 19 µm. The twin roll cast state reveals a typical basal texture and the annealed state shows a weakened texture, by spreading basal poles along the transverse direction. The twin roll cast Mg-2Zn-1Al-0.3Ca alloy offers a good ultimate tensile strength of 240 MPa. The course of the flow curves indicate that dynamic recrystallization occurs during hot deformation. For the validity range from 250 °C to 450 °C as well as equivalent logarithmic strain rates from 0.01 s^{-1} to 10 s^{-1} calculated model coefficients are shown. The average activation energy for plastic flow of the twin roll cast and annealed Mg-2Zn-1Al-0.3Ca alloy amounts to 180.5 kJ/mol. The processing map reveals one domain with flow instability at temperatures above 370 °C and strain rates ranging from 3 s^{-1} to 10 s^{-1}. Under these forming conditions, intergranular cracks arose and grew along the grain boundaries.

Keywords: twin roll casting; magnesium alloy; calcium; Mg-Zn-Al-Ca alloy; texture; flow curve; processing map

1. Introduction

Twin roll casting (TRC) of magnesium alloys becomes more important due to the economic and energy efficient process of manufacturing sheets and strips compared to conventional thin sheet production. Sheets of magnesium wrought alloys, produced via twin roll casting, exhibit good mechanical properties [1]; however, the main drawback concerning their application potential is their low formability at low temperatures [2]. The formability of magnesium wrought alloys is closely related to the resulting microstructure, texture and hcp lattice structure. Improving formability, through grain refinement and texture modification, can be obtained by different approaches: thermomechanical processing [3,4], severe plastic deformation such as equal-channel angular extrusion (ECAE) [5], differential speed rolling (DSR) [6] or cross-rolling (CR) [7] and chemical alloying. Processes of severe plastic deformation may be detrimental to cost development and limitations in possible shapes and dimensions. Studies about chemical alloying recommend rare earth (RE) elements [8,9] as well as Ca [10–12], Sr [13,14] or Li [15,16] being preferred to influence the texture modification, towards weaker basal or random texture development. Apart from their enhanced formability compared to conventionally available alloys, fabrication of novel alloys via twin roll casting has scarcely been published yet. Park et al. [2] demonstrate that twin roll casting of a ZAX400 alloy leads to the formation of an equiaxed dendrite microstructure with no occurrence of inverse segregations, which adversely affect the mechanical properties and surface quality of the strip. Wang et al. [17] observed columnar dendrites at the surface and a transition to equiaxed dendrites in the mid-thickness of an AM31 + 0.2 Ca alloy after twin roll casting. Main microstructural features are in accordance with those offered by other magnesium wrought alloys [1]. So far, most of previous works are focused on the conventional Mg alloys (AZ or ZK systems). However, systematic research on the hot deformation behavior of fine-grained Mg-Zn-(Al)-Ca alloy is rarely reported. In a study by Tong et al. [18] about the compressive deformation behavior of the as-extruded and as-ECAPed Mg–5.3Zn–0.6Ca (wt.%) alloys at 200 °C to 300 °C, the compression behaviors of both conditions were mainly dominated by climb-controlled dislocation creep through grain boundary diffusion. The as-ECAPed alloy presented a lower activation energy (100–108 kJ/mol instead of 160–172 kJ/mol for as-extruded alloy), which might be derived from non-equilibrium grain boundaries [18]. Kulyasova et al. [19] investigated the application of high-pressure torsion (HPT), which leads to the formation of ultrafine-grained structures in Mg-Zn-Ca with an average grain size of 150 nm. HPT processing and additional annealing at 200 °C results in a high ultimate tensile strength due to grain refinement and dispersion hardening; whereas, the retention of a good ductility (8.5%) is conditioned by the activation of dislocation slip in non-basal planes [19]. There are also publications dealing with RE elements addition in the Mg-Zn-Ca alloy. For example, Qi et al. [20] studied the hot compression behavior and processing characteristics of Mg-3Zn-0.3Ca-0.4La (wt.%). The results suggested that deformation parameters had significant effects on the deformation behaviour and dynamic recrystallization. The average activation energy of plastic deformation was calculated to be 188.9 kJ/mol and might be attributed to the existence of precipitates and a fine microstructure [20].

The magnesium alloy Mg-2Zn-1Al-0.3Ca (ZAX210) serves as investigation material of the present work. In recent years, the ZAX210 alloy has gained importance, due to its advantageous strength and formability because of Ca addition. Current studies are concerned, among other research, with casting and hot rolling of the ZAX210 alloy as well as the resulting microstructure, texture and mechanical properties [21–23]. Hoppe et al. [23] revealed a texture of the cast and hot rolled ZAX210 sheet exhibiting two peaks along the rolling direction as well as a basal pole spread in transverse direction. The authors assume that this texture evolution due to Ca is responsible for the improved formability. Letzig et al. [24] show microstructure, texture and mechanical properties of ZAX210 sheets after twin roll casting and hot rolling. The sheets offer a fine grained and homogenous microstructure and a texture similar to the RE-containing alloy ZE10. Twin roll casting and hot rolling of ZAX210 are also part of the studies of Ullmann et al. [25]. The authors conclude that the good formability of the ZAX210 sheets is due to the weakening of the texture, which originates from the recrystallization process during hot rolling were Ca induces particle stimulated nucleation.

The processing map technique based on the dynamic material model (DMM) has been widely used to understand the workability of materials [26]. With help of the map, the description of the hot forming behavior and the determination of the most reasonable thermal processing parameters is practicable.

The present work is focused on twin roll casting and the resulting microstructure, texture and mechanical properties as well as the hot deformation behavior of the ZAX210 alloy. These apparitions are associated with the hot workability (evaluated by processing maps) of the TRC and annealed state. The hot forming behavior is the first important component in the understanding of the entire process chain for the production of a good formable Mg alloy.

2. Materials and Methods

The magnesium alloy with a nominal composition (wt.%) ZAX210 was subjected to a twin roll caster (Institute of Metal Forming, Technische Universität Bergakademie Freiberg, Saxony, Germany) on industrial scale. The chemical composition of the investigated alloy is listed in Table 1. Ingots of the ZAX210 were melted in a steel crucible (Institute of Metal Forming, Technische Universität Bergakademie Freiberg, Saxony, Germany) at 730 °C under protective gas atmosphere, transferred into a preheated casting system and twin roll cast. The thickness of the TRC sheets was approximately 5.3 mm and is further referred to as the TRC state. Subsequently, the TRC sheets were annealed at different temperatures ranging from 340 °C to 460 °C for holding times ranging from 2 h to 12 h. After annealing, continuous plane strain compression tests at three different strain rates (0.1 s^{-1}, 1 s^{-1}, 10 s^{-1}) and temperatures (250 °C, 350 °C, 450 °C) were performed on a servo-hydraulic hot deformation simulator (Institute of Metal Forming, Technische Universität Bergakademie Freiberg, Saxony, Germany). The specimen size was 20 mm × 30 mm × 5.3 mm and deformed by a 6-mm-wide carbide tool insert perpendicular to the TRC direction up to equivalent logarithmic strains equal to 1. As lubricant, a liquid graphite-oil-mixture was used. Based on the recorded force-displacement data isothermal flow curves were calculated. Here, divergences due to friction at the tool interface as well as softening effects due to an increase in temperature from dissipated forming energy have been corrected numerically. To guarantee a sufficient degree of statistical certainty, five comparative samples of each forming condition were tested.

Table 1. Chemical composition of the ZAX210 alloy (wt.%).

Zn	Al	Ca	Mn	Cu	Fe	Ni	Others	Mg
2.290	0.920	<0.250	0.040	0.001	0.005	0.001	<0.045	Bal.

Samples of the TRC material were prepared for the characterization of the microstructure. Specimens for optical microscopy were prepared by conventional grinding and polishing with oxide polishing suspension (OPS), followed by etching in a solution of ethanol, glacial acetic acid, picric acid and distilled water. Optical and scanning electron microscopy (SEM, Institute of Material Science, Technische Universität Bergakademie Freiberg, Saxony, Germany) were used to study microstructural characteristics. The chemical composition of the individual structural constituents was measured by energy dispersive X-ray spectroscopy (EDX, Institute of Material Science, Technische Universität Bergakademie Freiberg, Saxony, Germany) and X-Ray Diffraction (XRD) analysis on a 10 mm × 5.3 mm cross section of the TRC strip. XRD was performed on a Seifert-FPM RD7 (Institute of Material Science, Technische Universität Bergakademie Freiberg, Saxony, Germany) using CuKα radiation (λ = 1.540598 Å). Diffraction patterns were recorded within the 2Θ-range of 20° to 150°. The step size was chosen at 0.02° and a step time of 25 s.

Tensile tests were conducted at room temperature. Samples were extracted in TRC direction (0°) according to DIN 50125. The specimen shape H with an initial measuring length of 80 mm was selected. Tensile tests were performed on the universal tensile testing machine AG-100 (Institute of Metal Forming, Technische Universität Bergakademie Freiberg, Saxony, Germany) with a crosshead speed of 2 mm/min.

Pole figures of the sheets in TRC as well as the annealed condition were calculated from electron backscatter diffraction (EBSD). The specimens were polished by ion beam for 2 h with a voltage of 4 kV and 2 mA beam current. The EBSD analysis was performed using a FEI Versa 3D scanning electron microscope (Academic Centre for Materials and Nanotechnology, University of Science and Technology, Krakow, Poland) equipped with a Hikari EBSD detector. The voltage used for acquisition of EBSD data was 15–20 kV, and the step size was 0.65 µm. All orientation maps were processed with the EDAX/TSL OIM Data collection software version 7 using a batch processing operation. For analysis of the EBSD data, calculation of ODF (orientation distribution function) and pole figures the free MTEX MATLAB toolbox [27] was used.

The preparation of the processing maps is based on thermodynamic analysis of the flow curves. Further information about processing maps based on DMM derivation can be found in [28]. The method considers the complementary relationship between the rate of heat generation induced by the forming process and the rate of energy dissipation associated with microstructural changes such as recovery and dynamic recrystallization (stability domains) and material damage (instability domains). To represent the power dissipation due to microstructural mechanism, a non-dimensional efficiency index η is used, see Equation (1) [26], where m is the strain rate sensitivity (function of forming temperature and strain rate). The changes of η on the temperature-strain rate field create the processing map. In addition to the η contours, an instability parameter $\xi(\dot{\varepsilon})$ given first by Prasad et al. [26] described by the Equation (2) is applied to delineate the temperature-strain rate regimes of flow instability on the processing map.

$$\eta = \frac{2m}{m+1} \tag{1}$$

$$\xi(\dot{\varepsilon}) = \frac{\partial \ln\left(\frac{m}{m+1}\right)}{\partial \ln \dot{\varepsilon}} + m \leq 0 \tag{2}$$

3. Results

3.1. Microstructure and Texture of the Twin Roll Cast ZAX210 Alloy

Twin roll casting of ZAX210 leads to the formation of the characteristic microstructure, which develops during TRC and is well investigated for the magnesium alloy AZ31 [1,29]. Due to the superimposed solidification and forming processes, the alloy exhibits an inhomogeneous microstructure [25]. The morphology of the microstructure is significantly influenced by the processing parameters. The inhomogeneous microstructure of the TRC strip can be divided into three zones (chill zone, columnar dendritic zone and central equiaxed zone) with different characteristics (Figure 1a). During solidification coarse columnar and severely deformed dendrites develop, which also tend to cause the occurrence of unwanted artificial twins in optical micrographs due to sample preparation. The dendrites grow in opposite direction to the heat flow of the melt, predominantly perpendicular to the surface of the rolls. The rolling leads to an inclination of the columnar dendrite with respect to the neutral axis of the TRC strip. Within the chill and the central equiaxed zone, fine grains occur because of constitutional undercooling or dynamic recrystallization during hot deformation.

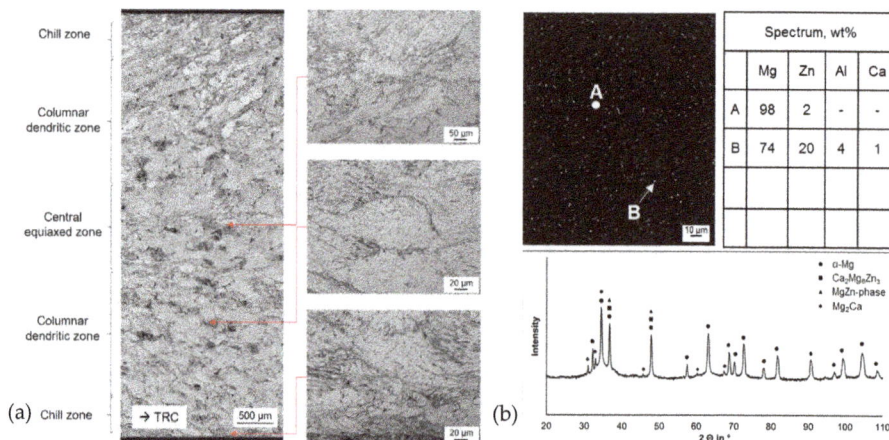

Figure 1. (a) Microstructure of the TRC (twin roll cast) state ZAX210 [25], divided into three characteristic sections through thickness cross-section and (b) scanning electron micrograph, results of the XRD analysis (X-ray diffraction) and chemical composition of structural constituents according to marking in SEM (scanning electron microscope) image determined via EDX analysis (energy dispersive X-ray spectroscopy) for the ZAX210 in TRC condition [30].

The interdendritic areas as well as a small segregation line in the middle of the strip reveal intermetallic compounds. Results of the XRD analysis indicate different kinds of intermetallic compounds (Mg_2Ca, MgZn-phases and $Ca_2Mg_6Zn_3$), but predominantly α-magnesium was detected (Figure 1b). Numerous measurements were carried out, nevertheless, due to the small size of the intermetallic compounds (<1 μm), further investigations with greater resolution are necessary to back these results as the surrounding α-magnesium matrix possibly influenced the measurement.

The pole figures of TRC state ZAX210 presented in Figure 2 indicate the formation of a basal texture with a slight spread along TD. $(10\bar{1}0)$ and $(10\bar{1}1)$ pole figures reveal preferential orientations of the pyramidal and prismatic planes. The peaks in the $(10\bar{1}0)$ and $(10\bar{1}1)$ pole figures indicates specific alignment of the unit cells around the c-axis. The angles of these peaks correspond to the a-axes angles of the unit cell as also recognizable in the illustrated ODF sections, and show a slight rotation about the c-axis. Just as in the basal pole figure, this rotation might be caused by a sloped specimen surface.

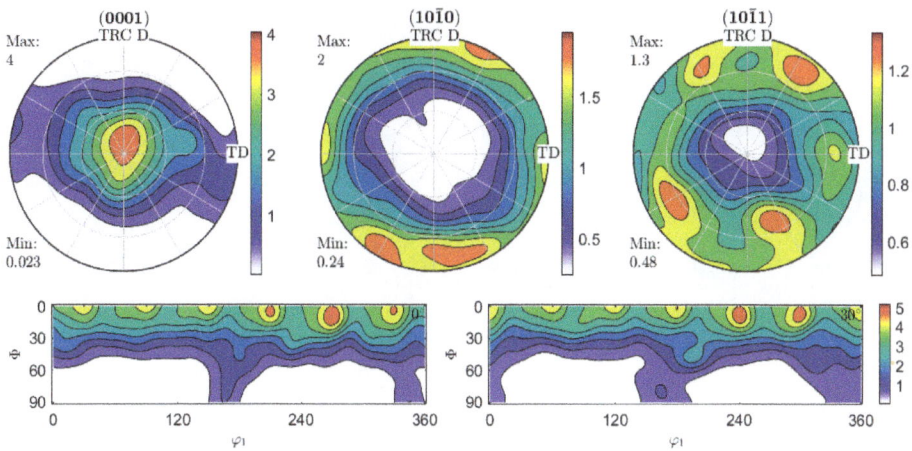

Figure 2. (0001)-basal, $(10\bar{1}0)$-prismatic and $(10\bar{1}1)$-pyramidal pole figures related to the TRC direction and transverse direction of the TRC strips denoted as TRC D and TD (transverse direction) respectively as well as calculated ODF sections at $\varphi_2 = 0°$ and $30°$.

3.2. Microstructure and Texture of the Annealed ZAX210 Alloy

Heat treatment was performed in order to ensure homogenization and the development of a microstructure with equiaxed grains. Resulting microstructures and grain size distributions measured by linear intercept method are summarized in Figure 3. During heat treatment of TRC strip, the rearrangement of the microstructure starts at 340 °C. However, areas of the initial microstructure remain surrounded by small recrystallized grains. With increasing temperature, the initial microstructure diminishes and recrystallization proceeds. Thus, heat treatment at 420 °C and a holding time of 2 h results in a rearranged microstructure offering fine equiaxed grains with an average grain size ranging from approximately 18 μm at the edge to approximately 20 μm in the middle of the strip (Figure 3, 420 °C, 2 h). An increasing temperature or holding time (Figure 3, 420 °C, 8 h and 460 °C, 8 h) leads to anomalous grain growth. Consequently, a bimodal grain size distribution arises with an average exceeding 100 μm.

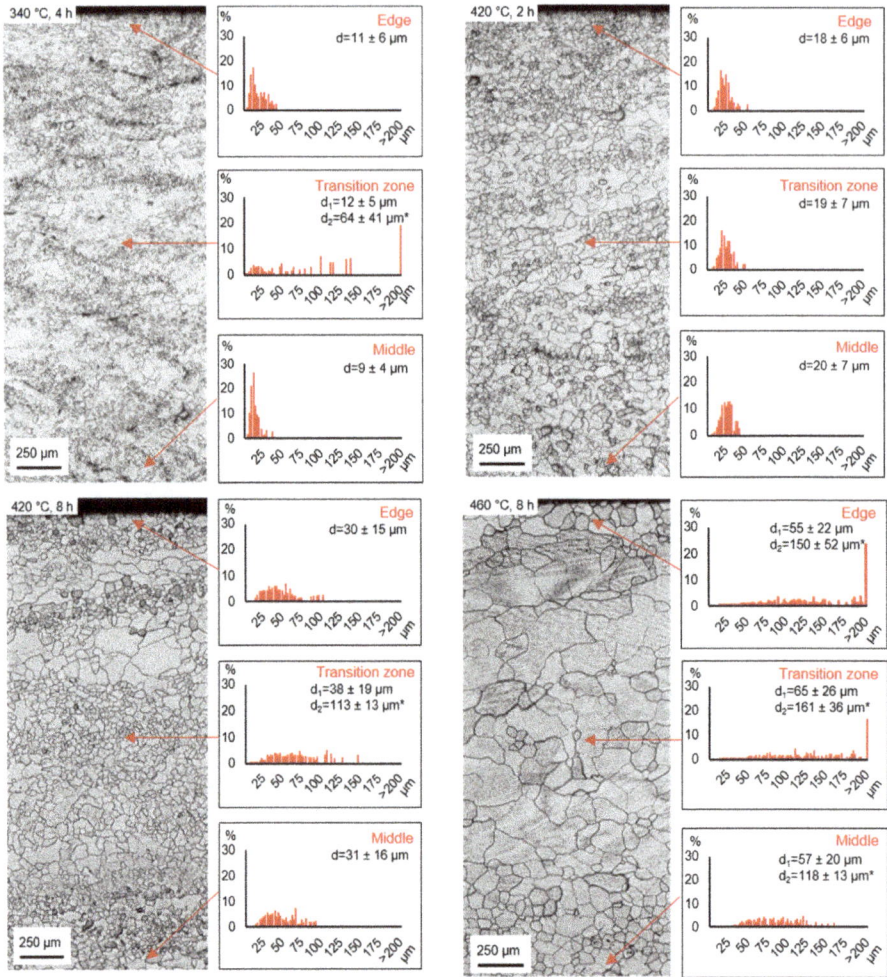

Figure 3. Microstructures of the annealed ZAX210 alloy with regard to the heat treatment conditions (340 °C, 4 h; 420 °C, 2 h; 420 °C, 8 h and 460 °C, 8 h) and associating grain size distribution depending on the characteristic section of the TRC strip (edge, transition zone and middle); diagrams show grain size class vs. percentage of the grain area in relation to the total area of all measured grains (→ TRC direction, * two maxima due to bimodal grain size distribution).

Figure 4 shows the texture of the annealed condition (420 °C, 2 h). Compared to the TRC condition, a strong spreading of the c-axis towards TD from ND occurred. As it can be seen from the $(10\bar{1}0)$ and $(10\bar{1}1)$ pole figures the prismatic and pyramidal poles reveal a similar alignment with TD. This prismatic and pyramidal pole spread originates from the basal pole spread, but also indicates a majority arrangement of the unit cells with parallel a-axes. As seen in the illustrated ODF sections, the pre-existing preference has remained, but here too the intensities have decreased because of the basal pole spread.

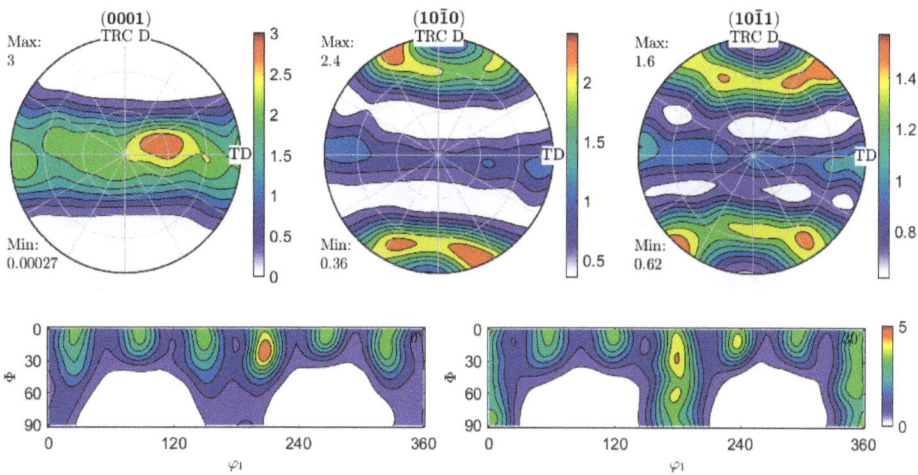

Figure 4. (0001)-basal, (10$\bar{1}$0)-prismatic and (10$\bar{1}$1)-pyramidal pole figures taken from TRC D–TD plane of the annealed (420 °C, 2 h) strip as well as calculated ODF sections at $\varphi_2 = 0°$ and 30°.

3.3. Mechanical Properties at Room Temperature

In TRC state, the ZAX210 alloy offers values of 205 MPa yield strength and 231 MPa ultimate tensile strength. Due to the inhomogeneous microstructure, which arises during twin roll casting because of solidification kinetics, the elongation at fracture is diminished. Other magnesium alloys, for example AZ31, AM50 or WE43, show similar results [1]. Heat treatment and the accompanying processes of static recrystallization lead to a significant improvement of the elongation at fracture. Depending on homogeneity of the microstructure as well as grain size and grain size distribution the elongation at fracture ranges from ~20% (460 °C, 8 h) to ~30% (420 °C, 2 h). The enhancement of the ductility is accompanied by a decrease in yield strength. During heat treatment dislocation climbing and annihilation takes place as well as precipitates are dissolved and consequently softening of the material occurs and the yield strength decreases (Figure 5). Increasing temperature or holding time of the annealing results in a slight decrease of the ultimate tensile strength.

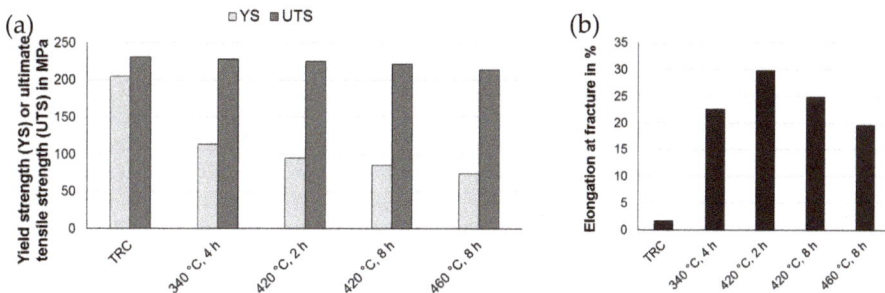

Figure 5. Mechanical properties at room temperature of the ZAX210 alloy after twin roll casting and different heat treatments: (**a**) yield strength and ultimate tensile strength and (**b**) elongation at fracture.

3.4. Hot forming Behavior of Twin Roll Cast and Annealed ZAX210 Alloy

Exemplary flow curves during plane strain compression at temperatures of 250 °C, 350 °C and 450 °C as well as strain rates of 0.1 s^{-1}, 1 s^{-1} and 10 s^{-1} are shown in Figure 6. The flow curves exhibit smooth hardening and softening phases, while the maximum flow stress value moves–together with the strain rate increase–towards greater equivalent logarithmic strain values. The flow stresses decrease when the strain rate is decreased or the temperature is increased. The flow stress initially increases because of strain hardening and reaches a peak value, subsequently decreasing with further strain. The course of the flow curves indicate that dynamic recrystallization occurs. The rates of strain hardening and strain softening vary with the deformation conditions. The initial hardening is associated with the increase in dislocation density. At lower strain rates, dislocation multiplication is less rapid and contributes to a lower strain hardening effect than that at higher strain rates. Deformation at higher temperatures or at lower strain rates result in higher dynamic recrystallization kinetics because of higher diffusivity at higher temperatures and more time for nucleation and growth of dynamically recrystallized grains.

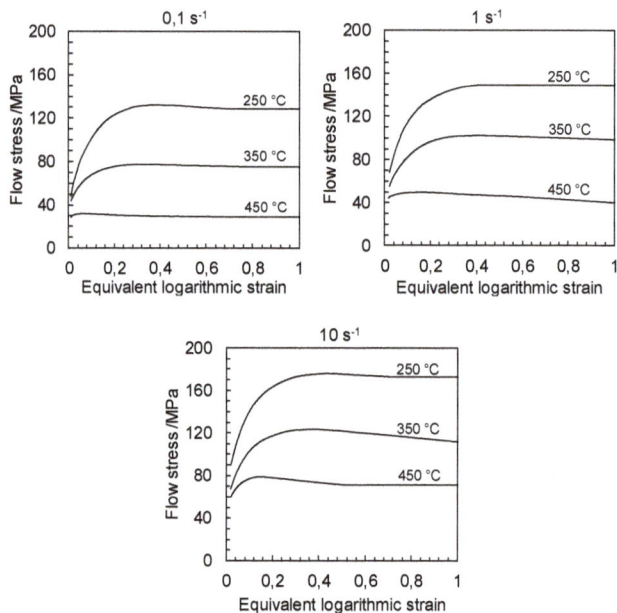

Figure 6. Exemplary flow curves of TRC and annealed ZAX210 alloy during plane strain compression at temperatures of 250 °C–450 °C and strain rates of 0.1 s^{-1}–10 s^{-1}.

4. Discussion

4.1. Texture Evolution of the Twin Roll Cast and Annealed ZAX210 Alloy

Twin roll casting results in the formation of a basal texture, where a slight tilt of the peak can be seen. Preferred orientations and strong basal textures usually arise during rolling, when slip mainly occurs on basal planes [31]. Typically, TRC strips (for example AZ31 alloy) exhibit a low basal texture due to the directional growth of columnar grains [32]. Results of the influence of a heat treatment on the resulting microstructure of a TRC state ZAX210 in comparison to other magnesium alloys are shown in [25]. The texture of the TRC strip originates from the rolling deformation, which is associated with the TRC process. Similar results were reported for TRC of AZ31 alloy in [32] and [33]. According to Soomro et al. [34], the cause of possible tilting is due to the formation of twins. It is reported that twinning as an important deformation mode in magnesium alloys can alter the orientation of original grains and consequently {10–12} tensile twins may be responsible for the basal pole tilt. Because of the low critical resolved shear stress (CRSS) of about 2.5 MPa [34] tensile twins can easily arise. Nevertheless, twinning was not observed in the present TRC strip. In contradiction to the literature the authors presume, that due to the high temperatures that occur, prismatic and pyramidal slip is also part of the dominating deformation mechanisms leaving little contribution due to twinning, as can be presumed from the alignment in the $(10\bar{1}0)$-prismatic and $(10\bar{1}1)$-pyramidal pole figures in annealed state. Further investigations are required in order to clarify the dominant mechanism for basal pole tilting during TRC. Because of the specific microstructure, which results from the combination of casting and rolling during TRC process, further heat treatment has an important effect on the microstructure as well.

During heat treatment of TRC strip the rearrangement of the microstructure is combined with softening processes, which already occur at low temperatures (Figure 3, 340 °C). The stored deformation energy, introduced during TRC, which is characterized by forming with minor reduction (log. strain < 0.1), is sufficient to initiate static recrystallization during heat treatment. In previous studies static recrystallization has been proposed to be related to particle-stimulated nucleation (PSN), shear band induced nucleation (SBIN) and deformation twin induced nucleation (DTIN). The XRD results of the TRC condition imply the occurrence of Mg_2Ca precipitates, which can act as nucleation sites for recrystallization. The addition of calcium contributes to the formation of the Mg_2Ca phase [35,36]. XRD results reveal, that precipitates have dissolved during heat treatment at 460 °C, because only α-magnesium matrix was detected. Very small particles are able to impede grain growth by inhibition of grain boundary movement. The heat treatment at temperatures above 420 °C may lead to their dissolution and consequently grain growth occurred [37,38]. Zimina et al. [39] revealed, that after a homogenization treatment of a TRC AZ31 strip tilting of the grains towards TRCD occurred and led to the balancing of the texture along the cross-section. The above-mentioned recrystallization mechanisms are known to result in a decrease of the basal texture intensity as well as in the development of more randomized textures [40,41]. Assuming, that PSN is the dominant recrystallization mechanism, randomly oriented nuclei were provided and a weaker recrystallization texture arose. Here are parallels to the displayed textures in TRC and annealed state recognizable. According to the results of Sandlöbes et al. [42] Ca addition is correlated with a significant decrease of the intrinsic stacking fault energy, which is connected with the ductility increase by the increased activity of pyramidal <c + a> dislocations.

4.2. Hot Deformation Behavior of Twin Roll Cast and Annealed ZAX210 Alloy

4.2.1. Analysis of Hot Compressive Deformation Behavior

By assessing the values of the power dissipation efficiency η and the instability parameter $\xi(\dot{\varepsilon})$ in dependency of temperature, logarithmic strain as well as strain rate in a processing map, the characteristics of plastic flow are evaluated. The power law relationship of Sellars and Tegart's [43,44] was used to derive the deformation mechanisms during hot deformation. The relationship is expressed as:

$$\dot{\varepsilon} = A \left[\sin h(\alpha\sigma)\right]^n \exp\left(-\frac{Q}{R \cdot T}\right) \tag{3}$$

where $\dot{\varepsilon}$ is the strain rate (s^{-1}), A is the material constant, n is the stress exponent, α is a fitting parameter, σ is the peak stress (MPa), Q is the average activation energy for plastic flow (kJ/mol), R is the gas constant (8.314 J/(mol·K)) and T is the thermodynamic deformation temperature (K). The parameters of this equation were obtained from the mean values of the slopes of graphs relating flow stress to strain rate and temperature in linear dependence, see Figures 7–9, and are as follows: $A = 3.463\cdot10^{15}, n = 4.8, \alpha = 0.0109$ MPa^{-1}, $Q = 180.5$ kJ/mol. The calculated model coefficients have validity for temperatures ranging from 250 °C to 450 °C as well as for equivalent logarithmic strain rates of 0.01 s^{-1} to 10 s^{-1}. The activation energy Q indicating deformation difficulty degree during hot deformation depends on concurrent dynamic precipitate, dislocation pinning effects and the second phase. The high Q value (higher than that for lattice diffusion of magnesium, which is 135 kJ/mol) might be attributed to the existence of precipitates and the fine microstructure. Precipitates could hinder dislocation movement during the deformation and exerted a resistant force for lattice self-diffusion. Additionally, the fine grains provide more grain boundaries acting as barriers to the dislocation movement and lattice self-diffusion. Besides, the addition of alloying elements presumably have an effect on the activation energy, due to the effect of solution strengthening.

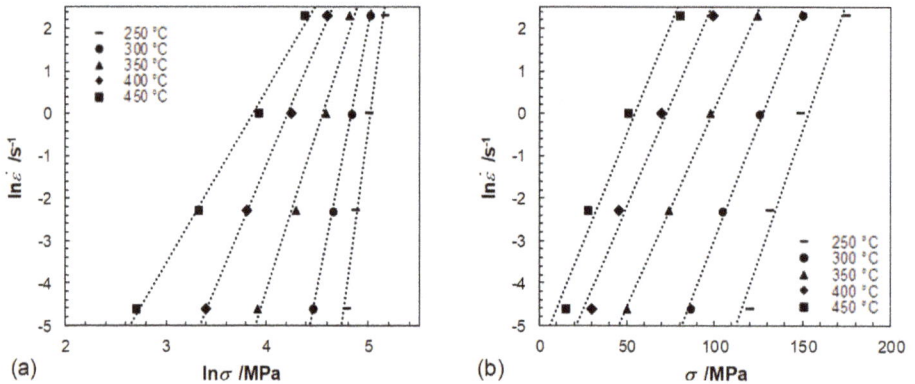

Figure 7. Relationship between ln$\dot{\varepsilon}$ and (a) lnσ (b) σ, temperature dependent.

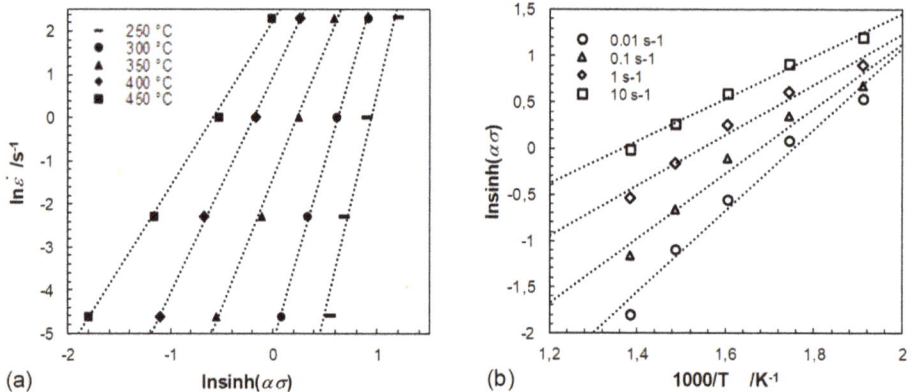

Figure 8. Relationship between (a) ln$\dot{\varepsilon}$−ln[sinh($\alpha\sigma$)] at different temperatures and (b) ln[sinh($\alpha\sigma$)]$^{-1}$/T as a function of the tested strain rates.

As is well known, the Zener–Hollomon parameter Z is determined by $Z = \dot{\varepsilon} \cdot \exp(Q/(RT))$. The dependence of lnZ to ln[sin h($\alpha\sigma$)] is shown in Figure 9. The interpretation of the power law relationship depends on the fitting of the relationship between the Zener–Hollomon parameter and ln[sin h($\alpha\sigma$)]. At low Z values linear regression was determined to an accuracy of $(R^2) = 0.95$, revealing a good fit of the hyperbolic sine function. Following the constitutive equation

$$Z = A \sin h^n(\alpha\sigma) \tag{4}$$

The stress exponent n at low Z values (low strain rates and high temperatures) is 4.8. Under consideration of the results from [45] that dislocation climb creep corresponds to n > 5 and solute drag creep to n = 3, the presented results let assume that dislocation climb creep is the dominant mechanism of plastic flow. Increasing lnZ values results in a deviation from linearity, due to the power law breakdown (PLB). PLB arises at high strain rates and low temperatures and is characterized by n = 10. It is recognized that shear banding occurs in the PLB regime. At high strain rates there is not enough time available for dislocations to rearrange. Consequently, dislocations are not able to sweep out excess vacancies which arise during plastic deformation [46]. Figure 9a shows the plot of

$lnZ - ln[sin h(\alpha\sigma)]$, corresponding to an activation energy $Q = 180.5$ kJ/mol and the fitting parameter $\alpha = 0.0109$ MPa^{-1}. In comparison to the curves in Figure 9b, an improved degree of linearity can be observed as expected from the characteristic hyperbolic sine function. However, with increasing Z values, linearity decreases resulting in the appearance of instability domains in the processing map (Figure 10). This result indicates that the data in the instability regime are not valid within the PLB regime, described by Equation (4).

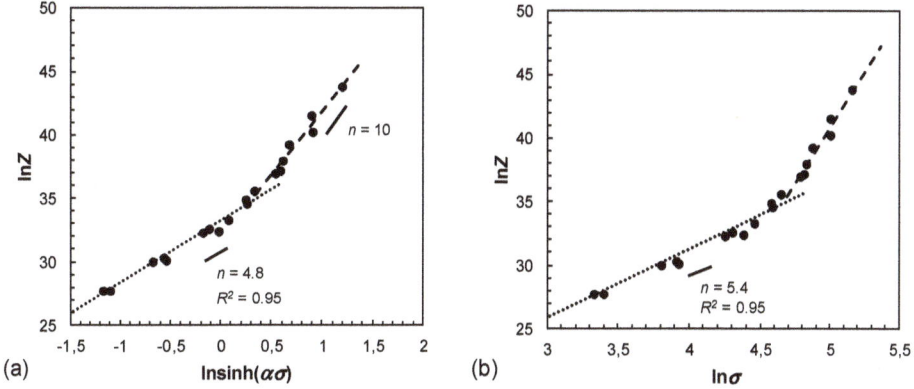

Figure 9. Relationship between Zener–Hollomon parameter and peak stress in hot compression test, (a) $lnZ - ln[sin h(\alpha\sigma)]$ and (b) $lnZ - ln\sigma$.

4.2.2. Processing Map

The processing map for TRC and annealed ZAX210 alloy at a logarithmic strain of 0.4, which is representative in a typical rolling pass, is shown in Figure 10a. The contour numbers denote percent power dissipation efficiency η and the shaded domains regions of flow instability $\xi(\dot{\varepsilon}) < 0$. The processing map exhibits one domain with flow instability at temperatures above 370 °C and strain rates between 3 s^{-1}–10 s^{-1}. The instability, which arises at high strain rates is attributed to localized shear [26]. The microstructure after deformation at 450 °C and 1 s^{-1}, which refers to the instability region of the processing map, is shown in Figure 10b. Under these forming conditions, intergranular cracks arose and grew along the grain boundaries. Consequently, the instability region is undesirable for processing and hence should be avoided. Although the resulting microstructure in Figure 10c has different grain sizes, it has corrugated grain boundaries with some fine-grained nucleation sites. These microstructural characteristics are suitable for hot forming processing such as rolling, extrusion and forging.

It should be noted that the processing maps are sensitive to the initial condition like chemical composition and processing history, resulting in different microstructures with different forming behavior. The hot workability of Ca-containing alloys like Mg-4Al-2Ba-2Ca [47], Mg-4Al-2Ba-2Ca [48], Mg-1Zn-1Ca [49] and Mg-3Sn-2Ca-0.4al-0.4Zn [50] has been investigated by developing processing maps, which indicate that the alloys are workable at temperatures between 300 °C and 500 °C at low strain rates. At strain rates above 0.1 s^{-1} and 1 s^{-1} instability regions occur, except for the area between 340 °C and 380 °C. Due to the different chemical composition and process history of the alloy described in this paper, various deviations may occur when comparing the microstructure, texture as well as the hot working behavior.

(a)

(b)

(c)

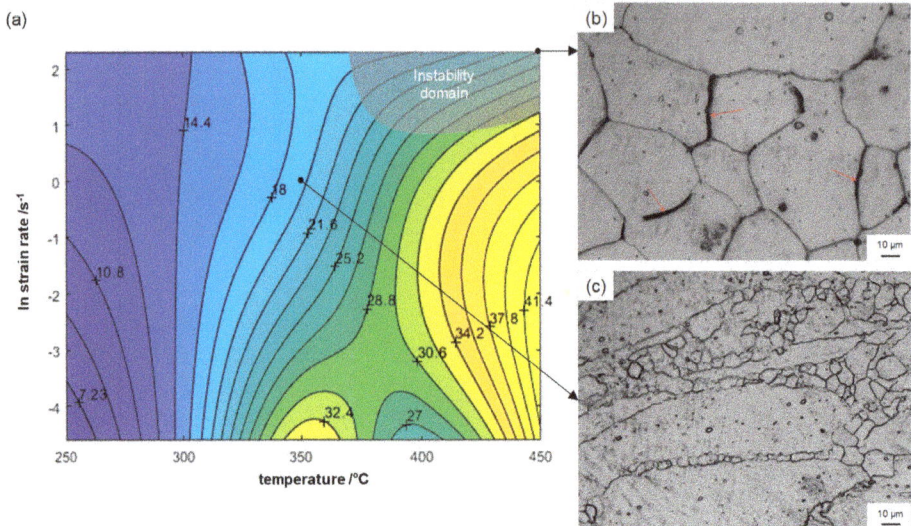

Figure 10. (**a**) Processing map for TRC and annealed ZAX210 at an equivalent logarithmic strain of 0.4 and (**b**) microstructure within the instability domain (450 °C; 10 s^{-1}) and (**c**) resulting microstructure in stable domain (350 °C; 1 s^{-1}).

5. Conclusions

The presented work focused on twin roll casting and the resulting microstructure and texture, mechanical properties as well as the hot deformation behavior of the Mg-2Zn-1Al-0.3Ca alloy. Findings obtained permit the following conclusions:

1. Twin roll casting leads to the formation of a characteristic microstructure, which can be divided into the three sections: chill zone, columnar dendritic zone and central equiaxed zone. Within the chill and the equiaxed zone fine grains occur as a result of constitutional undercooling or dynamic recrystallization during hot deformation.

2. Twin roll casting results in the formation of a weakly pronounced basal texture, which can be characterized by a slight asymmetric tilting of the basal pole. A preferential orientation by the prismatic or pyramidal poles is assumed in alignment with the twin roll casting direction.

3. During heat treatment of twin roll cast strip the rearrangement of the microstructure is combined with softening processes, which already occur at low temperatures (340 °C). The stored energy introduced during twin roll casting, which is characterized by forming with minor reduction (log. strain < 0.1), is sufficient to initiate static recrystallization during heat treatment. Heat treatment at 420 °C and a holding time of 2 h lead to a homogeneous microstructure consisting of fine equiaxed grains as a result of nucleation and grain growth. Due to the occurrence of Mg$_2$Ca phases in the TRC strip, particles stimulated nucleation is assumed being the dominant recrystallization mechanism.

4. After heat treatment a stronger spreading of the c-axis of towards TD occurred. Assuming, that PSN is the dominant recrystallization mechanism, randomly oriented nuclei were provided and a weaker recrystallization texture arose.

5. The features of the flow curves indicate that dynamic recrystallization occurs during hot deformation. At lower strain rates, dislocation multiplication is less rapid and contributes to a lower strain hardening effect than that at higher strain rates. Deformation at higher temperatures and/or at lower strain rates can result in higher dynamic recrystallization kinetics because of higher diffusivity at higher temperatures and more time for nucleation and growth of dynamic recrystallization. Thus, a lower strain hardening effect is observed at higher temperatures and lower strain rates.

6. The average activation energy for plastic flow amounts to 180.5 kJ/mol for the twin roll cast and annealed Mg-2Zn-1Al-0.3Ca alloy. The stress exponent *n* at low *Z* values (low strain rates and high temperatures) is 4.8, suggesting that dislocation climb creep dominates the plastic flow.

7. The processing map at an equivalent logarithmic strain of 0.4 exhibit one domain with flow instability at temperatures above 370 °C and strain rates between $3 \, s^{-1}$–$10 \, s^{-1}$. Under these forming conditions, intergranular cracks arose and grew along the grain boundaries. Consequently, the instability region is undesirable for processing and hence should be avoided.

Author Contributions: Conceptualization, K.K. and M.U.; Methodology, K.K.; Formal Analysis, M.U.; Investigation, K.K. and M.U.; Data Curation, M.U. and T.H.; Writing—Original Draft Preparation, K.K. and M.U.; Writing—Review & Editing, T.H.; Visualization, T.H.; Supervision, R.K. and U.P.; Project Administration, R.K.

Funding: This research was funded by the Federal Ministry of Education and Research Germany within the research project "SubSEEMag".

Conflicts of Interest: The authors declare no conflicts of interest.

References

1. Neh, K.; Ullmann, M.; Oswald, M.; Berge, F.; Kawalla, R. Twin roll casting and strip rolling of several magnesium alloys. *Mater. Today Proc.* **2015**, *2S*, S45–S52. [CrossRef]
2. Park, S.J.; Jung, H.C.; Shin, K.S. Deformation behaviors of twin roll cast Mg-Zn-X-Ca alloys for enhanced room-temperature formability. *Mater. Sci. Eng. A* **2017**, *679*, 329–339. [CrossRef]
3. Luo, X.; Fang, D.; Liu, B. Thermo-Mechanical Treatment on Microstructure and Mechanical Properties of Mg-5.0Sn-1.0Mn-0.4Zr alloy. *Adv. Mater. Phys. Chem.* **2018**, *8*, 44–50. [CrossRef]
4. Yamamoto, A.; Ikeda, M.; Tsubakino, H. Grain Refinement in AZ91E Magnesium Alloy by Thermo-mechanical Treatments. *Mater. Sci. Forum* **2005**, *475–479*, 493–496. [CrossRef]
5. Agnew, S.R.; Mehrotra, P.; Lillo, T.M.; Stoica, G.M.; Liaw, P.K. Crystalligraphic texture evolution of three wrought magnesium alloys during equal channel angular extrusion. *Mater. Sci. Eng. A* **2005**, *408*, 72–78. [CrossRef]
6. Gong, X.; Kang, S.B.; Cho, J.H.; Li, S. Effect of annealing on microstructure and mechanical properties of ZK60 magnesium alloy sheets processed by twin-roll cast and differential speed rolling. *Mater. Charact.* **2014**, *97*, 183–188. [CrossRef]
7. Catorceno, L.L.C.; de Abreu, H.F.G.; Padilha, A.F. Effects of cold and warm cross-rolling on microstructure and texture evolution of AZ31B magnesium alloy sheet. *J. Mag. Alloys* **2018**, *6*, 121–133. [CrossRef]
8. Sabat, R.K.; Mishra, R.K.; Sachdev, A.K.; Suwas, S. The deciding role of texture on ductility in a Ce containing Mg alloy. *Mater. Lett.* **2015**, *153*, 158–161. [CrossRef]
9. Bohlen, J.; Nürnberg, M.R.; Senn, J.W.; Letzig, D.; Agnew, S.R. The texture and anisotropy of magnesium-zinc-rare earth alloy sheets. *Acta Mater.* **2007**, *55*, 2101–2112. [CrossRef]
10. Serebryany, V.N.; Popov, M.V.; Gordeev, A.S.; Timofeev, V.N.; Rokhlin, L.L.; Estrin, Y.; Dobatkin, S.V. Effect of Texture and Microstructure on Ductility of a Mg-Al-Ca Alloy Processed by Equal Channel Angular Pressing. *Mater. Sci. Forum* **2008**, *584–586*, 375–379. [CrossRef]
11. Kyeong, J.S.; Kim, J.K.; Lee, M.J.; Park, Y.B.; Kim, W.T.; Kim, D.H. Texture Modification by Addition of Ca in Mg-Zn-Y Alloy. *Mater. Trans.* **2012**, *53*, 991–994. [CrossRef]
12. Wang, T.; Jiang, L.; Mishra, R.K.; Jonas, J.J. Effect of Ca Addition on the Intensity of the Rare Earth Texture Component in Extruded Magnesium Alloys. *Metall. Mater. Trans. A* **2014**, *45*, 4698–4709. [CrossRef]
13. Borkar, H.; Pekguleryuz, M. Microstructure and texture evolution in Mg-1%Mn-Sr alloys during extrusion. *J. Mater. Sci.* **2013**, *48*, 1436–1447. [CrossRef]
14. Borkar, H.; Hoseini, M.; Pekguleryuz, M. Effect of strontium on texture and mechanical properties of extruded Mg-2%Mn alloys. *Mater. Sci. Eng. A* **2012**, *549*, 168–175. [CrossRef]
15. Liu, W.; Liu, X.; Tang, C.; Yao, W.; Xiao, Y.; Liu, X. Microstructure and texture evolution in LZ91 magnesium alloy during cold rolling. *J. Mag. Alloys* **2018**, *6*, 77–82. [CrossRef]
16. Chang, T.; Wang, J.; Chu, C.; Lee, S. Mechanical properties and microstructures of various Mg-Li alloys. *Mater. Lett.* **2006**, *60*, 3272–3276. [CrossRef]

17. Wang, Y.; Kang, S.B.; Cho, J. Microstructure and mechanical properties of Mg-Al-Mn-Ca alloy sheet produced by twin roll casting and sequential warm rolling. *J. Alloys Compd.* **2011**, *509*, 704–711. [CrossRef]
18. Tong, L.B.; Zheng, M.Y.; Zhang, D.P.; Gan, W.M.; Brokmeier, H.G.; Meng, J.; Zhang, H.J. Compressive deformation behavior of Mg–Zn–Ca alloy at elevated temperature. *Mater. Sci. Eng. A* **2013**, *586*, 71–77. [CrossRef]
19. Kulyasova, O.B.; Islamgaliev, R.K.; Zhao, Y.; Valiev, R.Z. Enhancement of the Mechanical Properties of an Mg–Zn–Ca Alloy Using High-Pressure Torsion. *Adv. Eng. Mater.* **2015**, *17*, 1738–1741. [CrossRef]
20. Qi, J.; Du, Y.; Jiang, B.; Shen, M. Hot deformation behavior and microstructural evolution of Mg–Zn–Ca–La alloys. *J. Mater. Res.* **2018**, *33*, 2817–2826. [CrossRef]
21. Neh, K.; Ullmann, M.; Kawalla, R. Substitution of Rare Earth Elements in Hot Rolled Magnesium Alloys with Improved Mechanical Properties. *Mater. Sci. Forum* **2016**, *854*, 57–64. [CrossRef]
22. Kurz, G.; Petersen, T.; Gonzales, I.P.; Hoppe, R.; Bohlen, J.; Letzig, D. Substitution of Rare Earth Elements in Magnesium Alloys for the Sheet Production Via Twin Roll Casting. *Mag. Technol.* **2016**, 377–382. [CrossRef]
23. Hoppe, R.; Kurz, G.; Letzig, D. Substitution of Rare Earths in Magnesium Alloys. *Mater. Sci. Forum* **2016**, *854*, 51–56. [CrossRef]
24. Letzig, D.; Bohlen, J.; Kurz, G.; Victoria-Hernandez, J.; Hoppe, R.; Yi, S. Development of Magnesium Sheets. *Mag. Technol.* **2018**, 355–360. [CrossRef]
25. Ullmann, M.; Kittner, K.; Henseler, T.; Stöcker, A.; Prahl, U.; Kawalla, R. Development of new alloy systems and innovative processing technologies for the production of magnesium flat products with excellent property profile. *Procedia Manuf.* **2019**, *27*, 203–208. [CrossRef]
26. Prasad, Y.V.R.K.; Sasidhara, S. *Hot Working Guide: A Compendium on Processing Maps*; ASM International: Metals Park, OH, USA, 1997; ISBN 0871705982.
27. Klein, H.; Schwarzer, R.A. Texture analysis with MTEX—Free and Open Source Software Toolbox. *Solid State Phenom.* **2010**, *160*, 63–68. [CrossRef]
28. Ullmann, M.; Schmidtchen, M.; Kittner, K.; Henseler, T.; Kawalla, R.; Prahl, U. Hot Deformation Behaviour and Processing Maps of an as-cast Mg-6.8Y-2.5Zn-0.4Zr Alloy. *Mater. Sci. Forum* **2019**, *949*, 57–65. [CrossRef]
29. Ullmann, M.; Schmidtchen, M.; Kawalla, R. Dynamic recrystallization behaviour of twin roll cast AZ31 strips during hot deformation. *Key Eng. Mater.* **2014**, *622–623*, 569–574. [CrossRef]
30. Kittner, K.; Ullmann, M.; Henseler, T.; Kawalla, R.; Prahl, U. Microstructure, texture and mechanical properties of twin roll cast Mg-2Zn-1Al-0.3Ca alloy. In Proceedings of the MSSM2018, Paisley, UK, 7–10 August 2018.
31. Wang, Y.N.; Huang, J.C. Texture analysis in hexagonal materials. *Mater. Chem. Phys.* **2003**, *81*, 11–26. [CrossRef]
32. Masoumi, M.; Zarandi, F.; Pekguleryuz, M. Microstructure and texture studies on twin-roll cast AZ31 (Mg-3wt%Al-1wt%Zn) alloy and the effect of thermomechanical processing. *Mater. Sci. Eng. A* **2011**, *528*, 1268–1279. [CrossRef]
33. Tang, N.; Wang, M.P.; Lou, H.F.; Zhao, Y.Y.; Li, Z. Microstructure and texture of twin-roll cast Mg-3Al-1Zn-0.2Mn magnesium alloy. *Mater. Chem. Phys.* **2009**, *116*, 11–15. [CrossRef]
34. Soomro, M.W. Formability of Magnesium AZ80. Ph.D. Thesis, School of Engineering, Computer and Mathematical Sciences, Auckland, New Zealand, 2016.
35. Chaudry, U.M.; Kim, T.H.; Park, S.D.; Kim, Y.S.; Hamad, K.; Kim, J. On the high formability of AZ31-0.5Ca magnesium alloy. *Materials* **2018**, *11*, 2201. [CrossRef] [PubMed]
36. Yang, J.; Peng, J.; Li, M.; Nyberg, E.A.; Pan, F. Effects of Ca Addition on the Mechanical Properties and Corrosion Behavior of ZM21 Wrought Alloys. *Acta Metall. Sin.* **2017**, *30*, 53–65. [CrossRef]
37. Zeng, Z.R.; Zhu, Y.M.; Xu, S.W.; Bian, M.Z.; Davies, C.H.J.; Birbilis, N.; Nie, J.F. Texture evolution during static recrystallization of cold-rolled magnesium alloys. *Acta Mater.* **2016**, *105*, 479–494. [CrossRef]
38. Basu, I.; Al-Samman, T. Triggering rare earth texture modification in magnesium alloys by addition of zinc and zirconium. *Acta Mater.* **2014**, *67*, 116–133. [CrossRef]
39. Zimina, M.; Zimina, A.; Pokova, M.; Bohlen, J.; Letzig, D.; Kurz, G.; Knapek, M.; Cieslar, M. Influence of the heat treatment on the texture of twin-roll-cast AZ31 magnesium alloy. *Metal* **2014**. [CrossRef]
40. Griffiths, D. Explaining texture weakening and improved formability in magnesium rare earth alloys. *Mater. Sci. Technol.* **2015**, *31*, 10–24. [CrossRef]

41. Ball, E.A.; Prangnell, P.B. Tensile-compressive yield asymmetries in high strength wrought magnesium alloys. *Scr. Metall. Mater.* **1994**, *31*, 111–116. [CrossRef]

42. Sandlöbes, S.; Friak, M.; Korte-Kerzel, S.; Pei, Z.; Neugebauer, J.; Raabe, D. A rare-earth free magnesium alloy with improved intrinsic ductility. *Sci. Rep.* **2017**, *7*, 1–8. [CrossRef]

43. Garofalo. *Fundamentals of Creep and Creep-Rupture in Metals*; McMillan Series in Materials Science; McMillan: New York, NY, USA, 1965.

44. Weertman, J. Dislocation climb theory of steady-state creep. *ASM Trans. Quavt. Metal. Sci. J.* **1968**, *61*, 681–694. [CrossRef]

45. Sherby, O.D.; Taleff, E.M. Influence of grain size, solute atoms and second-phase particles on creep behaviour of polycrystalline solids. *Mater. Sci. Eng. A* **2002**, *322*, 89–99. [CrossRef]

46. Lesuer, D.R.; Syn, C.K.; Sherby, O.D. An evaluation of power law breakdown in metals, alloys, dispersion hardened materials, and compounds. In *Deformation, Processing and Properties of Structural Materials, Proceedings of the Honorary Symposium for Professor OD Sherby, Nashville, TN, USA, 16–20 March 2000*; Taleff, E.M., Syn, C.K., Lesuer, D.R., Eds.; TMS: Warrendale, PA, USA, 2000; pp. 81–194.

47. Rao, K.P.; Chalasani, D.; Suresh, K.; Prasad, Y.V.R.K.; Dieringa, H.; Hort, N. Connected Process Design for Hot Working of a Creep-Resistant Mg–4Al–2Ba–2Ca Alloy (ABaX422). *Metals* **2018**, *8*, 463. [CrossRef]

48. Suresh, K.; Rao, K.P.; Prasad, Y.V.R.K.; Wu, C.-M.L.; Hort, N.; Dieringa, H. Mechanism of Dynamic Recrystallization and Evolution of Texture in the Hot Working Domains of the Processing Map for Mg-4Al-2Ba-2Ca Alloy. *Metals* **2017**, *7*, 539. [CrossRef]

49. Rao, K.P.; Suresh, K.; Prasad, Y.V.R.K.; Dharmendra, C.; Hort, N.; Dieringa, H. High Temperature Strength and Hot Working Technology for As-Cast Mg–1Zn–1Ca (ZX11) Alloy. *Metals* **2017**, *7*, 405. [CrossRef]

50. Dharmendra, C.; Rao, K.P.; Suresh, K.; Hort, N. Hot Deformation Behavior and Processing Map of Mg-3Sn-2Ca-0.4Al-0.4Zn Alloy. *Metals* **2018**, *8*, 216. [CrossRef]

materials

MDPI

Article

Static and Dynamic Loading Behavior of Ti6Al4V Honeycomb Structures Manufactured by Laser Engineered Net Shaping (LENS™) Technology

Anna Antolak-Dudka [1],*, Paweł Płatek [2], Tomasz Durejko [1],*, Paweł Baranowski [3], Jerzy Małachowski [3], Marcin Sarzyński [2] and Tomasz Czujko [1]

[1]   Department of Advanced Materials and Technologies, Military University of Technology, Urbanowicza 2, Warsaw 00-908, Poland; tomasz.czujko@wat.edu.pl

[2]   Institute of Armament Technology, Military University of Technology, Urbanowicza 2, Warsaw 00-908, Poland; pawel.platek@wat.edu.pl (P.P.); marcin.sarzynski@wat.edu.pl (M.S.)

[3]   Department of Mechanics and Applied Computer Science, Military University of Technology, Urbanowicza 2, Warsaw 00-908, Poland; pawel.baranowski@wat.edu.pl (P.B.); jerzy.malachowski@wat.edu.pl (J.M.)

*   Correspondence: anna.dudka@wat.edu.pl (A.A.-D.); tomasz.durejko@wat.edu.pl (T.D.); Tel.: +48-261-839-445 (A.A.-D.); +48-261-837-135 (T.D.)

Received: 1 March 2019; Accepted: 9 April 2019; Published: 15 April 2019

Abstract: Laser Engineered Net Shaping (LENS™) is currently a promising and developing technique. It allows for shortening the time between the design stage and the manufacturing process. LENS is an alternative to classic metal manufacturing methods, such as casting and plastic working. Moreover, it enables the production of finished spatial structures using different types of metallic powders as starting materials. Using this technology, thin-walled honeycomb structures with four different cell sizes were obtained. The technological parameters of the manufacturing process were selected experimentally, and the initial powder was a spherical Ti6Al4V powder with a particle size of 45–105 µm. The dimensions of the specimens were approximately $40 \times 40 \times 10$ mm, and the wall thickness was approximately 0.7 mm. The geometrical quality and the surface roughness of the manufactured structures were investigated. Due to the high cooling rates occurring during the LENS process, the microstructure for this alloy consists only of the martensitic α' phase. In order to increase the mechanical parameters, it was necessary to apply post processing heat treatment leading to the creation of a two-phase $\alpha + \beta$ structure. The main aim of this investigation was to study the energy absorption of additively manufactured regular cellular structures with a honeycomb topology under static and dynamic loading conditions.

Keywords: honeycomb structure; additive manufacturing; laser engineered net shaping; LENS; Ti6Al4V alloy; energy absorption; dynamic tests

1. Introduction

Currently, the progress of civilization forces scientists and engineers to discover new technologies and materials necessary to optimize the products used in all areas of life. The automotive and aviation industries still require new solutions in terms of safety, where a strong impact is placed on the elements used so that they possess high energy absorption capacity during crash tests [1–4]. Moreover, these requirements are also essential in relation to military applications, such as passive protective systems [5,6] and various kinds of critical infrastructure elements [7]. There are many solutions based on lightweight cellular structures made from different materials, i.e., tubes, sandwiches, or honeycombs, which have been studied under static as well as dynamic loading conditions [8–13]. Particularly noteworthy is honeycomb structure topology, which is popular as an energy absorber; it also has received great attention in the biomedical field for applications such as 3D porous scaffolds for tissue engineering and its regeneration [14–17]. The phenomena of this structure result from its original properties. It is a combination of a low value of relative density and a high stiffness that derives from geometry and allows for the minimization of the amount of material used. Additionally, the periodic arrangement of cells plays a significant role. It has been observed that the crush strength depends on cell shape (hexagonal cells with different branch angles were studied) [18–22] and wall thickness [23–26]. The honeycomb structures were investigated with numerical analysis or experimentally in both in-plane [27–31] and out-of-plane directions [20,32–36] to obtain various parameters to provide answers about the mechanisms of failure.

Cellular structures exist as regular and stochastic objects. Stochastic objects are mostly manufactured by conventional methods such as vapor deposition, casting, sintering, and foaming polymer or metallic materials [37,38]. Unfortunately, there are some difficulties associated with these techniques, such as insufficient precision of cell projections, porosity of the produced structures and anisotropy of mechanical properties that derive from the stochastic arrangement of cells and differing unit cell sizes. Researchers have been looking for new technological solutions to overcome these limitations. One of the rapidly developing routes in manufacturing technologies is additive manufacturing (AM), which enables the production of regular cellular structures [39,40]. It is also called rapid prototyping (RP), which can be used as an original method for the fabrication of elements with periodically spaced cells. There are several different techniques that direct the fabrication of component parts by building them layer by layer with the use of various types of metal alloys. Generally, these methods are divided in two categories: Powder Bed Fusion (PBF) and DED (Direct Energy Deposition) [41]. The first one is represented by SLM (Selective Laser Melting), (DMLS) Direct Metal Laser Sintering, and EBM (Electron Beam Melting). Mentioned systems are very popular and allow for manufacturing regular structures with a complex geometry and a very low mass. The main idea of working this type of system is described in detail in following papers [40,42,43]. The other DED group of Metal Additive Manufacturing system is represented by the Laser Engineering Net Shaping system. It was discovered at Sandia National Laboratories and commercialized by Optomec, Inc. in 1997 [44]. It is a technique that allows for saving processing time by shortening the period between the design stage and the manufacturing process. The building of near-net shaped components can be fully controlled, and this is one of the most important benefits. The amount of given powder, the feed of the work table, the laser power and the focus of the laser beam can be precisely selected depending on the needs. The completed 3D parts are made by building them layer-by-layer from powder applied directly to the place the laser beam affects. Components manufactured by the LENS technique can be made from easily accessible engineering materials such as stainless steel, Ni-based super alloys or titanium alloys. The mechanical properties of parts made by the LENS technique have been presented in following scientific papers [37,45–47]. The microstructure, mechanical properties and corrosion resistance of 316L stainless steel samples manufactured by the LENS technique were investigated by Ziętala et al. [48]. These samples were characterized by unique dual-phase microstructures unprecedented in stainless steel fabricated by conventional methods, which is the reason for the improvement of the mechanical and corrosion properties. There are also a few works concerning the components made from titanium alloys using the LENS technique. The relationship between the influence of building parameters and deposition of Ti6Al4V samples was investigated by Kummailil [49]. Furthermore, Zhai et al. conducted small and long fatigue crack growth tests [50]. In the results, it was shown that the lamellar structure that is created during post-LENS heat treatment offers higher fracture toughness and ductility. Blackwell et al. examined the possibility of the production of alpha-beta Ti6Al2Sn4Zr6Mo (Ti-6246) titanium alloy samples [46]. Additionally, materials with a chemical gradient obtained by the LENS technique are becoming more focused. There are many different combinations of materials used, i.e., TiC/Ti composite with compositions changing from pure Ti to 95 vol.% TiC [45] or thin wall tubes with a $Fe_3Al/SS316L$ graded structure [51].

On the basis of a conducted literature review, the authors have spotted that there is a limited number of papers considering the possibility of using the DED LENS system in the manufacturing process of regular cellular structures with high energy absorption capacity. Results of investigations presented in many papers are limited to simple geometrical shapes of absorbers like cylindrical and rectangular tubes. Moreover, many of them were realized with the use of stainless-steel powder alloys which give good-quality manufactured models. Titanium alloy seems to be the perfect candidate because of its significant properties of high strength-to-weight ratio, excellent corrosion resistance and high melting point. All these advantages make this alloy very attractive for applications in the medical, aviation, and aerospace industries. The Ti6Al4V chemical composition is the most popular among titanium-based alloys, and it is very popular as a material used in additive manufacturing techniques. The mechanical properties of this alloy closely depend on the structure and size as well as the morphology of the grains. At equilibrium, Ti6Al4V is an alloy consisting of two phases: α and β phase. However, additive manufacturing techniques are characterized by the fact that during production of the details, the degree of cooling rates is so high whereby a martensitic α' phase creates. The presence of this phase increases the yield strength and tensile strength with a simultaneous decrease in the plasticity of the built samples. In order to improve the ductility or the fracture toughness a post-processing heat treatment is required and provides a two-phases structure [43,52,53]. The heat treatment applied for the components manufactured using additive techniques from various materials can also improve the microstructure homogeneity or removes stresses created during the building process [54,55].

Taking into account the mechanical properties of Ti6Al4V titanium alloy and the technological possibilities of the LENS system, the authors proposed investigations to examine the possibility of obtaining Ti6Al4V regular structures with hexagonal cells by the LENS technique. Mapping quality, metallurgical quality and microstructure were tested for the structures before and after applying the appropriate heat treatment. This paper is a continuation of the authors' previous paper [56] in which a methodology investigation of the deformation process of honeycomb cellular structures manufactured using LENS was discussed. Additionally, the mechanical response of manufactured specimens of structures were evaluated not only in static, but also under dynamic loading conditions.

2. Materials and Methods

2.1. LENS^TM Technique

The components presented in this work were manufactured by the Optomec LENS MR-7 (Albuquerque, USA). The scheme of the LENS system operation is shown in Figure 1. In general, the principle of the device's operation is based on the selective deposition of metallic or ceramic powders on a prepared substrate or on the previously built layer and melting of them with 500 W of high-power fiber laser at the same time.

The LENS machine has an operating system that allows the operator to design and prepare the manufacturing process. The building of components starts with a 3D CAD solid model, which is sliced by the PartPrep program into layers of an assumed thickness. In the next step the LENS control software selects and determines production parameters, such as the powder flow rate, acceleration, and deceleration of the working table or laser power.

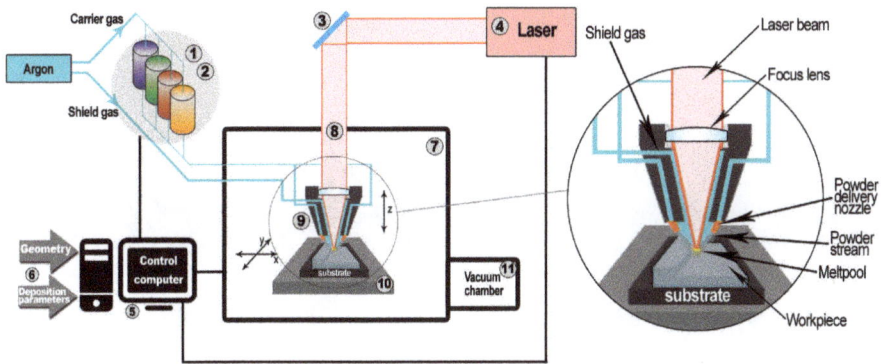

Figure 1. Scheme of the laser engineered net shaping (LENS) system [56]: **1.** Powder supply; **2.** pneumatic vibrating system; **3.** optical system; **4.** IPG fiber laser; **5.** controlling computers; **6.** input data; **7.** working chamber; **8.** optical path of the laser; **9.** working head with four nozzles; **10.** numerically controlled working table (movement in the *X-Y* plane); **11.** antechamber.

Powders used in this technique should be of high purity and chemically homogenous with a spherical shape. The recommended particle size distribution ranges from 45–150 µm [57]. Meeting these requirements will ensure good metallurgical quality of the produced components. Four chemically different powders can be used simultaneously during one technological process since the device is equipped with four powder feeders. In addition, the feeding of powders from nozzles placed in the laser head is very precise, even if the quantity of the powder's flow is small. This allows for creating gradient structures or structures reinforced with the strengthening phase. The substrate on which the previously designed element will be built should have the same chemical composition as the batch powder. This will avoid the occurrence of very high stresses that can be generated during the process.

The LENS process is carried out under an argon gas atmosphere; therefore, the amount of oxygen molecules in the chamber is below 10 parts per million. For this reason, powders with high reactivity can be used in the production process.

2.2. Modelling of Thin-Walled Honeycomb Structure

The aim of this work was to produce four variants of thin-walled honeycomb structures differing in the size of the unit cells. The selection of the cell sizes was a result of the structure's geometrical optimization and the technological capabilities of the device. It was determined that the smallest diameter of a circle described on a hexagonal cell possibly obtained by the LENS technique is 3 mm (Figure 2). The other three cells are 4, 5, and 6 mm in diameter. The assumed dimensions of the obtained structures were approximately 40 × 40 × 10 mm, and the wall thickness should be approximately 0.7 ± 0.1 mm. The value of the relative density obtained for the specimens is presented in Table 1.

Table 1. Relative density of the developed honeycomb structures.

Specimen	Honeycomb_3	Honeycomb_4	Honeycomb_5	Honeycomb_6
	No. 1	No. 2	No. 3	No. 4
Unit cell size (mm)	3	4	5	6
Relative density (−)	0.36	0.31	0.3	0.23

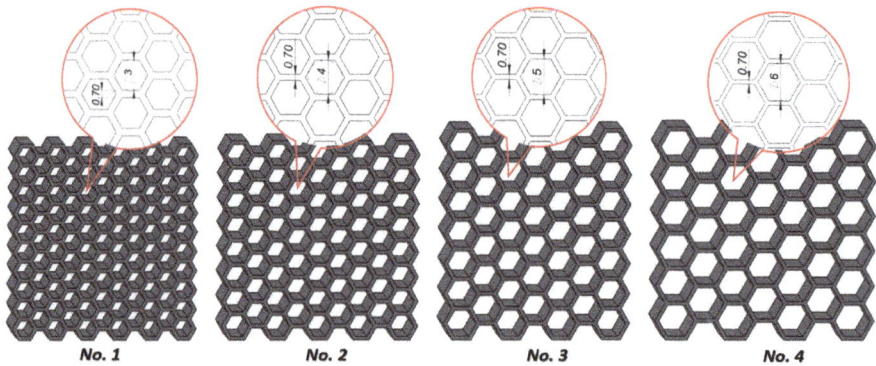

Figure 2. Four variants with of thin-walled honeycomb structures differing in the size of the cells (No. 1–No. 4, dimensions in mm).

For manufacturing the above-presented structures, a commercial Ti6Al4V powder was used. The powder was produced by an argon atomization method and was delivered by TLS Technik GmbH & Co. The size of the powder particles ranges from 45–105 μm with a spherical shape. The morphology and microstructure of the powders are given in Figure 3. It has been confirmed that the particles have a spherical shape and mostly smooth surface with microsatellites. In the cross-section, some pores with micrometer size were observed inside the particles.

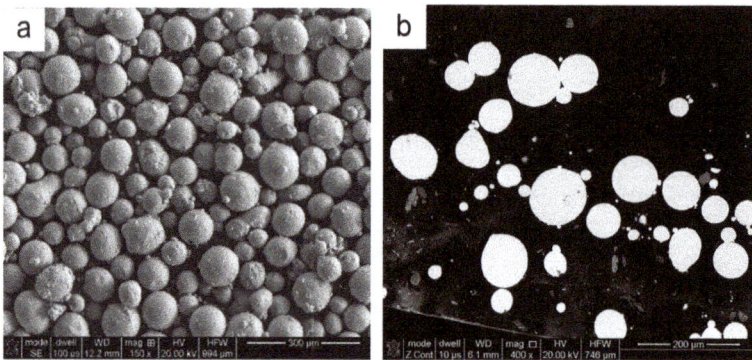

Figure 3. The morphology (**a**) and microstructure (**b**) of the Ti6Al4V powder used for manufacturing of honeycomb structures by the laser engineered net shaping (LENS^{TM}) technique.

The honeycomb structures were built using the laser engineered net shaping (LENS^{TM}) technique on a Ti6Al4V substrate that was previously skimmed with acetone and sandblasted. The process was conducted in an argon atmosphere, and the oxygen content in the working chamber was approximately 2 ppm. The parameters of the manufacturing process were selected experimentally, and they are presented in Table 2.

Table 2. Basic parameters used for manufacturing honeycomb structures by the LENS technique.

Laser Power (W)	Powder Feedrate (rpm)	Layer Thickness [mm]	Powder (material)	Substrate (material)
180	9.8	0.3	Ti6Al4V	Ti6Al4V

The obtained thin-walled structures presented in Figure 4 were examined in terms of their geometrical and microstructural properties.

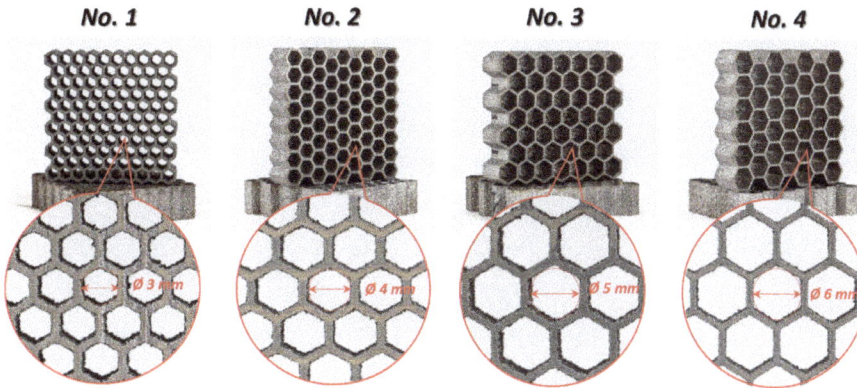

Figure 4. Four variants of Ti6Al4V thin-walled honeycomb structures with different the cell size (No. 1–No. 4) manufactured by the laser engineered net shaping (LENS^TM) technique.

3. Results and Discussion

3.1. Geometrical Assessment

The geometrical quality of the manufactured structure specimens was evaluated based on the data gathered with the application of computer tomography Metrology XTH 225 (Nikon, Leuven, Belgium) and photographs made with the use of optical microscopy (VHX-6000, Keyence International NV/SA, Mechelen, Belgium). The results of the measurements are presented in Figure 5, and they were partially presented in [56]. The obtained wall thickness of specimens differ in comparison to the values assumed during the preparation of the 3D CAD model. It is caused mainly due to the adopted regime of the structure manufacturing process. The wall thickness was defined as a single route of the working head.

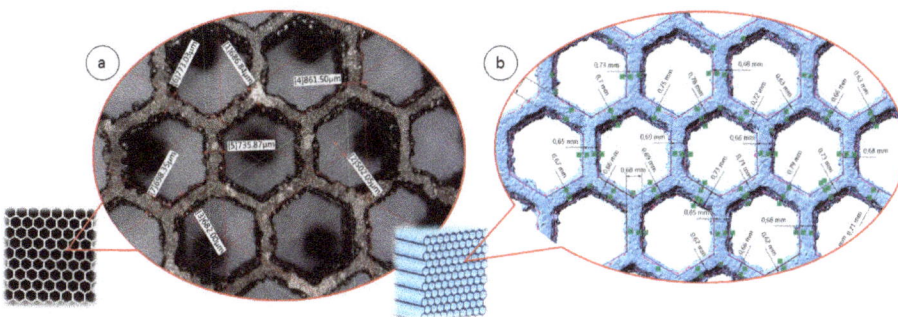

Figure 5. Geometrical evaluation of structure specimens: (**a**) with the application of optical microscopy, (**b**) based on a 3D model reconstructed from Computed Tomography (CT) data [56].

The honeycomb structures were subjected to heat treatment just after manufacturing, which was performed at 1050 °C for 2 h. The process was undertaken in a Nabertherm R80/b750/12-B170 tubular furnace (Nabertherm GmbH, Lilienthal, Germany) with a low vacuum. The furnace chamber was purged with argon before the heating processes started. The specimens were heated and cooled down with the furnace. The choice of heat treatment parameters was determined by the analysis of literature [58,59].

The influence of the adopted heat treatment process on structure geometrical quality was evaluated based on the analysis of CT data collected before and after the process. Based on the comparison of the obtained results, spatial maps of the geometrical deviation for all specimens were defined (Figures 6–9).

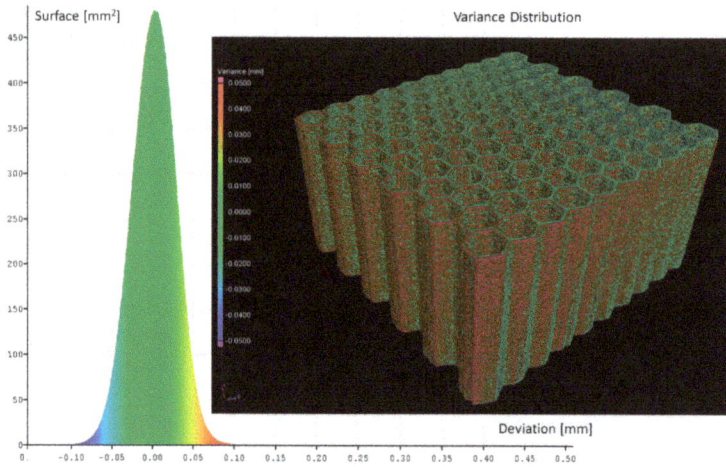

Figure 6. Geometrical quality control of specimen No. 1 before versus after heat treatment.

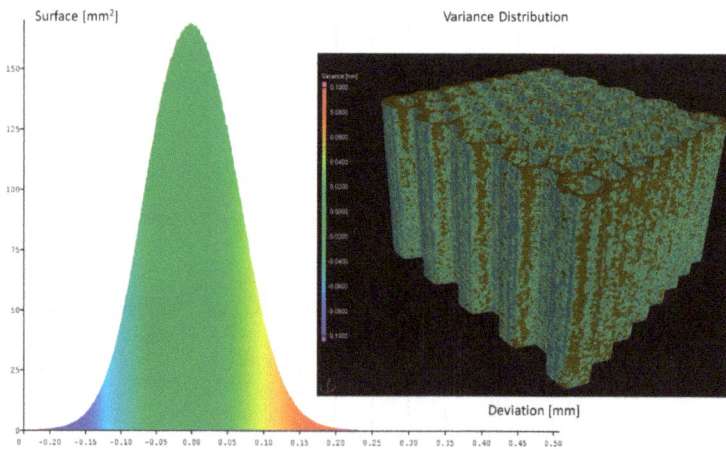

Figure 7. Geometrical quality control of specimen No. 2 before versus after heat treatment.

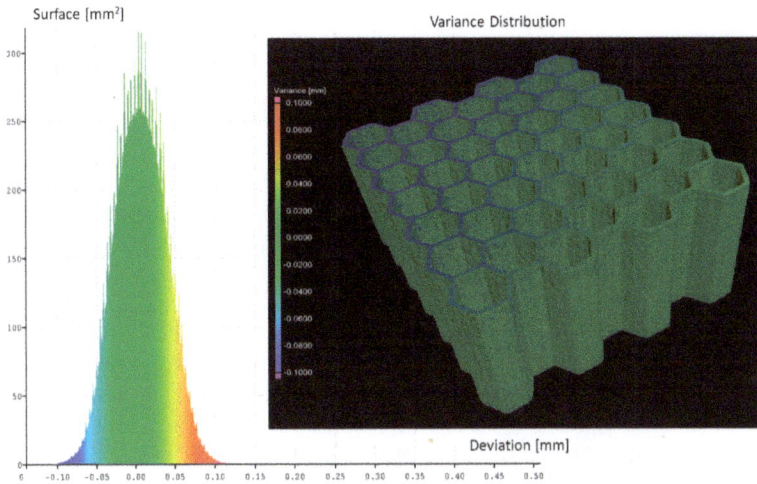

Figure 8. Geometrical quality control of specimen No. 3 before verses after heat treatment.

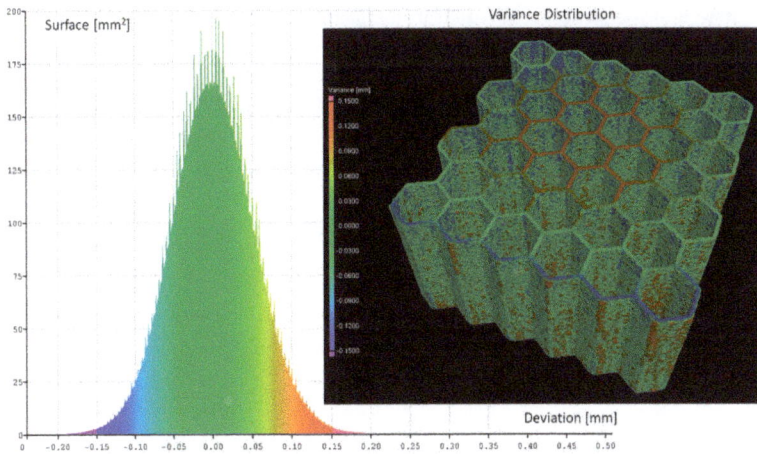

Figure 9. Geometrical quality control of specimen No. 4 before versus after heat treatment.

Based on the presented figures, it could be stated that the additional applied heat treatment process strongly influenced the stress relief process. The lowest value of the elementary unit cell size and the highest value of the relative density of specimen No. 1 indicated the highest surface dimensional deviation (Figure 6). It was mainly caused by the poor heat dissipation process, which also affected the structure of the material. Implementation of a larger value of unit cell size enabled better geometrical quality. Additionally, due to the lower value of relative density, the thermal conditions of heat dissipation were better and caused the lowest deviation of structure dimensions (Figures 7 and 9).

Application of a modular contact profilometer enabled the determination of the surface roughness of the specimens, which are presented in Table 3. On the basis of the presented results, it could be stated that the roughness of the samples after the sandblasting process is lower than those before sandblasting. This is the effect of the smoothing surface of the honeycomb structures, which remove the unmelted powder particles adhesively bonded with them. Independent of the cell size, all sandblasted samples are characterized by a statistically comparable level of roughness (Table 3).

147

Table 3. Results of the roughness tests of samples manufactured by the LENS technique without heat treatment.

Sandblasting	Honeycomb Structure	Ra (μm)	St. dev.	Rz (μm)	St. dev.
Before	No. 4	28.66	2.62	170.59	14.83
	No. 3	29.19	1.89	173.97	15.56
	No. 2	27.89	1.98	163.71	17.48
	No. 1	28.26	1.32	166.45	16.78
After	No. 4	22.19	2.74	133.76	7.29
	No. 3	24.52	1.77	140.58	10.42
	No. 2	21.34	2.59	127.77	15.77
	No. 1	22.99	0.70	143.95	13.43

3.2. Evaluation of the Microstructure and Mechanical Properties of Ti6Al4V

The next stage of the investigations was related to the evaluation of the microstructure of the Ti6Al4V material. For this purpose, the surface of metallographic samples cut from honeycomb components with or without heat treatment (1050 °C/2 h) was prepared by grinding, polishing and etching with Kroll's reagent. The microstructure of the samples was examined by a FEI Quanta 3D (FEI company, Hillsboro, USA) scanning electron microscope (SEM) equipped with energy dispersive spectroscopy (EDS). Based on compared SEM photographs presented in Figure 10, it is possible to state that the structure revealed in the sample without annealing consists only of the martensitic phase, which is a Ti-based solid solution (non-equilibrium phase). This phase is formed in titanium alloys during very fast cooling from the temperature area above the transformation $\alpha\rightarrow\beta$ temperature. Such conditions of high supercooling occur during the LENS building process. Whereas the microstructure of sample manufactured by LENS technique and heat treated consists of two phases $\alpha + \beta$, which were described and confirmed in a previous work [56].

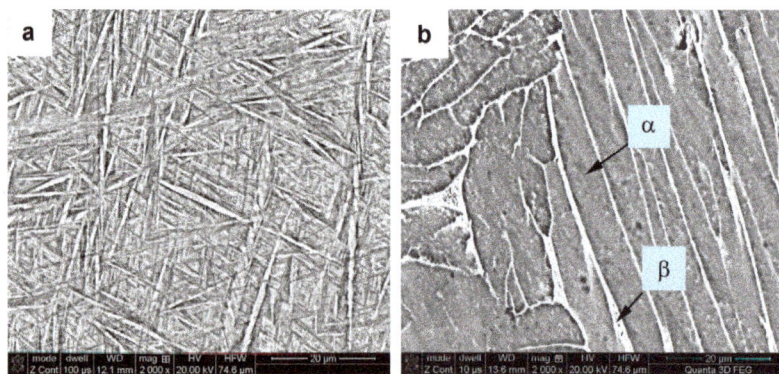

Figure 10. The SEM micrographs of honeycomb components microstructure before (**a**) and after (**b**) heat treatment (1050 °C/2 h).

The mechanical properties of additively manufactured Ti6Al4V materials were determined during a uniaxial tensile test under quasi-static loading conditions. The typical dog bone specimens with a thickness of 0.7 ± 0.1 mm were manufactured with the same technological parameters as the structure specimens. Tensile tests were performed on a MTS Criterion C45 testing machine (MTS Systems Corporation, Eden Prairie, USA), which gives a strain rate of the magnitude $1.7 \times 10 \ s^{-1}$. The entire process was monitored and recorded using TW-Elite software (ver. 2.3.1, MTS Systems Corporation, Eden Prairie, USA). Figure 11 presents the results obtained for the origin (NHT) and modified (HT) specimens during the heat treatment process. The characteristic mechanical properties of the Ti6Al4V material are presented in Table 4.

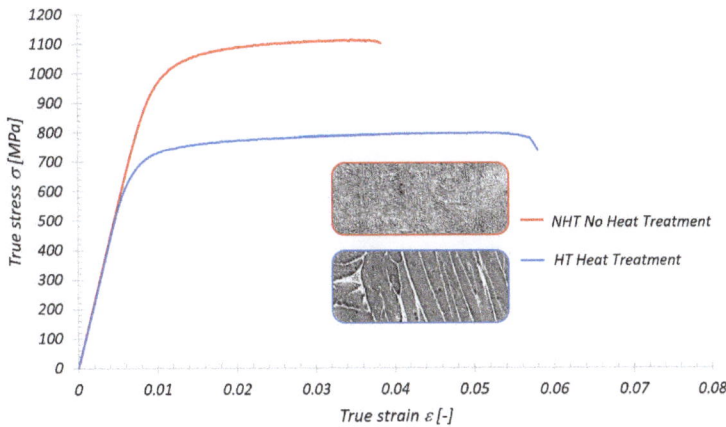

Figure 11. An example of stretching curves for samples cut from the Ti6Al4V thin walls obtained using the LENS technique without (NHT) and with (HT) additional heat treatment process [56].

Table 4. Strength parameters determined in a tensile test for samples cut from Ti6Al4V thin walls obtained by the LENS technique (before and after the heat treatment process).

Specimen Sample	R $_{0.2}$ (MPa)	Rm (MPa)	A (%)	E (GPa)
NHT	988	1110	3.7	110
HT	705	794	5.6	108

3.3. Compression Tests under Quasi-Static Loading Conditions

The mechanical response of regular cellular structures with honeycomb topology was determined during experimental uniaxial compression tests performed under quasi-static and dynamic loading conditions. In both cases, the study structures were placed in an in-plane direction. The obtained results are presented below.

The first stage of the investigations was conducted with the use of an MTS Criterion C45 universal tensile test machine and TW-Elite software, which recorded the history of the deformation process. The specimens were placed in the in-plane direction and compressed with 1 mm/s velocity. On the basis of the conducted tests, plots of the deformation force and deformation energy were defined (Figures 12–15). From these plots, it was possible to define the relationship between the structure's relative density when referring to the energy absorption capacity (Figure 16).

Analyzing the history of the deformation force plots, it could be stated that specimens No. 1 (honeycomb_3) (Figure 12) and No. 2 (honeycomb_4) (Figure 13) demonstrate a high value of the maximum deformation force due to the high geometrical stiffness of the structures of the specimens. The slope of the first part of the historical plots connected with the local buckling effect is similar and caused due to the mechanical properties of the applied Ti6Al4V material and the friction between the surfaces of the specimen and the grip of the testing machine. Depending on the applied elementary unit cell size, the maximum value of the respective deformation force is different. Structures with the lowest unit cell size and highest relative density indicated the highest value of the deformation force. After the buckling effects, a failure mechanism was very quickly achieved. It was mainly caused by the shearing and cracking of the structure walls. The higher value of a unit cell size is represented by specimen No. 3 (honeycomb_5) (Figure 14) and specimen No. 4 (honeycomb_6) (Figure 15), which affects the lower geometrical stiffness of the structures. The maximum value of the deformation force is relatively lower in comparison to specimens No. 1 and No. 2. Moreover, the buckling effect and loss of structural stability appeared later, which caused a delay in the failure mechanism. The plot of the force deformation is milder in this case. The number of the local maximum deformation force is lowest and average (Figure 15). Additionally, the chart presented in Figure 16 demonstrates the relationship between the deformation energy of the specimen referring to the deformation. It could be observed that the higher geometrical stiffness of the specimen caused an increasing maximum value of the deformation energy and caused the reduction of the specimen shortening. Application of topology with a higher value of unit cell size (with lower value of relative density) shows that the maximum value of deformation energy is significantly lower, but the range of the specimen shortening is relatively higher. The value of the deformation energy obtained for the same shortening of all structure specimens (the value of shortening was 20 mm) is presented in Table 5. Based on the presented results, it could be seen that the deformation energy depends on the value of the relative density. Nevertheless, this relationship is not linear, which means that it could also be dependent on the friction process between collaborating elements and the geometrical quality of the additively manufactured structure specimens. Due to having the lowest value of unit cell size, specimen No. 1 was rougher, and the dimensional deviations were higher in comparison to the other specimens.

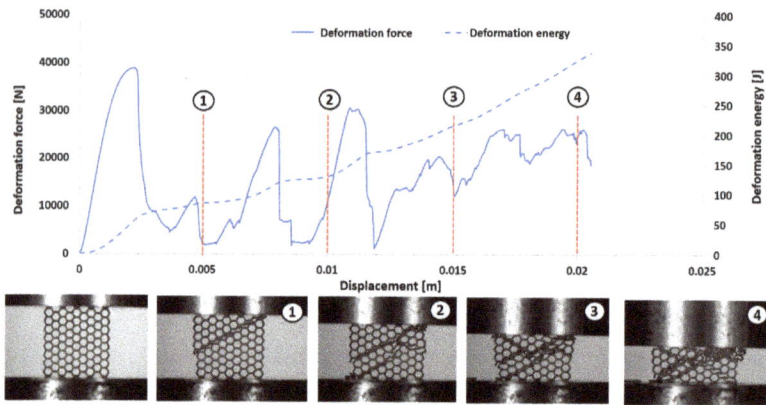

Figure 12. Deformation process of specimen No. 1 manufactured additively with LENS (1–4 stages of deformation during static compression test).

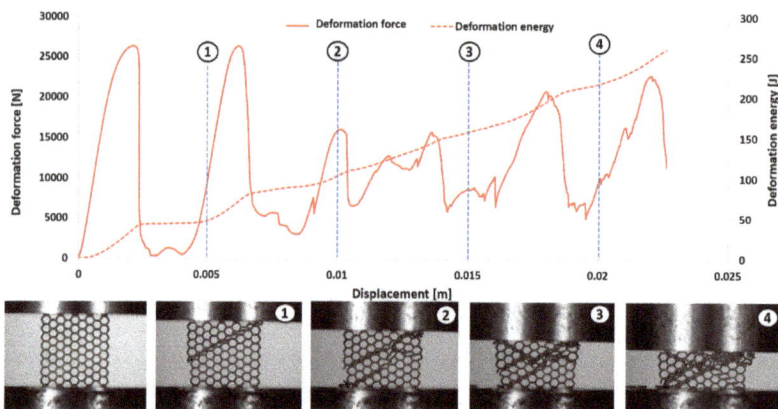

Figure 13. Deformation process of specimen No. 2 manufactured additively with LENS (1–4 stages of deformation during static compression test).

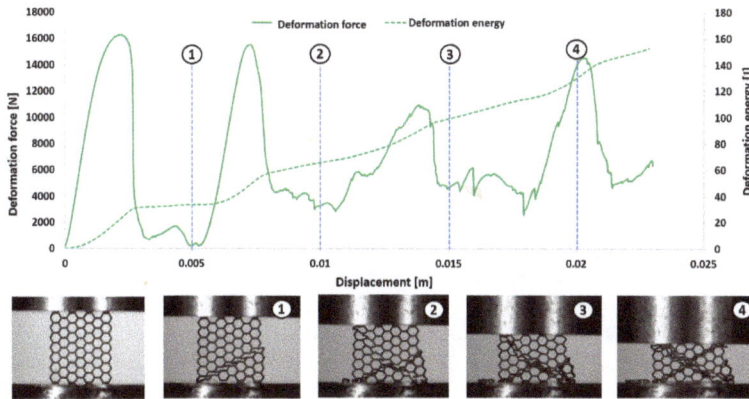

Figure 14. Deformation process of specimen No. 3 manufactured additively with LENS (1–4 stages of deformation during static compression test).

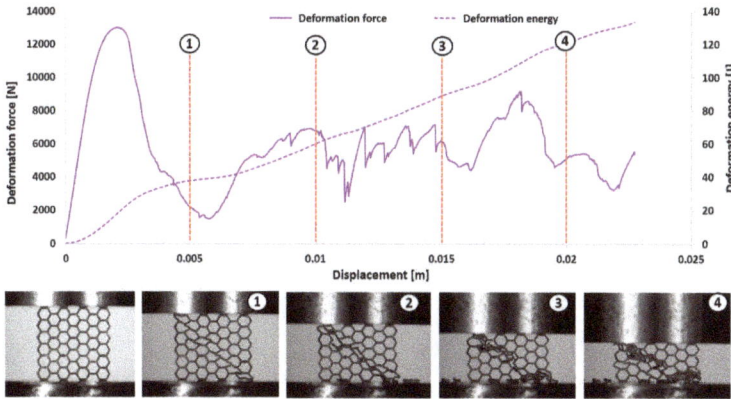

Figure 15. Deformation process of specimen No. 4 manufactured additively with LENS (1–4 stages of deformation during static compression test).

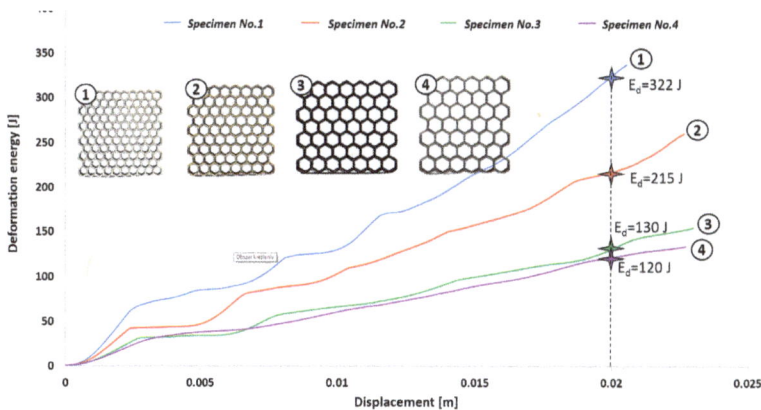

Figure 16. Comparison of deformation energy curves related to structure unit cell size (1–4—the thin-walled honeycomb structures with the different cell size).

151

Table 5. The comparison of maximum values of absorbed energy referring to honeycomb structure specimens.

	No. 1	No. 2	No. 3	No. 4
Relative density (−)	0.36	0.31	0.3	0.23
Max. value of absorbed energy (J)	322	215	130	120

3.4. Compression Tests under Dynamic Loading Conditions

The dynamic tests were the other stage of investigations conducted on the mechanical response of additively manufactured Ti6Al4V regular cellular structures with honeycomb topology under dynamic loading conditions. They were carried out with the use of the universal column impact test machine Instron Dynatup 9250 HV with an additional system of data acquisition and a high-speed camera. Adopted initial loading conditions were defined based on the mass of the impactor and its initial velocity. The following initial boundary conditions were used for the tests: impactor mass was 8 kg, and its velocity was 20 m/s. On the basis of the conducted compression tests, the results presented in Figures 17–20 were obtained. Analyzing the presented deformation energy plots, it could be stated that the maximum value of the absorbed energy is related to the relative density of the structure of the specimens. The structure of specimen No. 1, with the smallest elementary unit cell size, indicates a higher value of relative density and geometrical stiffness. These features caused the value of the absorbed energy to be significantly higher in comparison to other specimens. Additionally, due to the considerably greater number of cells in the structural arrays, the number of the local maximum deformation force arrived with more frequency. The adopted dynamic initial boundary conditions caused the mechanism of the structural densification to appear for testing cases.

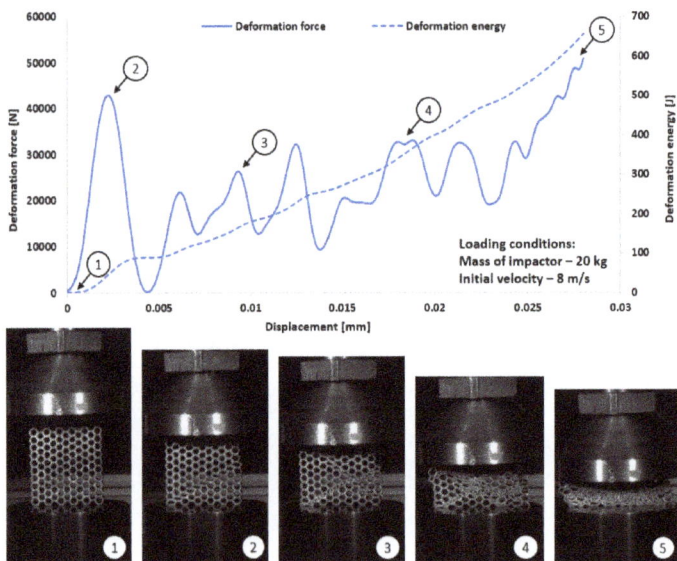

Figure 17. The results of dynamic tests obtained for honeycomb specimen No. 1 (1–5 stages of deformation during dynamic compression test).

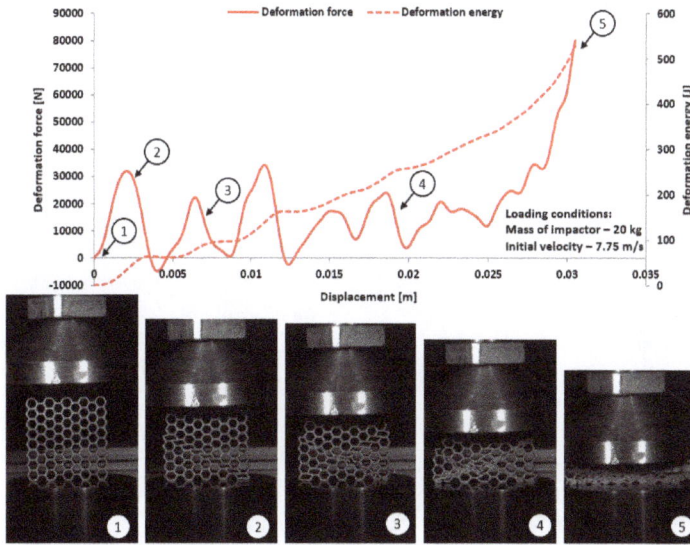

Figure 18. The results of dynamic tests obtained for honeycomb specimen No. 2 (1–5 stages of deformation during dynamic compression test).

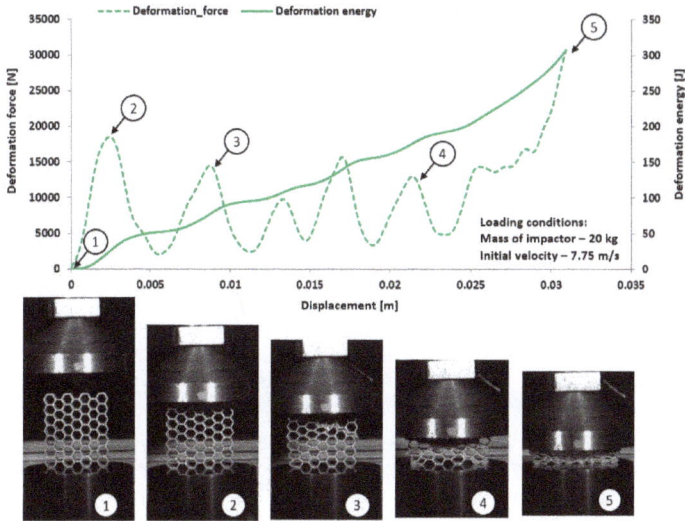

Figure 19. The results of dynamic tests obtained for honeycomb specimen No. 3 (1–5 stages of deformation during dynamic compression test).

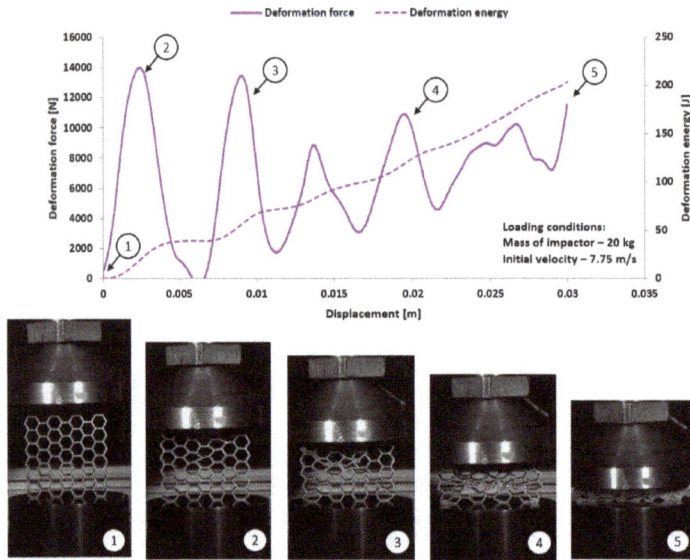

Figure 20. The results of dynamic tests obtained for honeycomb specimen No. 4 (1–5 stages of deformation during dynamic compression test).

Comparing the maximum values of the deformation plots in the first stage of the deformation process before the buckling effect, it is possible to observe that the values are relatively higher for quasi-static loading conditions, and they are also contingent on the relative density. The highest value of the deformation force was achieved for specimen No. 1, and the lowest value was achieved for specimen No. 4. Presented in Figure 18 are the results obtained for specimen No. 2, which indicate a negative value of the deformation force after being the first maximum. This phenomenon suggests that after the densification of the first row of structure cells, the rapture damage mechanism caused a spring-back effect of the impactor mass. This evidence was observed in specimens No. 2 and No. 4. The main reason to justify this situation is the presence of local material defects such as microcracks or pores. The presence of the maximum local deformation force could be related to the number of elementary cell rows in the structure. Specimens with the lowest unit cell size demonstrate a more variable plot of deformation force history, which is contrary to a specimen with a larger value of unit cell size. Moreover, the process of structural densification is smoother because the damaged rows of the array do not affect the force deformation history, as in the case of specimens No. 1 or No. 2.

Considering the dynamic character of the interaction between the specimen and the impactor, it could be observed that the structures with the highest value of the relative density are stiffer and the range of deformation is lower, even when the same impact loading conditions were applied.

The other aspect worth discussing is the deformation rate sensitivity. Comparing the quasi-static and dynamic results, it can be observed that specimens No. 1 and No. 2 demonstrate a high deformation rate sensitivity. Due to the high mass of the structure, the values of the deformation energy are significantly higher in comparison to the results of the quasi-static tests. The low values of relative density and the lowest structural stiffness caused specimens No. 3 and No. 4 to be less and almost insensitive to deformation rate effects. The comparison between the results is presented in Figure 21 and in Table 6. The results obtained in the quasi-static compression tests are marked by dotted lines, and those for the dynamic tests are defined by continuous lines.

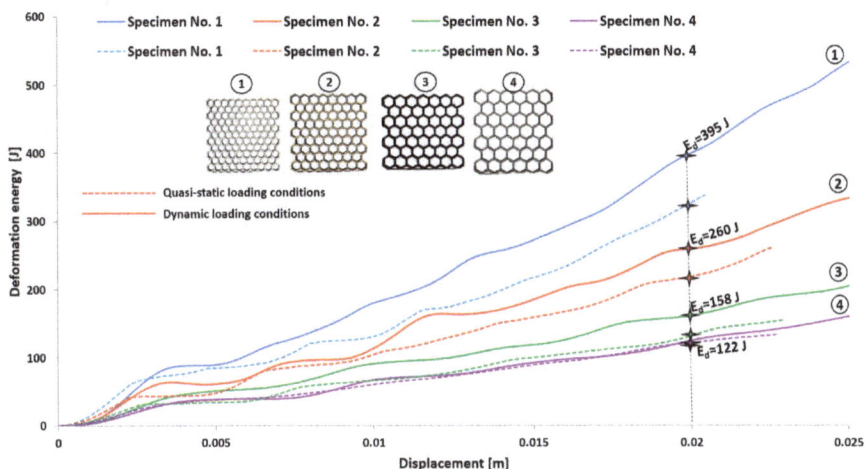

Figure 21. The comparison of deformation energy plots between dynamic and quasi-static results (1–4—the thin-walled honeycomb structures with the different cell size).

Table 6. The comparison of the maximum values of absorbed energy referring to honeycomb structure specimens.

	No. 1	No. 2	No. 3	No. 4
Relative density (−)	0.36	0.31	0.3	0.23
Max. value of absorbed energy in quasi-static (J)	322	215	130	120
Max. value of absorbed energy in dynamic (J)	395	260	158	122
Average increase of absorbed energy	22.6%	20.1%	21.5%	1.6%

Figure 22 presents the deformation rate sensitivity of honeycomb specimens versus the various values of relative density. Based on the obtained data, the value of the deformation energy depends on the relative density which is very important from the application point of view. It determines the mass of object and also the effects on its geometrical stiffness. Application of structures with the higher value of relative density causes increasing value of the impact force in the initial stage of structure deformation. Moreover, afterwards the process of structure deformation is more rapid and more destructive due to arriving of the damage mechanism (cracking). Structures with the lower value of the relative density indicate a more smooth deformation history plot. This feature is mostly caused due to buckling and bending mechanisms which preceded the cracking mechanism. Considering the dynamic response of the structure specimen it could be stated that it strongly depends on the value of relative density. Higher values of relative density (specimen No. 1 and No. 2) cause the increasing value of the deformation energy. This phenomenon is generally caused due to the higher mass of specimens and results from inertia effects. Application of lower values of the relative density (Specimen No. 3 and No. 4) allows a reduction in the mass of the object and enables minimization of the effects of impact force in the initial stage of the deformation process. Considering the possibility of honeycomb structures application as a proposal dedicated to energy absorption solutions it is recommended to use the specimens with the lower value of relative density merged with other diverse solid materials. This proposal could be used in civilian (automotive, railway) as well as military (passive protective systems) applications.

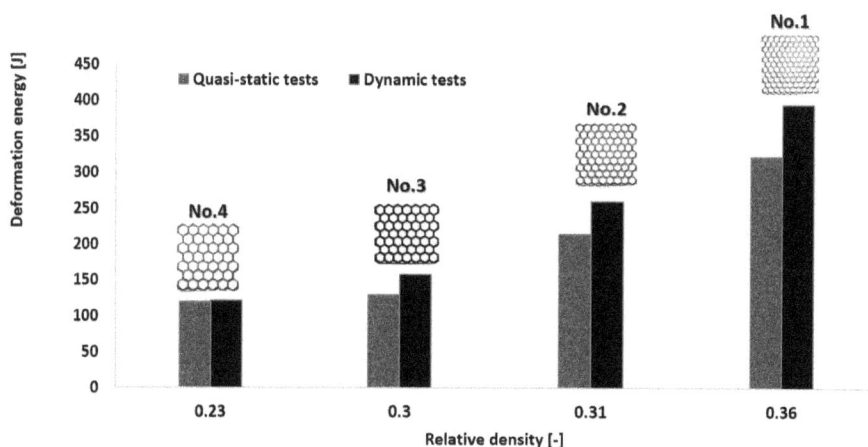

Figure 22. The influence of relative density on structural deformation process under static and dynamic loading conditions.

4. Conclusions

The main aim of this investigation was to analyze the mechanical response of additively manufactured regular cellular structures with a honeycomb topology manufactured additively from a Ti6Al4V titanium alloy with the use of a Laser Engineering Net Shaping system under static and dynamic loading conditions. Based on obtained experimental results following conlcusions are listed:

1. Geometrical assumptions were adopted during the specimen design process that considered the technological possibilities of the used additive manufacturing system. Moreover, the specimens were designed as cuboid elements with similar dimensions and wall thicknesses.
2. Based on the additional heat treatment process that was conducted, it was revealed that the applied Ti6Al4V titanium alloy materials require a heat treatment process in order to improve the mechanical properties of the material (increase of ductility) and stress relief annealing. As a result, the higher range of Ti6Al4V titanium alloy plastic deformation allowed for increasing the structure specimen's energy absorption capacity. Furthermore, it enabled reduction of the destructive effect of material brittle damage, which is essential referring to safety issues.
3. Uniaxial tests of structural specimens were performed under both static and dynamic loading conditions, which allowed for the evaluation of the specimens' energy absorption capacity and the sensitivity of the developed specimens on the strain rate. Based on the obtained results, it could be stated that an increasing value of relative density causes a growing sensitivity of the structure for strain rate effects.

Author Contributions: A.A.-D. performed the LENS manufacturing process and prepared documentation. P.P. designed static/dynamic tests and prepared the final manuscript. T.D. designed the LENS manufacturing process, revised the manuscript and approved the final version. P.B. discussed the data J.M. discussed the data and revised the manuscript. M.S. edited the data. T.C. analyzed the data and approved the final version. All authors have read and approved the content of the manuscript.

Funding: The research was supported by the National Science Centre under research grant No. 2015/17/B/ST8/00825.

Acknowledgments: Additionally, the authors would like to express gratitude to Piotr Dziewit from the Military University of Technology and Tadeusz Szymczak from the Motor Transport Institute for help performing the drop weight impact tests.

Conflicts of Interest: The authors declare no conflict of interest.

References

1. Kedzierski, P.; Gieleta, R.; Morka, A.; Niezgoda, T.; Surma, Z. Experimental study of hybrid soft ballistic structures. *Compos. Struct.* **2016**, *153*, 204–211.
2. Slawinski, G.; Niezgoda, T. Protection of Occupants Military Vehicles Against Mine Threats and Improvised Explosive Devices (IED) | Ochrona Zalogi Pojazdu Wojskowego Przed Wybuchem Min I Improwizow Anych Urzadzen Wybuchowych (IED). *J. Konbin* **2015**, *33*, 123–134. [CrossRef]
3. Arkusz, K.; Klekiel, T.; Sławiński, G.; Będziński, R. Influence of energy absorbers on Malgaigne fracture mechanism in lumbar-pelvic system under vertical impact load. *Comput. Methods Biomech. Biomed. Eng.* **2019**, 1–11. [CrossRef] [PubMed]
4. Bocian, M.; Jamroziak, K.; Kosobudzki, M. The analysis of energy consumption of a ballistic shields in simulation of mobile cellular automata. *Adv. Mater. Res.* **2014**, *1036*, 680–685. [CrossRef]
5. Baranowski, P.; Malachowski, J. Numerical study of selected military vehicle chassis subjected to blast loading in terms of tire strength improving. *Bull. Pol. Acad. Sci. Techn. Sci.* **2015**, *63*, 867–878.
6. Mayer, P.; Pyka, D.; Jamroziak, K.; Pach, J.; Bocian, M. Experimental and Numerical Studies on Ballistic Laminates on the Polyethylene and Polypropylene Matrix. *J. Mech.* **2017**, 1–11. [CrossRef]
7. Mazurkiewicz, .; Małachowski, J.; Baranowski, P. Optimization of protective panel for critical supporting elements. *Compos. Struct.* **2015**, *134*, 493–505. [CrossRef]
8. Wang, Z.; Liu, J.; Hui, D. Mechanical behaviors of inclined cell honeycomb structure subjected to compression. *Compos. Part B Eng.* **2017**, *110*, 307–314.
9. Kotkunde, N.; Deole, A.D.; Gupta, A.K.; Singh, S.K. Comparative study of constitutive modeling for Ti-6Al-4V alloy at low strain rates and elevated temperatures. *Mater. Des.* **2014**, *55*, 999–1005. [CrossRef]
10. Huang, W.; Zhang, W.; Li, D.; Ye, N.; Xie, W.; Ren, P. Dynamic failure of honeycomb-core sandwich structures subjected to underwater impulsive loads. *Eur. J. Mech. A/Solids* **2016**, *60*, 39–51. [CrossRef]
11. Rajaneesh, A.; Sridhar, I.; Rajendran, S. Relative performance of metal and polymeric foam sandwich plates under low velocity impact. *Int. J. Impact Eng.* **2014**, *65*, 126–136. [CrossRef]
12. Crupi, V.; Epasto, G.; Guglielmino, E. Comparison of aluminium sandwiches for lightweight ship structures: Honeycomb vs. foam. *Marine Struct.* **2013**, *30*, 74–96. [CrossRef]
13. Zhang, B.; Hu, S.; Fan, Z. Anisotropic Compressive Behavior of Functionally Density Graded Aluminum Foam Prepared by Controlled Melt Foaming Process. *Materials* **2018**, *11*, 2470. [CrossRef] [PubMed]
14. Zhang, Q.; Yang, X.; Li, P.; Huang, G.; Feng, S.; Shen, C.; Han, B.; Zhang, X.; Jin, F.; Xu, F.; et al. Bioinspired engineering of honeycomb structure—Using nature to inspire human innovation. *Prog. Mater. Sci.* **2015**, *74*, 332–400. [CrossRef]
15. Wang, X.; Xu, S.; Zhou, S.; Xu, W.; Leary, M.; Choong, P.; Qian, M.; Brandt, M.; Xie, Y.M. Topological design and additive manufacturing of porous metals for bone scaffolds and orthopaedic implants: A review. *Biomaterials* **2016**, *83*, 127–141. [CrossRef] [PubMed]
16. Ataee, A.; Li, Y.; Fraser, D.; Song, G.; Wen, C. Anisotropic Ti-6Al-4V gyroid scaffolds manufactured by electron beam melting (EBM) for bone implant applications. *Mater. Des.* **2018**, *137*, 345–354. [CrossRef]
17. Engelmayr, G.C.; Cheng, M.; Bettinger, C.J.; Borenstein, J.T.; Langer, R.; Freed, L.E. Accordion-like honeycombs for tissue engineering of cardiac anisotropy. *Nat. Mater.* **2008**, *7*, 1003–1010. [CrossRef]
18. Zhang, X.; Zhang, H.; Wen, Z. Experimental and numerical studies on the crush resistance of aluminum honeycombs with various cell configurations. *Int. J. Impact Eng.* **2014**, *66*, 48–59. [CrossRef]
19. Xiao, L.; Song, W.; Wang, C.; Liu, H.; Tang, H.; Wang, J. Mechanical behavior of open-cell rhombic dodecahedron Ti-6Al-4V lattice structure. *Mater. Sci. Eng. A* **2015**, *640*, 375–384. [CrossRef]
20. Restrepo, D.; Mankame, N.D.; Zavattieri, P.D. Programmable materials based on periodic cellular solids. Part I: Experiments. *Int. J. Solids Struct.* **2016**, *100–101*, 485–504. [CrossRef]
21. Chen, Y.; Li, T.; Jia, Z.; Scarpa, F.; Yao, C.W.; Wang, L. 3D printed hierarchical honeycombs with shape integrity under large compressive deformations. *Mater. Des.* **2018**, *137*, 226–234. [CrossRef]
22. Hedayati, R.; Sadighi, M.; Mohammadi Aghdam, M.; Zadpoor, A.A. Mechanical properties of additively manufactured thick honeycombs. *Materials* **2016**, *9*, 613. [CrossRef]
23. Hu, L.L.; Yu, T.X. Dynamic crushing strength of hexagonal honeycombs. *Int. J. Impact Eng.* **2010**, *37*, 467–474. [CrossRef]

24. Chen, S.; Yu, H.; Fang, J. A novel multi-cell tubal structure with circular corners for crashworthiness. *Thin-Walled Struct.* **2018**, *122*, 329–343. [CrossRef]

25. Yamashita, M.; Gotoh, M. Impact behavior of honeycomb structures with various cell specifications— Numerical simulation and experiment. *Int. J. Impact Eng.* **2006**, *32*, 618–630. [CrossRef]

26. Lin, T.C.; Yang, M.Y.; Huang, J.S. Effects of solid distribution on the out-of-plane elastic properties of hexagonal honeycombs. *Compos. Struct.* **2013**, *100*, 436–442. [CrossRef]

27. Liu, W.; Wang, N.; Luo, T.; Lin, Z. In-plane dynamic crushing of re-entrant auxetic cellular structure. *Mater. Des.* **2016**, *100*, 84–91. [CrossRef]

28. Fu, M.H.; Chen, Y.; Hu, L.L. Bilinear elastic characteristic of enhanced auxetic honeycombs. *Compos. Struct.* **2017**, *175*, 101–110. [CrossRef]

29. Huang, J.; Gong, X.; Zhang, Q.; Scarpa, F.; Liu, Y.; Leng, J. In-plane mechanics of a novel zero Poisson's ratio honeycomb core. *Compos. Part B Eng.* **2016**, *89*, 67–76. [CrossRef]

30. Zhu, H.X.; Mills, N.J. The in-plane non-linear compression of regular honeycombs. *Int. J. Solids Struct.* **2000**, *37*, 1931–1949. [CrossRef]

31. Kucewicz, M.; Baranowski, P.; Małachowski, J.; Popławski, A.; Płatek, P. Modelling, and characterization of 3D printed cellular structures. *Mater. Des.* **2018**, *142*, 177–189. [CrossRef]

32. Sun, G.; Jiang, H.; Fang, J.; Li, G.; Li, Q. Crashworthiness of vertex based hierarchical honeycombs in out-of-plane impact. *Mater. Des.* **2016**, *110*, 705–719. [CrossRef]

33. Zhang, Y.; Liu, T.; Ren, H.; Maskery, I.; Ashcroft, I. Dynamic compressive response of additively manufactured AlSi10Mg alloy hierarchical honeycomb structures. *Compos. Struct.* **2018**, *195*, 45–59. [CrossRef]

34. Zhang, X.; An, L.; Ding, H. Dynamic crushing behavior and energy absorption of honeycombs with density gradient. *J. Sandw. Struct. Mater.* **2014**, *16*, 125–147. [CrossRef]

35. Zhang, Y.; Lu, M.; Wang, C.H.; Sun, G.; Li, G. Out-of-plane crashworthiness of bio-inspired self-similar regular hierarchical honeycombs. *Compos. Struct.* **2016**, *144*, 1–13. [CrossRef]

36. Tao, Y.; Chen, M.; Chen, H.; Pei, Y.; Fang, D. Strain rate effect on the out-of-plane dynamic compressive behavior of metallic honeycombs: Experiment and theory. *Compos. Struct.* **2015**, *132*, 644–651. [CrossRef]

37. Chen, W.; Xia, L.; Yang, J.; Huang, X.; Haghpanah, B.; Papadopoulos, J.; Vaziri, A.; Felbrich, B.; Wulle, F.; Allgaier, C.; et al. The topological design of multifunctional cellular metals. *Mech. Mater.* **2018**, *25*, 309–327.

38. Cui, S.; Gong, B.; Ding, Q.; Sun, Y.; Ren, F.; Liu, X.; Yan, Q.; Yang, H.; Wang, X.; Song, B. Mechanical metamaterials foams with tunable negative poisson's ratio for enhanced energy absorption and damage resistance. *Materials* **2018**, *11*, 1869. [CrossRef]

39. Bandyopadhyay, A.; Heer, B. Additive manufacturing of multi-material structures. *Mater. Sci. Eng. R Rep.* **2018**, *129*, 1–16. [CrossRef]

40. Singh, S.; Ramakrishna, S.; Singh, R. Material issues in additive manufacturing: A review. *J. Manuf. Process.* **2017**, *25*, 185–200. [CrossRef]

41. ASTM F2792-12a—Standard Terminology for Additive Manufacturing Technologies. ASTM Int'l, Fri Oct 25 07:45:56 EDT 2013.

42. Ngo, T.D.; Kashani, A.; Imbalzano, G.; Nguyen, K.T.Q.; Hui, D. Additive manufacturing (3D printing): A review of materials, methods, applications and challenges. *Compos. Part B* **2018**, *143*, 172–196. [CrossRef]

43. DebRoy, T.; Wei, H.L.; Zuback, J.S.; Mukherjee, T.; Elmer, J.W.; Milewski, J.O.; Beese, A.M.; Wilson-Heid, A.; De, A.; Zhang, W. Additive manufacturing of metallic components—Process, structure and properties. *Prog. Mater. Sci.* **2018**, *92*, 112–224. [CrossRef]

44. Mudge, R.R.P.; Wald, N.N.R. Laser engineered net shaping advances additive manufacturing and repair. *Weld. J. N. Y.* **2007**, *86*, 44–48.

45. Liu, W.; DuPont, J.N. Fabrication of functionally graded TiC/Ti composites by laser engineered net shaping. *Scr. Mater.* **2003**, *48*, 1337–1342. [CrossRef]

46. Blackwell, P.L.; Wisbey, A. Laser-aided manufacturing technologies; their application to the near-net shape forming of a high-strength titanium alloy. *J. Mater. Process. Technol.* **2005**, *170*, 268–276. [CrossRef]

47. Polanski, M.; Kwiatkowska, M.; Kunce, I.; Bystrzycki, J. Combinatorial synthesis of alloy libraries with a progressive composition gradient using laser engineered net shaping (LENS): Hydrogen storage alloys. *Int. J. Hydrog. Energy* **2013**, *38*, 12159–12171. [CrossRef]

48. Ziętala, M.; Durejko, T.; Polański, M.; Kunce, I.; Płociński, T.; Zieliński, W.; azińska, M.; Stępniowski, W.; Czujko, T.; Kurzydłowski, K.J.; et al. The microstructure, mechanical properties and corrosion resistance of 316 L stainless steel fabricated using laser engineered net shaping. *Mater. Sci. Eng. A* **2016**, *677*, 1–10. [CrossRef]

49. Kummailil, J.; Sammarco, C.; Skinner, D.; Brown, C.A.; Rong, K. Effect of Select LENS^{TM} Processing Parameters on the Deposition of Ti-6Al-4V. *J. Manuf. Process.* **2005**, *7*, 42–50. [CrossRef]

50. Zhai, Y.; Lados, D.A.; Brown, E.J.; Vigilante, G.N. Fatigue crack growth behavior and microstructural mechanisms in Ti-6Al-4V manufactured by laser engineered net shaping. *Int. J. Fatigue* **2016**, *93*, 51–63. [CrossRef]

51. Durejko, T.; Zietala, M.; Polkowski, W.; Czujko, T. Thin wall tubes with Fe3Al/SS316L graded structure obtained by using laser engineered net shaping technology. *Mater. Des.* **2014**, *63*, 766–774. [CrossRef]

52. Galarraga, H.; Warren, R.J.; Lados, D.A.; Dehoff, R.R.; Kirka, M.M.; Nandwana, P. Effects of heat treatments on microstructure nad properties of Ti-6Al-4V ELI alloy fabricated by electron beam melting (EBM). *Mater. Sci. Eng. A* **2017**, *685*, 417–428. [CrossRef]

53. Kumar, P.; Ramamurty, U. Microstructural optimization through heat treatment for enhancing the fracture toughness and fatigue crack growth resistance of selective laser melted Ti–6Al–4V alloy. *Acta Mater.* **2019**, *169*, 45–59. [CrossRef]

54. Schneider, J.; Lund, B.; Fullen, M. Effect of heat treatment variations on the mechanical properties of Inconel 718 selective laser melted specimens. *Addit. Manuf.* **2018**, *21*, 248–254. [CrossRef]

55. Maamoun, A.H.; Elbestawi, M.; Dosbaeva, G.K.; Veldhuis, S.C. Thermal post-processing of AlSi10Mg parts produced by Selective Laser Melting using recycled powder. *Addit. Manuf.* **2018**, *21*, 234–247. [CrossRef]

56. Baranowski, P.; Płatek, P.; Antolak-Dudka, A.; Sarzyński, M.; Kucewicz, M.; Durejko, T.; Małachowski, J.; Janiszewski, J.; Czujko, T. Deformation of honeycomb cellular structures manufactured with Laser Engineered Net Shaping (LENS) technology under quasi-static loading: Experimental testing and simulation. *Addit. Manuf.* **2019**, *25*, 307–316. [CrossRef]

57. OPTOMEC–LENS. *MR-7 PSU System Manual*; Made in the USA; OPTOMEC: Albuquerque, NM, USA, 2009.

58. Donachie, M.J. *Titanium: A Technical Guide 2000*; ASM International: Metals Park, OH, USA, 2000.

59. Vrancken, B.; Thijs, L.; Kruth, J.P.; Van Humbeeck, J. Heat treatment of Ti6Al4V produced by Selective Laser Melting: Microstructure and mechanical properties. *J. Alloy. Compound.* **2012**, *541*, 177–185. [CrossRef]

materials

MDPI

Article

Effects of Thermal Variables of Solidification on the Microstructure, Hardness, and Microhardness of Cu-Al-Ni-Fe Alloys

Maurício Silva Nascimento [1,*], Givanildo Alves dos Santos [1], Rogério Teram [1], Vinícius Torres dos Santos [2,3], Márcio Rodrigues da Silva [2,3] and Antonio Augusto Couto [4,5]

[1] Department of Mechanics, Federal Institute of Education, Science and Technology of São Paulo, 01109-010 São Paulo, Brazil; givanildo@ifsp.edu.br (G.A.d.S.); rogerioteram@ifsp.edu.br (R.T.)
[2] Salvador Arena Foundation Educational Center, 09850-550 São Bernardo do Campo, Brazil; vinicius.santos@termomecanica.com.br (V.T.d.S.); marcio.rdrgs.slv@gmail.com (M.R.d.S.)
[3] Department of Research and Development, Termomecanica São Paulo S.A., 09612-000 São Bernardo do Campo, Brazil
[4] Nuclear and Energy Research Institute, Center for Materials Science and Technology, IPEN, 05508-000 São Paulo, Brazil; aacouto@ipen.br
[5] Department of Engineering, Mackenzie Presbyterian University, UPM, 01302-907 São Paulo, Brazil
* Correspondence: mauricio.nascimento@ifsp.edu.br; Tel.: +55-11-2763-7536

Received: 28 February 2019; Accepted: 15 April 2019; Published: 18 April 2019

Abstract: Aluminum bronze is a complex group of copper-based alloys that may include up to 14% aluminum, but lower amounts of nickel and iron are also added, as they differently affect alloy characteristics such as strength, ductility, and corrosion resistance. The phase transformations of nickel aluminum–bronze alloys have been the subject of many studies due to the formations of intermetallics promoted by slow cooling. In the present investigation, quaternary systems of aluminum bronze alloys, specifically Cu–10wt%Al–5wt%Ni–5wt%Fe (hypoeutectoid bronze) and Cu–14wt%Al–5wt%Ni–5wi%Fe (hypereutectoid bronze), were directionally solidified upward under transient heat flow conditions. The experimental parameters measured included solidification thermal parameters such as the tip growth rate (V_L) and cooling rate (T_R), optical microscopy, scanning electron microscopy (SEM) analysis, hardness, and microhardness. We observed that the hardness and microhardness values vary according to the thermal parameters and solidification. We also observed that the Cu–14wt%Al–5wt%Ni–5wi%Fe alloy presented higher hardness values and a more refined structure than the Cu–10wt%Al–5wt%Ni–5wt%Fe alloy. SEM analysis proved the presence of specific intermetallics for each alloy.

Keywords: solidification thermal parameters; Cu-Al-Ni-Fe bronze alloys; hardness; microhardness; specific intermetallics

1. Introduction

Cast copper alloys are used in applications that require metals with superior corrosion resistance, high electrical and thermal conductivity, good surface quality for bearings, and other special properties. Among the full range of copper alloys, aluminum bronzes are the best available material for fulfilling these requirements [1,2]. Aluminum bronzes are copper-based alloys that may include up to 14% aluminum, but lower amount of nickel and iron are also added to produce different alloy strength, ductility, and corrosion resistance [3–6]. In the maritime field, nickel aluminum–bronze alloys are known as "propeller bronze", representing their application in the manufacturing of propellers of ships and submarines [7].

In the Cu-Al-Ni-Fe alloys, the aluminum component is the main alloying element, with a content normally varying between 8% and 13%. Greater contents are used for obtaining high hardness and reduce the ductility of the alloy. However, high levels of aluminum provide the appearance of γ2 phase, which is detrimental to its mechanical resistance and corrosion. Some elements such as Ni and Fe combine with Al to form complex phases called κ, avoiding the emergence of the γ2. Nickel is added in amounts ranging from 1% to 7% and its presence improves corrosion resistance, increases mechanical strength, and contributes to increased erosion resistance in environments with high water flow velocity. Iron is present in nickel aluminum–bronze to refine the structure and increase the toughness. The low solubility of iron at low temperatures in these alloys is the main reason for the appearance of precipitates rich in iron, which can be combined to produce the required mechanical properties [6].

The phase transformations of aluminum–bronze have been the subject of many studies due to the formations of intermetallics promoted by slow cooling [4,8–11]. The phase diagram of the Cu-Al system shows the different microstructures that arise in the cooling of the investigated alloys (Figure 1).

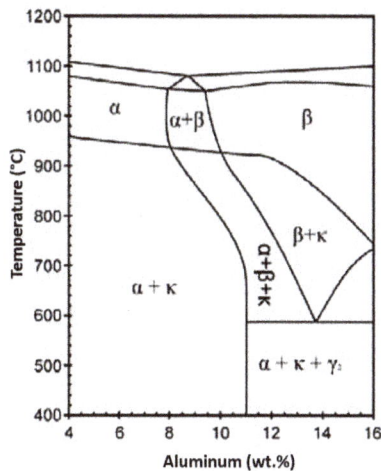

Figure 1. Phase diagram of the Cu-Al system with addition of 5 wt% nickel and 5 wt% iron [12].

The β phase is a solid solution phase at high temperatures in nickel aluminum–bronze and presents disordered BCC (body-centered cubic) crystalline structure. The β phase is present mainly at high temperatures and is considered the first solid generated in the transformation of the liquid state to the solid state; later, part of the β phase becomes the α phase [7]. The α phase represents a solid equilibrium solution or matrix with a FCC (face-centered cubic) crystalline structure. The α phase is formed from the β phase around 1030 °C and exhibits a Widmastätten structure [6,9,13]. In addition to the α phase, the aluminum–bronze alloy also exhibits a β phase that originates from three main types of intermetallics present in these alloys, labeled κ, which is formed via slow cooling: Kappa II (κII), Kappa III (κIII), and Kappa IV (κIV), shown in Figure 2 [4,6–11,13].

We aimed to study the microstructure resulting from Cu-10wt%Al-5wt%Ni-5wt%Fe (hypoeutectoid bronze, cited as Cu10Al alloy) and Cu-14wt%Al-5wt%Ni-5wi%Fe (hypereutectoid bronze, cited as Cu14Al alloy) alloys after undergoing a directional solidification process. Directional solidification allows different microstructures to be obtained in the length of the molten ingot, influencing the alloy properties. The effects of the manufacturing processes on the microstructure and properties of engineering materials have been highlighted in various studies [14–20]. Thermal parameters of solidification, as tip growth rate (V_L) and cooling rate (T_R), were correlated with hardness and microhardness values for both alloys studied. Optical microscopy and scanning electron microcopy (SEM) images were obtained from various positions in the ingot for both alloys.

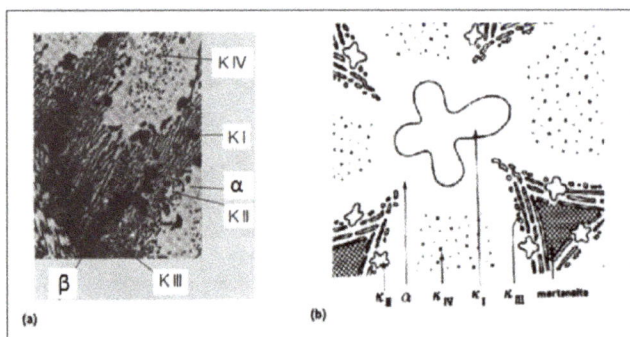

Figure 2. Distribution of the different phases and intermetallic components of the nickel aluminum–bronze cooled slowly: (**a**) optical microscopy [4] and (**b**) schematic representation [8].

2. Materials and Methods

The directional solidification apparatus has a cylindrical shape (Figure 3), covered with refractory bricks and externally coated with steel plate. The heat required to keep the liquid metal heated before the cooling process was created by electrical resistors controlled with an external control panel. Two support tubes supported the ingot, the outer one being composed of SAE 1020 steel, and the internal tube was stainless steel AISI 304. Refractory cement was placed between these two tubes to increase the insulation of the internal space of the furnace. A tube inside the two support tubes directed the water jet into a plate responsible for the removal of heat from the molten metal. This plate was composed of SAE 1020 steel and was 5 mm thick. The upper surface of the sheet, which remained in contact with the liquid metal, was sanded with 1200 mesh sandpaper. The ingot mold was composed of stainless steel AISI 304, with a height of 160 mm and internal and external diameters of 60 and 76 mm, respectively. For the acquisition of temperature data, type K thermocouples were used, with distances of 4, 8, 12, 16, 35, 53, and 73 mm relative to the position of the upper surface of the heat exchange plate. These thermocouples were connected to National Instruments NI 9212 (National Instruments, Debrecen, Hungary) and NI cDAQ 9171 data acquisition devices (National Instruments, Debrecen, Hungary), responsible for sending the collected data to a computer via a USB cable. The temperature data obtained by the thermocouples were provided at the frequency of one per second.

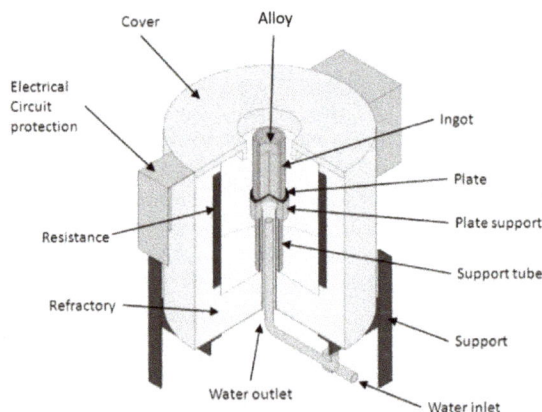

Figure 3. Schematic illustration of the upward unidirectional solidification furnace [14].

The alloys were cast in a Fortelab muffle-type electric furnace (Fortelab, São Carlos, SP, Brazil) in a Salamander SIC AS2 graphite crucible. The chemical composition of the alloys was analyzed using X-ray spectrometry (XRS) using a Panalitycal Magix Fast Spectrometer (Panalitycal, Almelo, The Netherlands) (Table 1). The alloys were heated to temperatures above their liquid temperature. After this, the crucible was removed from the furnace and the liquid metal was poured into the ingot mold in the unidirectional solidification furnace. Cooling of the liquid metal inside the ingot started when the water jet was connected at a flow rate of 18 L/min.

Table 1. Chemical composition of ingots in weight%.

Alloy	Al	Ni	Fe	Others	Cu
Cu–10wt%Al–5wt%Ni–5wt%Fe	10.79	4.42	3.67	0.051	Remaining
Cu–14wt%Al–5wt%Ni–5wt%Fe	14.23	5.44	5.39	0.340	Remaining

The tip growth rate (V_L) was calculated by deriving the function $P = f(t)$. This function is the relationship between the position of the thermocouple (P) and the time interval between the start of the alloy cooling and the time at which the liquidus temperature (T_L) is observed in each thermocouple. With this, V_L corresponds to the velocity of the solidification front passage in each thermocouple. The cooling rate (T_R) values for each position on the thermocouple were obtained experimentally from the temperature variation values as a function of time, at a temperature before and after the *liquidus* temperature ($\Delta T/\Delta t$). For metallographic analysis, samples of cross-sections of the molten ingot were selected. The analyzed surfaces of the samples were selected from different positions (P) in relation to the heat exchange surface. These distances were 4, 8, 12, 16, 26, 35 and 53 mm. Each sample was embedded in Bakelite, sanded with sands of different granulations, and polished with 3–6 μm diamond paste. The etchant used to reveal the microstructure consisted of a solution of 10.7% HCl, 3.4% Fe_3Cl, and 85.9% distilled water. The reaction time was 25 s. A Zeiss AxioVert A1 microscope (Carl Zeiss, Gottingen, Germany) was used to obtain optical images of the microstructure. Samples were analyzed by scanning electron microscopy (SEM) using Phenom Pro X and Jeol JSM 6510 equipment (Jeol, Tokyo, Japan) for checking the phases and intermetallics morphology. The mechanical characteristics were evaluated by the hardness test, according to ASTM E10-2012 [21] standard in a Wilson UH-930 hardness tester (Boehler, Lake Bluff, IL, USA) using a load of 62.5 kgf and a sphere 2.5 mm in diameter. The hardness test was performed at five points of each position on the thermocouple. The microhardness was tested according to ASTM E92-2003 [22] standard in a Boehler VH1102 microhardness tester (Boehler, Lake Bluff, IL, USA) at five different points of each position on the thermocouple using force of 1 kgf.

3. Results and Discussion

Figure 4 presents the thermal parameters V_L and T_R experimentally obtained as a function of the distance to the heat exchange surface (P). For both alloys, V_L values decreased with higher P values. The Cu14Al alloy, which has a higher amount of Al in its composition, had higher initial V_L values than the Cu10Al alloy (Figure 4A). The values of T_R, similar to V_L, decreased as the distance from the heat exchange surface (P) increased. We observed that the Cu10Al alloy had values slightly larger than those for the Cu14Al alloy (Figure 4B). Analyzing the results obtained for both alloys, we observed that the Al content influences the values of V_L and T_R.

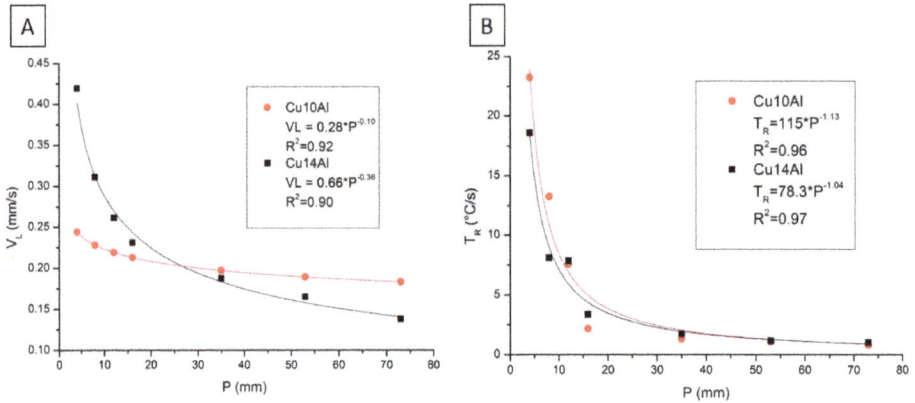

Figure 4. Solidification thermal parameters: (**A**) correlation between tip growth rate (V_L) and distance from heat extraction surface (P); and (**B**) correlation between cooling rate (T_R) and distance from heat extraction surface (P). The error bars represent the standard deviation of the measurements obtained.

The data obtained in the hardness test are presented in Figure 5. The experimental equations that correlate the hardness values (HB) with the distance of the heat exchange surface (P) values and with the values of T_R in the graphs were obtained by the least square method using Origin software. The linear fit of the data suggests that the hardness values (HB) decrease with increasing distance of the heat exchange surface (P). For T_R, the adjustment indicates the opposite: the values of hardness increase with the increase in T_R. This is important because it shows that it is possible to predict the hardness performance of both alloys by changing the cooling conditions. Comparing the two alloys studied, the Cu14Al alloy has higher hardness values than the Cu10Al alloy. This suggests that the increase in Al content influences this property. The linear adjustment also suggests that there are maximum hardness values. If P = 0 mm, we can define these values. For the Cu10Al and Cu14Al alloys, the values were 196 and 284 HB, respectively.

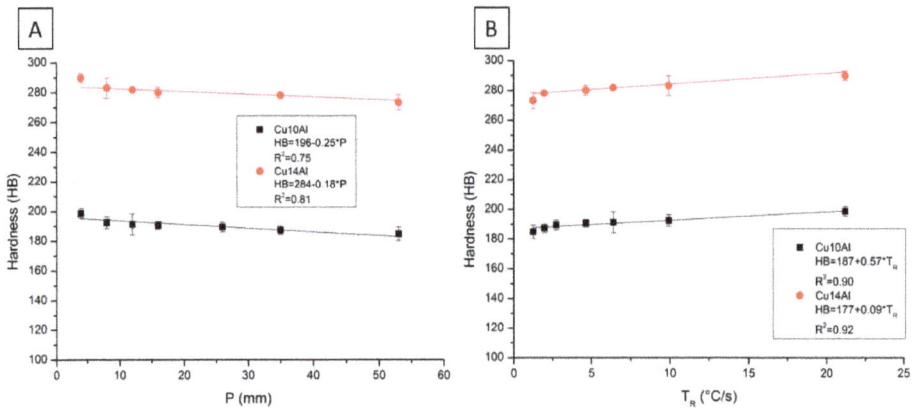

Figure 5. (**A**) Correlation between hardness (HB) and distance from heat extraction surface (P); and (**B**) correlation between HB and cooling rate (T_R). The error bars represent the standard deviation of the measurements obtained.

The data obtained in the microhardness test are presented in Figure 6. The experimental equations correlate the microhardness values (HV1) with the distance of the heat exchange surface (P). The values of T_R presented in the graphs were obtained by the least square method using Origin software. The linear fit of the data suggests that HV1 increases with the increase in the distance of the heat exchange surface (P). For T_R, the adjustment indicates the opposite: the values of hardness decrease with the increase in T_R.

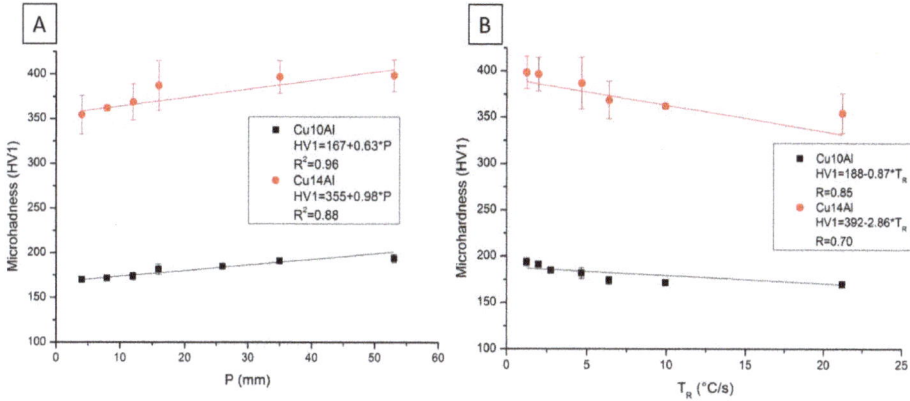

Figure 6. (**A**) Correlation between microhardness (HV1) and distance from heat extraction surface (P); and (**B**) correlation between microhardness (HV1) and cooling rate (T_R). The error bars represent the standard deviation of the measurements obtained.

Figure 7 depicts the transverse micrographs of the two alloys studied at positions 4, 8, 12, 16, 35, and 53 mm with respect to the heat extraction surface (P). Comparing both alloys, the Cu10Al alloy presents the α phase in Widmastätten morphology, whereas the Cu14Al alloy presents a diffuse morphology with small microstructures inside the grain. At the position P = 4 mm, we observed that the grains of the Cu14Al alloy have smaller dimensions than for the Cu10Al alloy. At position P = 53 mm, the dendritic arms were observed in dark color for the Cu14Al alloy. The dendritic arms being in positions of higher values of P and not in smaller values, show that the grain size increases as the value of P increases.

The images obtained by SEM for both alloys are shown in Figure 8. Hasan et al. [8] studied the morphology, crystallography, and composition of the phases present in Cu10Al alloy, determining the characteristics of each phase. Jahanafrooz et al. [9] studied the mechanism of phase formation in Cu10Al alloy during solidification. Pisarek [11] proposed a crystallization model for Cu-Al-Ni-Fe alloys. The microconstituents in the SEM images obtained in this work were identified based on the similarity of the SEM images presented by the authors mentioned above. We observed that the Cu14Al alloy had a larger number of microconstituents. The Cu10Al alloy more prominently presents the α phase.

The Cu14Al alloy had higher hardness values, a structure with smaller grains, and more microconstituents evidenced by the SEM analysis than the Cu10Al alloy, suggesting that the higher Al content influences these properties. It should be noted that the Cu14Al alloy presents the γ2 phase, characteristic of the high aluminum content in the alloy. This phase is detrimental because it reduces the performance of the alloy for corrosion resistance. The fact that this alloy has in its composition Fe and Ni contents, this phase appears in smaller quantity, since these elements bind to Al forming the microconstituents κmentioned above. It is also observed the appearance of the retained beta phase caused by the high rate of cooling. This phase is martensitic giving higher hardness values for the alloy.

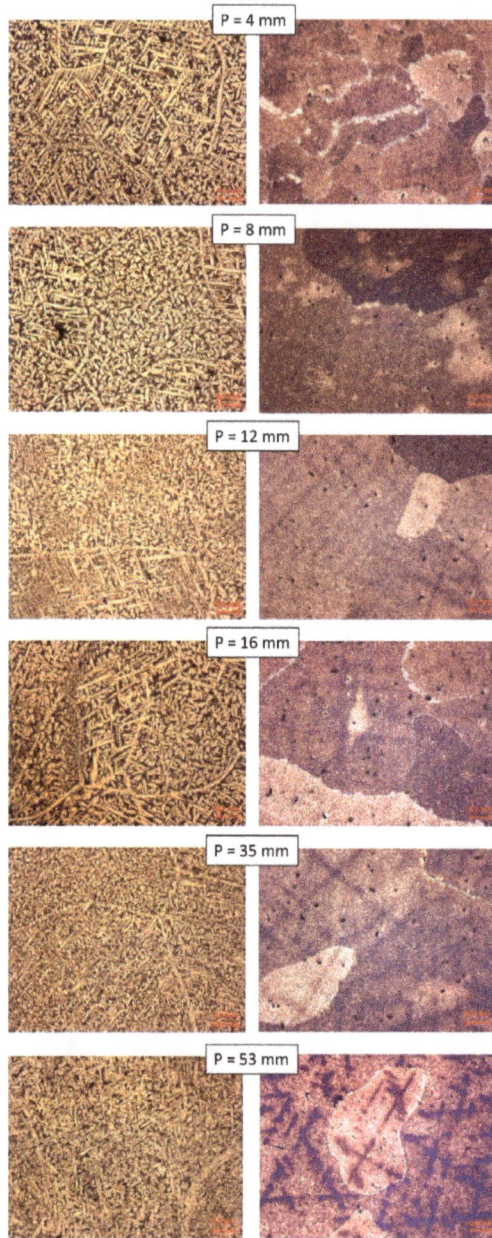

Figure 7. Micrograph of the various samples taken from different positions (P) as a function of the heat exchange surface: (**left**) Cu10Al alloy; and (**right**) Cu14Al alloy.

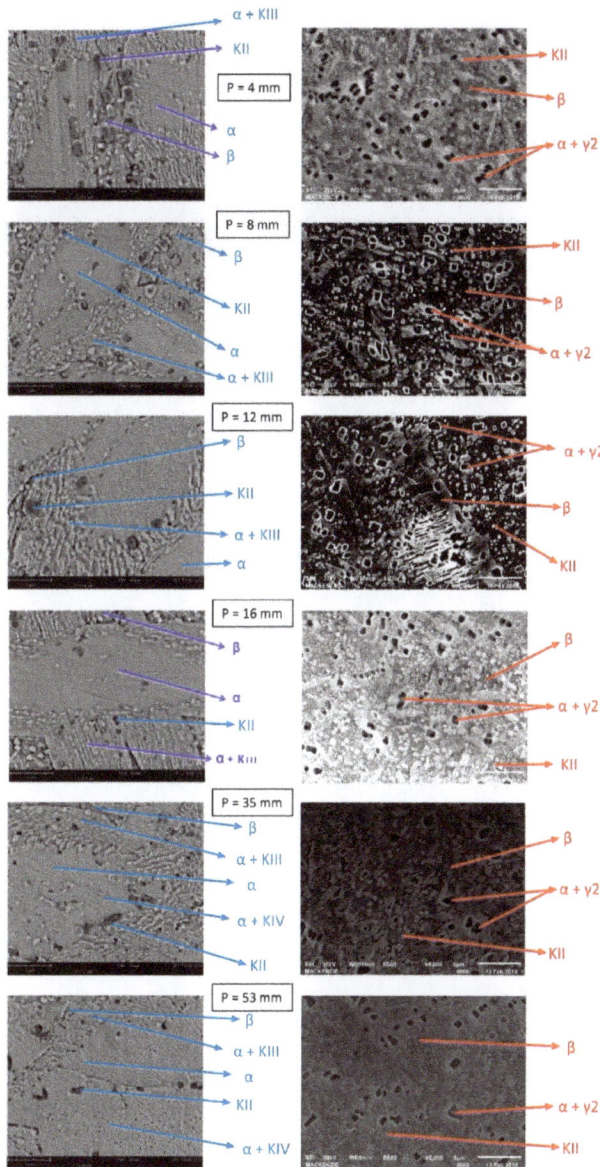

Figure 8. Scanning electron microscopy (SEM) analysis of the various samples taken from different positions (P) as a function of the heat exchange surface: (**left**) Cu10Al alloy; and (**right**) Cu14Al alloy.

4. Conclusions

The values of the solidification thermal parameters V_L and T_R decrease for larger distances from the heat exchange surface. The Cu14Al alloy, which has the highest amount of Al in its composition, has higher initial V_L higher values than the Cu10Al alloy. The linear fit of the data suggests that the hardness values (HB) decrease with increasing distance from the heat exchange surface (P). For T_R, the adjustment indicates the opposite: the values of hardness increase with increasing T_R values. The linear fit of the data also suggests that the microhardness values (HV1) increase with increasing distance from the heat exchange surface (P). For T_R, the adjustment indicates the opposite: the values of hardness decrease with increasing T_R.

Comparing the transverse optical micrographs for both alloys, the Cu10Al alloy presents the α phase in Widmastätten morphology, whereas the Cu14Al alloy presents a diffuse morphology with small microstructures inside the grain. At position P = 4 mm, we observed that the grains of the Cu14Al alloy were smaller than those in the Cu10Al alloy. At position P = 53 mm, the dendritic arms were observed to have a dark color for the Cu14Al alloy. The ability to observe the dendritic arms at positions of higher value of P and not in smaller values shows that the size of the grain increases as the value of P increases.

In the SEM images for both alloys, we observed that the Cu14Al alloy has more microconstituents. The Cu10Al alloy presented the most prominent α phase. The Cu14Al alloy has higher hardness values, a structure with smaller grains, and more microconstituents, as evidenced by the SEM analysis, than the Cu10Al alloy, suggesting that the higher Al content influences these properties.

Author Contributions: M.S.N., V.T.d.S., and R.T. performed the experiments; V.T.d.S. and M.R.d.S. contributed materials and analysis tools; M.S.N., V.T.d.S., G.A.d.S., and A.A.C. contributed results analysis; and M.S.N. and G.A.d.S. wrote the paper.

Funding: Materials for this research were donation by Termomecanica São Paulo S.A.

Acknowledgments: The authors are grateful for the financial support provided by PROEQUIPAMENTOS 2014-CAPES, Termomecanica São Paulo S.A. for the donation of materials for analysis and assignment of laboratories for analysis, Mackenzie Presbyterian University for support in obtaining MEV images and the Federal Institute of São Paulo (IFSP) for supporting research activities.

Conflicts of Interest: The authors declare no conflict of interest.

References

1. ASM Metals Handbook. *Properties and Selection: Nonferrous Alloy and Special-Purpose Materials*; ASM International: Novelty, OH, USA, 1992; Volume 2.
2. Wharton, J.A.; Barik, R.C.; Kear, G.; Wood, R.J.K.; Stokes, K.R.; Walsh, F.C. The Corrosion of nickel–aluminum bronze in seawater. *Corros. Sci.* **2005**, *47*, 3336–3367. [CrossRef]
3. Vaidyanath, L.R. *The Manufacture of Aluminum-Bronze Casting*; Indian Copper Information Center: West Calcutta, India, 1968; Paper n°38.
4. Culpan, E.A.; Rose, G. Microstructural characterization of cast nickel aluminum bronze. *J. Mater. Sci.* **1978**, *13*, 1647–1656. [CrossRef]
5. Meigh, H.J. *Cast and Wrought Aluminum Bronzes Properties, Processes and Structure*; IOM Communications: Detroit, MI, USA, 2000.
6. Richardson, I. *Guide to Nickel Aluminum Bronze for Engineers*; Copper Development Association: Birmingham, MI, USA, 2010; Volume 222.
7. Pierce, F.A. The Isothermal Deformation of Nickel Aluminum Bronze in Relation to Friction Stir Processing. Master's Thesis, Naval Postgraduate School, Monterey, CA, USA, June 2004.
8. Hasan, F.; Jahanafrooz, A.; Lorimer, G.W.; Ridley, N. The Morphology, Crystallography, and Chemistry of Phases in As-Cast Nickel Aluminum Bronze. *Met. Trans. A* **1982**, *13*, 1337–1345. [CrossRef]
9. Jahanafrooz, A.; Hasan, F.; Lorimer, G.W.; Ridkey, N. Microstructural Development in Complex Nickel-Aluminum Bronze. *Met. Trans A* **1983**, *14a*, 1951–1956. [CrossRef]
10. Howell, P.R. *On the Phases Microconstituents in Nickel-Aluminum Bronzes*; Copper Development Association: Birmingham, MI, USA, 2000.
11. Pisarek, B. Model of Cu–Al-Fe-Ni Bronze Crystallization. *Arch. Foundry Eng.* **2013**, *13*, 72–79. [CrossRef]
12. Sláma, P.; Dlouhy, J.; Kover, M. Influence of heat treatment on the microstructure and mechanical properties of aluminum bronze. *Mater. Technol.* **2018**, *48*, 599–604.
13. Faires, K.B. Characterization of Microstructure and Microtexture in Longitudinal Sections from Friction Stir Processed Nickel-Aluminum Bronze. Master's Thesis, Naval Postgraduate School, Monterey, CA, USA, June 2003.

14. Nascimento, M.S.; Franco, A.T.R.; Frajuca, C.; Nakamoto, F.Y.; Santos, G.A.; Couto, A.A. An Experimental Study of the Solidification Thermal Parameters Influence upon Microstructure and Mechanical Properties of Al-Si-Cu Alloys. *Mater. Res.* **2018**, *21*, e20170864. [CrossRef]

15. Miranda, F.; Rodrigues, D.; Nakamoto, F.Y.; Frajuca, C.; Santos, G.A. The Influence of the Sintering Temperature on the Grain Growth of Tungsten Carbide in the Composite WC-8Ni. *Mater. Sci. Forum* **2017**, *899*, 424–430. [CrossRef]

16. Miranda, F.; Rodrigues, D.; Nakamoto, F.Y.; Frajuca, C.; Santos, G.A.; Couto, A.A. Microstructural Evolution of Composite 8 WC-(Co, Ni): Effect of the Addition of SiC. *Defect Diffus. Forum* **2017**, *371*, 78–85. [CrossRef]

17. Santos, G.A.; Goulart, P.R.; Couto, A.A.; Garcia, A. Primary Dendrite Arm Spacing Effects upon Mechanical Properties of an Al-3w%Cu-1w%Li Alloy. In *Properties and Characterization of Modern Materials. Advanced Structured Materials*; Springer: Berlin, Germany, 2017; Volume 33.

18. Osorio, W.R.; Goulart, P.R.; Garcia, A.; Santos, G.A.; Neto, C.M. Effect of dendritic arm spacing on mechanical properties and corrosion resistance of Al 9 Wt Pct Si and Zn 27 Wt Pct Al alloys. *Metall. Mater. Trans A* **2006**, *37*, 2525–2538. [CrossRef]

19. Santos, G.A.; Neto, C.M.; Osório, W.R.; Garcia, A. Design of mechanical properties of a Zn27Al alloy based on microstructure dendritic array spacing. *Mater. Des.* **2007**, *28*, 2425–2430. [CrossRef]

20. Nascimento, M.S.; Frajuca, C.; Nakamoto, F.Y.; Santos, G.A.; Couto, A.A. Correlação entre variáveis térmicas de solidificação, microestrutura e resistência mecânica da liga Al-10%Si-2%Cu. *Matéria* **2017**, *22*, e11774. [CrossRef]

21. ASTM E10. *Standard Test Method for Brinell Hardness of Metallic Materials*; ASTM International: West Conshohocken, PA, USA, 2012.

22. ASTM E92. *Standard Test Method for Vickers Hardness and Knoop Hardness of Metallic Materials*; ASTM International: West Conshohocken, PA, USA, 2003.

materials

MDPI

Article

Structure and Deformation Behavior of Ti-SiC Composites Made by Mechanical Alloying and Spark Plasma Sintering

Dariusz Garbiec [1,*], Volf Leshchynsky [1], Alberto Colella [2], Paolo Matteazzi [2] and Piotr Siwak [3]

[1]   Metal Forming Institute, 14 Jana Pawla II St., 61-139 Poznan, Poland; leshchynsky@inop.poznan.pl
[2]   MBN Nanomaterialia, 42 Via G. Bortolan, 31050 Vascon Di Carbonera, Italy;
     alberto.colella@matres.org (A.C.); matteazzi@mbn.it (P.M.)
[3]   Poznan University of Technology, 5 Marii Sklodowskiej-Curie Square, 60-965 Poznan, Poland;
     piotr.siwak@put.poznan.pl
*   Correspondence: dariusz.garbiec@inop.poznan.pl; Tel.: +48-61-657-05-55

Received: 28 February 2019; Accepted: 15 April 2019; Published: 18 April 2019

Abstract: Combining high energy ball milling and spark plasma sintering is one of the most promising technologies in materials science. The mechanical alloying process enables the production of nanostructured composite powders that can be successfully spark plasma sintered in a very short time, while preserving the nanostructure and enhancing the mechanical properties of the composite. Composites with MAX phases are among the most promising materials. In this study, Ti/SiC composite powder was produced by high energy ball milling and then consolidated by spark plasma sintering. During both processes, Ti_3SiC_2, TiC and Ti_5Si_3 phases were formed. Scanning electron microscopy, energy-dispersive X-ray spectroscopy and X-ray diffraction study showed that the phase composition of the spark plasma sintered composites consists mainly of Ti_3SiC_2 and a mixture of TiC and Ti_5Si_3 phases which have a different indentation size effect. The influence of the sintering temperature on the Ti-SiC composite structure and properties is defined. The effect of the Ti_3SiC_2 MAX phase grain growth was found at a sintering temperature of 1400–1450 °C. The indentation size effect at the nanoscale for Ti_3SiC_2, $TiC+Ti_5Si_3$ and SiC-Ti phases is analyzed on the basis of the strain gradient plasticity theory and the equation constants were defined.

Keywords: MAX phase; Ti_3SiC_2; composite; high energy ball milling; spark plasma sintering; structure; mechanical properties; deformation behavior

1. Introduction

During the last two decades, nanocrystalline materials with grain sizes below 100 nm prepared using various methods have been widely studied due to their enhanced properties when compared with coarse-grained polycrystalline materials. Mechanical milling/alloying (MM/MA) is an effective method to prepare nanostructured materials such as pure element nanocrystalline metals, supersaturated solid solutions, intermetallic compounds, dispersion strengthened alloys and amorphous alloys. Mechanochemical synthesis—chemical reactions induced by high energy ball milling (HEBM)—is one of the promising routes for the synthesis of different materials and composites such as $M_{n+1}AX_n$ ternary layered compounds, where n=1, 2 or 3, M is an early transition metal, A is an A-group element (mostly groups 13 and 14), and X is C or N. Mechanochemical reactions may be divided into by two categories [1], specifically: (i) those which occur during the mechanical activation process, and (ii) those which occur during subsequent thermal treatment [2]. Indeed, the material grain size influences the mechanochemical reactions of both categories. Applying the spark plasma sintering (SPS) route has proved [3] to be extremely important to retaining the nanostructure of composites and to achieve their unique properties due to the formation of various phases during the sintering process [4].

Despite the fact that Ti and its alloys are widely used in many areas of engineering, the mechanical properties of these materials could be insufficient in some structural applications, especially at elevated temperatures [5]. One of the opportunities to improve the mechanical properties of Ti and its alloys could be to reinforce them with ceramics [6]. The most frequently used ceramic reinforcements are TiC [7] and SiC [6]. SiC ceramic reinforcement may be in the form of fibers, which provides high specific strength and stiffness, but only in the direction of the fibers [6,8]. The main limitations of the fiber-type reinforcement of Ti-based composites are the high cost, non-formability and difficultly of machining [9]. A better choice is to use SiC particles. The SiC ceramic can react with Ti during powder metallurgy processing routes. One of the results of chemical reactions between Ti and SiC during sintering is the formation of MAX phases.

MAX phases have attracted much attention because of their unique combination of both metal- and ceramic-like properties [10], such as high fracture toughness, high Young's moduli, high thermal and electrical conductivities, easy machinability, excellent thermal shock resistance, high damage tolerance, and microscale ductility [11]. In particular, the fact that MAX phases exhibit properties, which are typical both of metals and ceramics became known during the last two decades thanks to the studies of Prof. Barsoum et al. (Drexel University, USA) [12]. This group of researchers found that MAX phases are natural nanolaminates and have high electrical and heat conductivity, and lower coefficients of friction as compared to known hard materials, high rigidity in combination with a low density and high fracture toughness [13]. To synthesize bulk MAX phases, various methods have been used, e.g., both cold compacting and pressureless sintering [14], hot pressing (HP) [15], hot isostatic pressing (HIP) [16] and the field assisted sintering technology (FAST) [17]. In some cases, the self-propagating high-temperature synthesis (SHS) reaction also occurred [18]. However, the structure of MAX phase-based composites usually consists of various phases, and the influence of these phases on the mechanical properties of composites has not been studied in detail.

The objective of this work is to analyze the Ti_3SiC_2 MAX phase-based composite deformation behavior based on Vickers nanohardness and microhardness measurements of SPSed composites. The sintering process of an HEBMed Ti/SiC nanostructured composite powder, the influence of the sintering temperature on the Ti_3SiC_2 MAX phase-based composite structure and the mechanical properties (nanohardness of various phases, microhardness and fracture toughness) are analyzed.

2. Materials and Methods

HEBMed Ti/SiC (50 vol%) nanostructured composite powder was developed and supplied by MBN Nanomaterialia and then SPSed in vacuum using an HP D 25-3 furnace (FCT Systeme, Rauenstein, Germany). To avoid particle oxidation, the powder was transferred from the manufacturing facility in a specially sealed parcel under vacuum, and the parcel was opened directly before filling the graphite die. The sintering temperatures of 1350, 1400 and 1450 °C were applied at a heating rate of 200 °C/min. The holding time was 10 min. Afterwards, the SPSed compacts were cooled to ambient temperature. The compacting pressure of 50 MPa was applied throughout all the sintering stages. Discs of a diameter of 20 mm and thickness of 5 mm were produced.

The agglomerate size distribution was measured by laser diffraction using a Mastersizer 2000 analyzer (Malvern Panalytical, Malvern, UK). X-ray structural studies of the powder and the SPSed compacts were conducted using a Kristalloflex 4 diffractometer (Siemens, Berlin, Germany) using MoK$_\alpha$ radiation with 2Θ 15–50°. Microscopic observations and elemental microanalysis were performed using two scanning electron microscopes: an Inspect S (FEI, Hillsboro, TX, USA) and a MIRA3 (TESCAN, Brno, Czech Republic) equipped with an energy-dispersive X-ray spectroscope (EDS) (EDAX, Mahwah, NJ, USA). The effective density measurements were made by the Archimedes method. The Vickers hardness measurements were carried out using an FV-800 hardness tester (Future-Tech, Kawasaki, Japan). The applied loads were 0.4903, 4.903 and 19.61 N and the holding time was 15 s. The fracture toughness was calculated based on the crack length measured from the corner of the indentation made by a Vickers indenter using Equation (1) [19]:

$$K_{1c} = 0.15 \sqrt{\frac{HV_{30}}{\sum l}} \tag{1}$$

where HV_{30} is the Vickers hardness measured under the load of 294.2 N and l is the total length of cracks initiated from the corners of the indenter. The Vickers indentation nanohardness measurements were performed using a Picodentor HM500 nanoindenter (Fisher, Windsor, SC, USA). The applied load was 0.05 N and the holding time was 5 s.

It is well known today that materials exhibit properties on the sub-micro and nanoscale, which are entirely different from those exhibited on the macroscale. One of the main effects exhibiting this difference is the increase in hardness with a decrease in indentation load or depth, known as the indentation size effect (ISE). Various nano- and microstructural features, such as lattice defects, grain boundaries, interfacial effects, phase structure and composition, etc., are responsible for such behavior of a material. That is why the ISE examination of Ti_3SiC_2 MAX phase-based composites is extremely important to understand the scientific issues involved in the genesis of size effects similar to work [20].

The nanoindentation process is described in Figure 1 by the dependence of load F applied to the nanoindenter tip to depth h of tip penetration in a given sample (Figure 1a). In the present work, a Vickers tip was used. Loading started at point I and continued along the loading arrow up to the point J where F became F_{max}. The maximum depth of penetration is shown as h_{max}. It was identified by the perpendicular JL drawn on the depth h axis. The unloading curve defined the path of unloading up to K where the load was reduced to zero while the final depth was elastically recovered from h_{max} at L to h_f at K (Figure 1a). Calculation of plastic deformation depth $h_{pl} = h_x - h_{el}$ was made on the basis of determining square root h_{el} of the equation describing the unloading curve (Figure 1b).

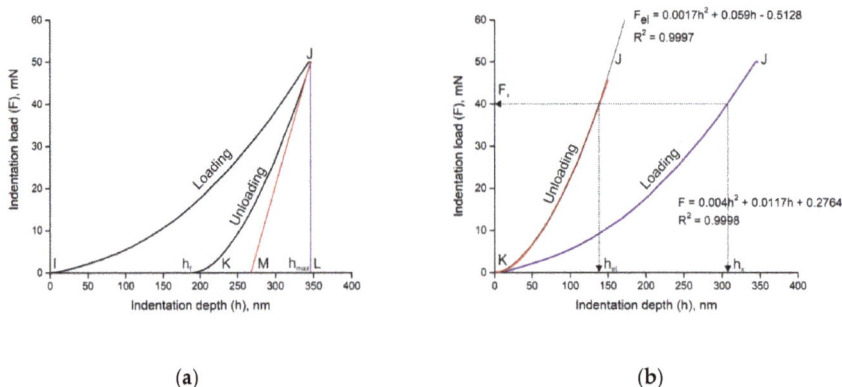

(a)   (b)

Figure 1. Indentation curve schematics: indentation curve of Ti_3SiC_2 MAX phase of specimen SPSed at 1350 °C (**a**) and approximation and calculation of indentation parameters (**b**).

3. Results and Discussion

3.1. Reactions of Sintering Process

Figure 2 presents the morphology and structure of the Ti/SiC HEBM powder. This powder consists of agglomerates with both, flaky and granular shaped particles (Figure 2a). The agglomerate size distribution of the HEBMed powder is presented in Table 1. The average size of the agglomerates is 42.2 μm, but some of the agglomerate sizes are close to 100 μm. The agglomerate structure analysis reveals that the SiC particles are randomly distributed in the Ti matrix and their sizes vary in a wide range (Figure 2b). Areas with a high content of SiC particles are adjacent to pure Ti. These Ti/SiC HEBMed powder feedstock structure features influence the further Ti-SiC reaction regime during sintering.

(a)   (b)

Figure 2. Morphology (**a**) and internal/cross cut structure (**b**) of Ti/SiC HEBMed powder agglomerates.

Table 1. Agglomerate size distribution of Ti/SiC HEBM powder.

D_{10}, µm	D_{50}, µm	D_{90}, µm
17.6	42.2	82.2

The X-ray spectra of the HEBM Ti/SiC powder (Figure 3) demonstrate that the powder feedstock consists of only Ti and SiC. Therefore, the chosen HEBM regime of the Ti/SiC powder did not lead to Ti-SiC reactions, and this powder feedstock was used for further SPS.

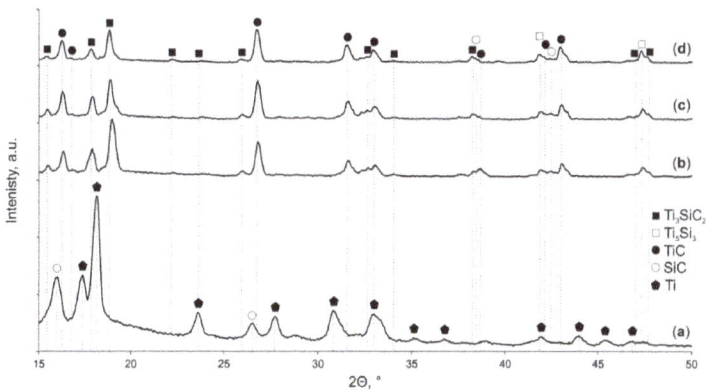

Figure 3. X-ray diffraction spectra of: Ti/SiC HEBMed powder (**a**); composite SPSed at 1350 °C (**b**); composite SPSed at 1400 °C (**c**) and composite SPSed at 1450 °C (**d**).

The densification behavior of the Ti-SiC composite during SPS was estimated based on analysis of the punch displacement and temperature dependences on the sintering time shown in Figure 4. The following process stages are seen: (i) air removal, (ii) pressing, (iii) heating, and (iv) holding at the sintering temperatures of 1350, 1400 and 1450 °C, and, subsequently, (v) cooling (Figure 4a). The punch displacement diagram for sintering at 1350 °C and 2.5 min (Figure 4b) demonstrates an

SHS-type reaction, which is characterized by a considerable increase in the densification rate (punch displacement rate) and temperature peak at ~700 °C and diffusion reactions during holding at the temperature of 1300–1450 °C.

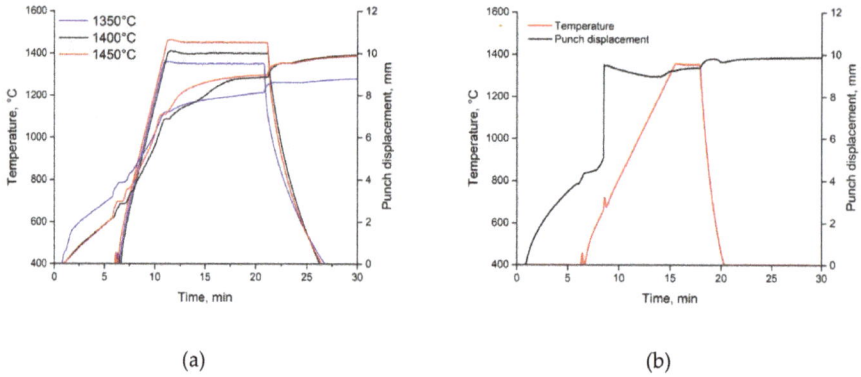

(a)   (b)

Figure 4. Punch displacement and temperature change during SPS of Ti/SiC HEBM powder at 1350, 1400 and 1450 °C and 10 min (**a**) and at 1350 °C and 2.5 min (**b**).

The X-ray diffraction analysis results (Figure 3) reveal the formation of a three-phase structure. It was found that the phase composition of the SPSed composites consists of dominant of Ti_3SiC_2 and TiC and Ti_5Si_3 phases. The results of the structure examination of the SPSed composites are shown in Figures 5 and 6. The data reveal the following specific features of the sintering process: (i) the dissolution of small SiC particles formed due to HEBM and the formation of new crystals during sintering, (ii) the presence of relatively large SiC-based particles that demonstrates their stability at the chosen sintering temperatures, (iii) the complexity of the chemical composition of the sintered phases due to the multiple diffusion of Ti, Si and C atoms, and (iv) grain growth of the MAX-based phase at high sintering temperatures.

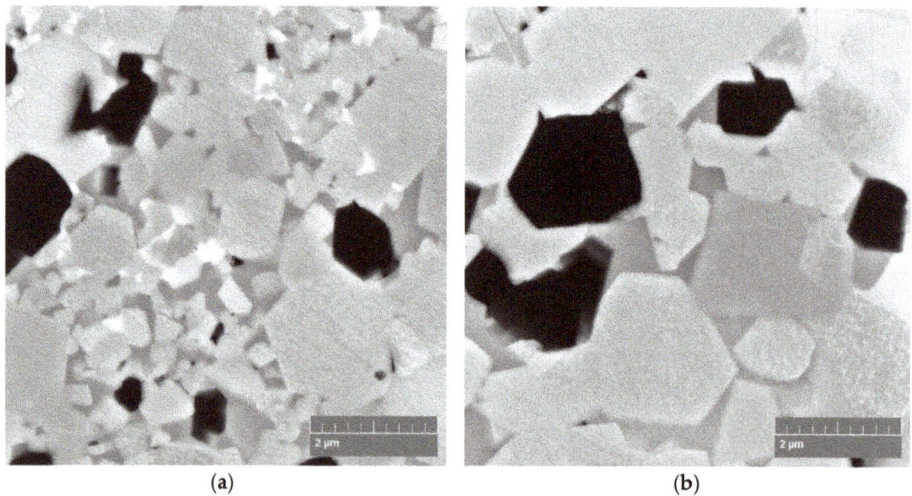

(a)   (b)

Figure 5. *Cont.*

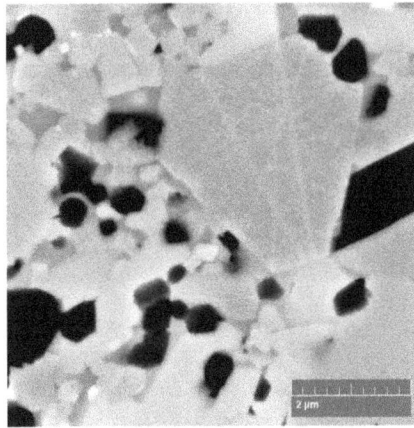

(c)

Figure 5. SEM micrographs of composites SPSed at: 1350 °C (**a**); 1400 °C (**b**) and 1450 °C (**c**).

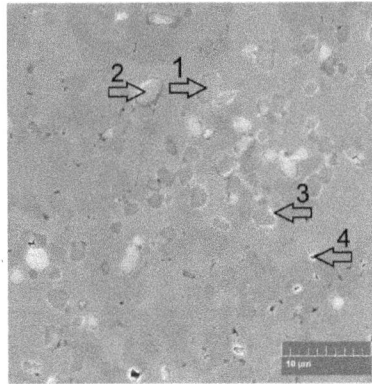

(a)

(b)

Figure 6. SEM micrographs of composite SPSed at 1400 °C: in-beam secondary electrons with magnification 5kx, arrow #1—dark phase, arrow #2—grey phase, arrow #3—bright phase and arrow #4—pore area (**a**) and comparison of in-beam secondary electrons and backscattered electrons micrographs (**b**).

175

The reactions between a Ti-based matrix and SiC particles were determined in works [21–23]. Gottselig et al. [21] studied the reaction behavior of Ti sputtered on SiC. It was found that mainly the ternary Ti_3SiC_2 MAX phase (>90%) is formed in the sintering temperature range between 1250 and 1500 °C. In experiments joining SiC with Ti layers at 1450 °C and at compacting pressures between 5 and 30 MPa, the formation of the ternary MAX phase leads to a high joining strength. A high sintering temperature results in fast diffusion of the Si and C atoms from the SiC particles into the Ti matrix formed due to severe plastic deformation of the Ti particles during the HEBM process. The milled SiC particles of a small size have an SHS reaction with Ti similar to those described in [24]. This SPS structure formation process is similar to the SPS joining described by Zhao et al. [23]. The diffusion layer at the Ti/SiC interface is shown [23] to be stratified at 1400 °C. This layer was enriched with Ti and C, while depleted in Si. These results suggested the formation of a TiC layer at the interface via interdiffusion between the interlayer and the SiC particles.

The EDS analysis results shown in Table 2 demonstrate similar diffusion processes which results in reactions described in detail by Gotman et al. [22]. The main possible chemical reactions during the sintering process are shown below [23]:

(i) SiC particle surface reactions and SiC dissolution:

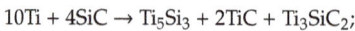

$8Ti + 3SiC \rightarrow 3TiC + Ti_5Si_3$;

$10Ti + 4SiC \rightarrow Ti_5Si_3 + 2TiC + Ti_3SiC_2$;

$2Ti + SiC + TiC \rightarrow Ti_3SiC_2$;

(ii) Reactions of Si and C atoms diffusing into Ti matrix:

$5Ti + 3Si \rightarrow Ti_5Si_3$;

$Ti + C \rightarrow TiC$.

Table 2. SEM-EDS analysis results of composites SPSed at 1350, 1400 and 1450 °C.

Phase Type	Sintering Temperature, °C	Phase Color	Area Number	Element Content, at%		
				Ti	Si	C
SiC particles	1350	Dark	1.1	07.63	51.82	40.55
		Gray	1.2	23.12	38.30	38.58
	1400	Dark	1.1	12.05	52.13	36.41
		Gray	1.2	34.09	29.50	33.45
	1450	Dark	1.1	09.99	56.18	29.56
		Gray	1.2	17.97	48.72	32.69
Ti matrix	1350	Gray	1.2	47.91	19.40	32.69
		Bright	1.3	43.47	18.84	37.70
	1400	Gray	1.2	47.79	25.77	26.45
		Bright	1.3	51.65	11.24	37.11
	1450	Gray	1.2	48.15	14.99	31.49
		Bright	1.3	47.68	39.32	13.00

Gotman et al. [22] clearly showed that all the above-listed reactions are thermodynamically favored, which means that the interaction between SiC and the Ti matrix can start with the formation of both TiC and Ti_5Si_3. The thermodynamic calculations conducted in work [22] show that the reaction layer at the surface of the SiC particles can contain both TiC and Ti_5Si_3. The micrographs in Figures 5 and 6 shows the bright contrast of these areas (see, for example, arrow #3 in Figure 6a) which reveals the presence of TiC and Ti_5Si_3 phases and the formation of a diffusion barrier layer. However, it can be seen that this barrier layer is not continuous. It is interesting to note that Ti diffusion into SiC does occur and results in obtaining a gray contrast of some SiC particles (see, for example, arrow #2 in Figure 6a), despite the presence of a TiC-Ti_5Si_3 barrier layer. Hence, some large dark SiC particles become gray due to a change in their chemical composition (an increase in Ti content, see Table 2). It is known that some errors occurred when determining the Ti and C content in former SiC particles by EDS. For this reason, additional EDS analysis was conducted on the fractured surfaces and showed similar results. The structure analysis results reveal that the large SiC particles cannot be completely decomposed during reaction sintering. Therefore, only particles of a size below the critical may be dissolved. Thus, the HEBM regime needs to be chosen to avoid the presence of large particles.

The main Ti_3SiC_2 MAX phase is believed to be formed due to the diffusion of Si and C atoms during the dissolution of small SiC particles. However, this process is not uniform, because of the extreme difference of SiC particle content in the various fields of the HEBMed powder particles (Figure 1b). Areas of small crystals are seen in the SEM micrograph of the composite SPSed at 1400 °C (Figure 6a). Some of them have a bright contrast, which indicates TiC and Ti_5Si_3. The same phases are seen in the light microscope micrograph of the composite SPSed at 1350 °C (Figure 7) etched with an HF-HNO$_3$-H$_2$O solution with the ratio HF:HNO$_3$:H$_2$O–1:1:5: Ti_3SiC_2 MAX phase—platelet bright crystals (arrow #1), areas of small crystals with the presence of TiC and Ti_5Si_3 (arrow #2), and former SiC particles (arrow #3).

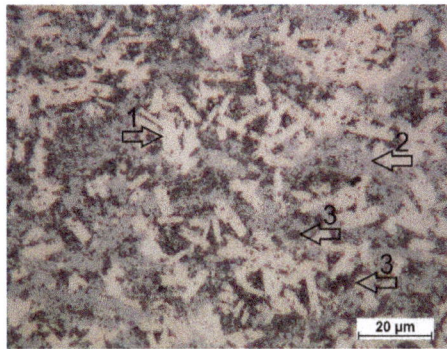

Figure 7. LM micrograph of composite SPSed at 1350 °C.

The structure of the composites SPSed at the temperature ranging between 1350 and 1450 °C presented in Figure 5 demonstrates the effect of the sintering temperature on the grain size of the Ti_3SiC_2 MAX phase. The Ti_3SiC_2 MAX phase recrystallization effect is seen in the composites SPSed at 1400 and 1450 °C. Comparison of the SEM in-beam secondary electrons and backscattered electrons micrographs (Figure 6b) allows the ratio between Si, C and Ti to be roughly evaluated and the conclusion to be drawn that both the structure formation of the SiC particles and the Ti matrix are governed by the diffusion of Si, C into Ti and Ti into the SiC particles. Therefore, the mechanical properties of the composite are controlled by these factors.

3.2. Mechanical Properties of Composites

The results of the effective density, hardness and fracture toughness measurements of composites SPSed at 1350, 1400 and 1450 °C are summarized in Table 3. It can be clearly seen that increasing the sintering temperature affected an increase in the density from 4.33 to 4.42 g/cm^3. The density of the SPSed composites varied only in the range of 0.09 g/cm^3. For this reason, the influence of porosity on the diffusion rates of the synthesis reactions was not taken into account. The hardness also increased with raising the sintering temperature. The specimens SPSed at 1400 and 1450 °C exhibit an increase in both the Vickers hardness and fracture toughness calculated from the indentation tests. The fracture toughness of the composites reached the maximum for the specimen SPSed at 1450 °C (K_{1c} = 6.06 MPa·m$^{1/2}$), whereas the fracture toughness of pure TiC is about 3.70 MPa·m$^{1/2}$ [25] and the maximal fracture toughness of the Ti_3SiC_2 MAX phase is about 11.50 MPa·m$^{1/2}$ [10,12].

Table 3.   Effective density, hardness and fracture toughness of composites SPSed at 1350, 1400 and 1450 °C.

Sintering temperature, °C	Effective density, g/cm^3	Hardness (HV$_{0.05}$), GPa	Hardness (HV$_{0.5}$), GPa	Hardness (HV$_2$), GPa	Fracture toughness (K$_{1c}$), MPa·m$^{1/2}$
1350	4.33 ± 0.02	12.71 ± 0.39	11.54 ± 0.85	10.57 ± 0.7	n/a
1400	4.36 ± 0.01	14.06 ± 1.79	12.49 ± 1.85	11.55 ± 0.77	5.42 ± 0.47
1450	4.42 ± 0.01	16.34 ± 2.38	13.08 ± 0.96	12.83 ± 0.95	6.06 ± 0.02

The increase in fracture toughness with the sintering temperature may be attributed to deflection of the cracks due to an increase in the Ti$_3$SiC$_2$ MAX phase content and a possible decrease in the Ti$_5$Si$_3$ brittle phase concentration (Figure 8).

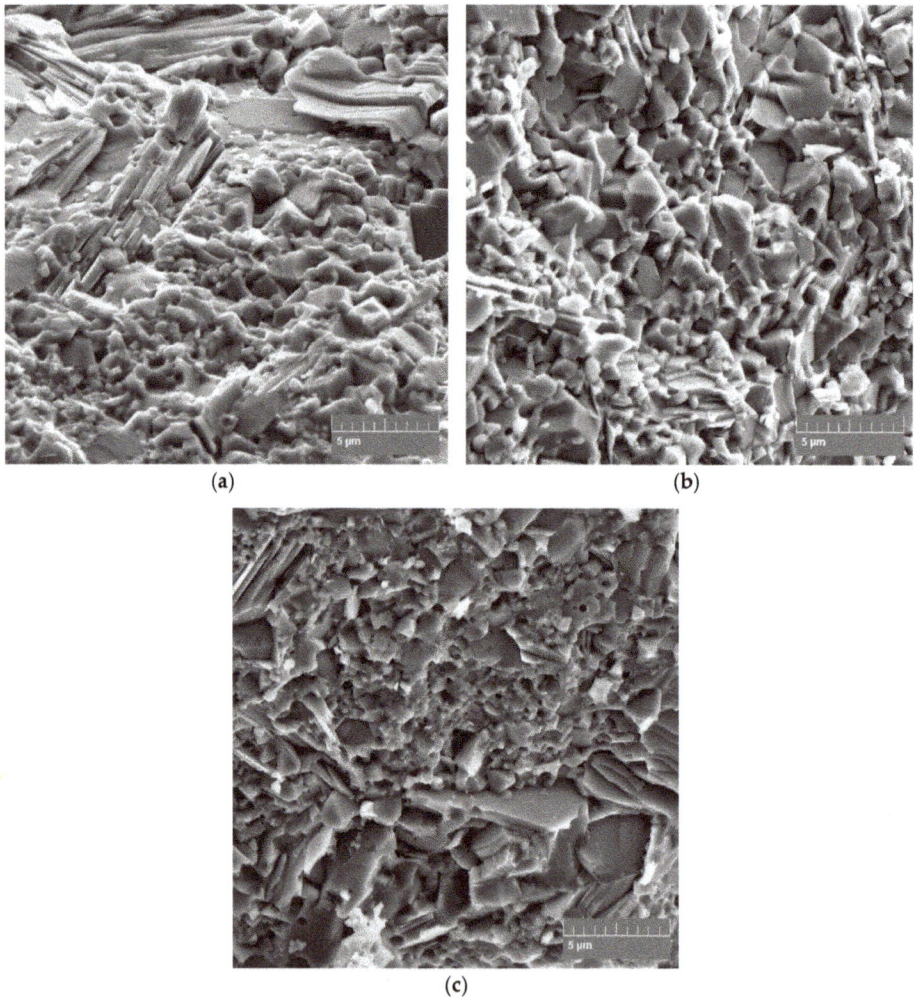

(a)

(b)

(c)

Figure 8. SEM micrographs of fracture surface topography of composites SPSed at: 1350 °C (**a**); 1400 °C (**b**) and 1450 °C (**c**).

It is well known that the physical, structural, mechanical and other properties change drastically on the nanoscale, thereby attracting attention to the size effects in various materials, especially ceramics. Therefore, it is extremely important to clarify the basic effects of size effects in materials, particularly ceramics, which are characteristically brittle in nature. Hence, examining the ISE of nanostructured ceramic-based composites is very important from the viewpoint of synthesizing new materials. Examining the nanomechanical behavior of composite phases will allow further understanding of their deformation and fracture mechanisms and the real area of its application to be determined. Thus, the major objectives of the nanohardness analysis of the composites is: (i) to study the nanoindentation response of SPSed composites sintered at various temperatures, and (ii) to examine the strain gradient plasticity (SGP) model to explain the ISE of various phases of sintered composites. Therefore, nanohardness measurements of the three phases were performed.

The variation of the nanohardness of the SPSed composites as a function of load and depth is shown in Figure 1, as an example, and in Figure 9b for composites SPSed at 1400 °C. It is important to note that the phases of the SPSed composites exhibit the presence of a very strong ISE which has not yet been studied in detail. The validity of the SGP model of the ISE explanation and characterization for the SPSed composites may be evaluated by comparison of the indentation and SGP equation parameters. The data presented in Figure 9a show that in the SPSed composites the ratio of elastic to plastic depths $(h_{max} - h_f)/h_f$ increases for the SiC particles and decreases for the Ti_3SiC_2 MAX phase and the $TiC+Ti_5Si_3$ phase. Clearly, the lower the $(h_{max} - h_f)/h_f$ ratio is, the higher the final depth h_f of penetration (Figure 9a), which indicates higher plastic strain. The data presented in Figure 9a demonstrate a similar deformation ability of the Ti_3SiC_2 MAX phase and $TiC+Ti_5Si_3$ phase. However, the deformation mechanisms of these phases are not clear, and the load-indentation depth diagrams of the $TiC+Ti_5Si_3$ phase contains the areas of crack generation (Figure 9b). The ISE mechanisms were analyzed by Maiti et al. [20], and the validity of various of phenomenological behaviors of nano-mechanical models (elastic recovery, proportional specimen resistance and Nix and Gao) to the experimental data of a ZrO_2 ceramic was determined. However, only the Nix and Gao model has a proper physical background—the SGP theory developed by Fleck et al. [26] and Nix and Gao [27]. It allows some physical-based parameters of composite plastic deformation to be estimated. Their estimation on the basis of the nanohardness measurement of the Ti_3SiC_2 MAX phase, $TiC+Ti_5Si_3$ and SiC phases will allow the real deformation mechanisms to be determined.

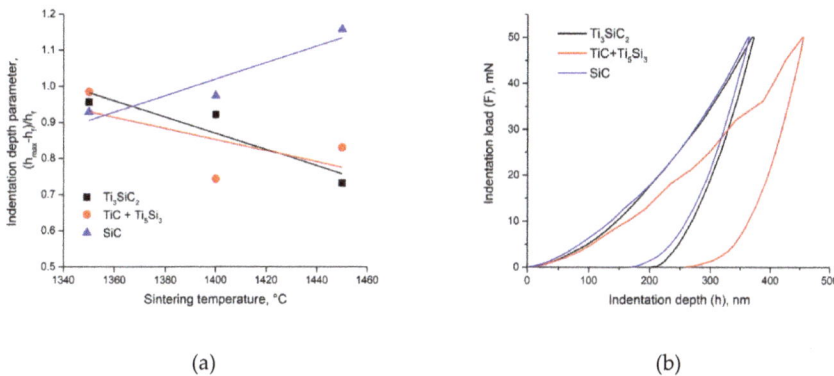

(a)   (b)

Figure 9. Indentation parameter $(h_{max} - h_f)/h_f$ dependence on sintering temperature (**a**) and load-penetration depth curves of phases of composite SPSed at 1400 °C (**b**).

In general, in SGP theory, it is assumed [27] that the indentation is accommodated by circular loops of geometrically necessary dislocations with Burgers vectors normal to the plane of the surface. As the indenter is forced into the surface of a single crystal, a certain density of geometrically necessary (GND) dislocations ρ_G is required to account for the permanent shape change at the surface. Other dislocations, called statistically stored dislocations by Ashby [28], would also be created and they would contribute to the deformation resistance. The geometry of indentation contact is characterized by the following parameters: (i) the angle between the surface of the conical or pyramidal indenter and the plane of the surface Θ, (ii) the contact radius a, and (iii) the depth of indentation h.

To estimate the shear strength, the authors of [27] use the Taylor relation:

$$\tau = \alpha \mu b \sqrt{\rho_T} = \alpha \mu b \sqrt{\rho_G + \rho_S} \tag{2}$$

where ρ_T is the total dislocation density in the indentation, ρ_S is the density of statistically stored dislocations, μ is the shear modulus, b is the Burgers vector and α is a constant (0.5). ρ_S is not expected to depend on the depth of indentation.

Taking into account the fact that a Tabor's [29] factor of 3 can be used to convert the equivalent flow stress to hardness, the von Mises yield criterion may be presented as:

$$\sigma = \sqrt{3}\tau; H = 3\sigma \triangleright H = 3\sqrt{3}\alpha \mu b \sqrt{(\rho_G + \rho_S)} \tag{3}$$

and the hardness equation can be shown as:

$$\frac{H}{H_0} = \sqrt{1 + \left(\frac{h^*}{h}\right)} \triangleright \left(\frac{H}{H_0}\right)^2 = h^*\left(\frac{1}{h}\right) + 1 \tag{4}$$

where H_0 is the hardness that would arise from the statistically stored dislocations alone, in the absence of any geometrically necessary dislocations, and:

$$h^* = \frac{81}{2}b\alpha^2 tg^2\theta\left(\frac{\mu}{H_0}\right)^2 \tag{5}$$

where h^* is a length that characterizes the depth dependence of the hardness. It must be noted that h^* is not a constant for a given material or indenter geometry. It depends on the statistically stored dislocation density through H_0.

The validity of Nix and Gao model to the nanohardness experimental data is defined by calculating the linear approximation $(H/H_0)^2 = f(1/(h - h_f))$ of the nanohardness measurement results (Figure 10). The results demonstrate that only the Ti$_3$SiC$_2$ MAX phase exhibits dependence $(H/H_0)=f(1/(h - h_f))$ close to the Nix and Gao equation [27] (Equation (4), the imulation veracity is $R^2 = 0.8680-0.9885$). The fitting of the experimentally measured F-h data of the TiC+Ti$_5$Si$_3$ phase gives Nix and Gao parameters with a goodness of fit (R^2) in the range of 0.7357–0.8552 (Figure 10). Slightly better results are for the SiC-based particles—$R^2 = 0.9281-0.9680$. This means that the calculated strain gradient parameters of only the Ti$_3$SiC$_2$ MAX phase reflect the real dislocation mechanisms of deformation.

The ISE is observed in both the nanohardness and microhardness of various metals [30]. While the explanations of SGP theory shown above seem to be valid for nanoindentation [27], the mechanism of ISE for the microindentation of SPSed composites needs to be corrected taking into account the specific structure features. Some experimental results of determining the nanohardness and microhardness ISE are shown in Table 4.

The values of true hardness H_0 and characteristic depth h^* are shown in Table 4 for all the examined phases. The H_0 values show that the true hardness parameter of all the phases is higher than that of the ZrO$_2$ ceramic presented in paper by Maiti et al. [20], because it is calculated for indentation depth $h_x = h - h_{el}$. The characteristic depth h^* of the Ti$_3$SiC$_2$ MAX phase varies in the range of $h^* = 0.07 - 0.11$, which is in the range of h^* defined for different metals–Ni, Ag, Au and Cu [31]. Parameter h^* defined for other phases of the SPSed composite varies in a wide range 0.03–0.19 which indicates the weak effectiveness of the Nix and Gao deformation model for TiC+Ti$_5$Si$_3$ and SiC phases.

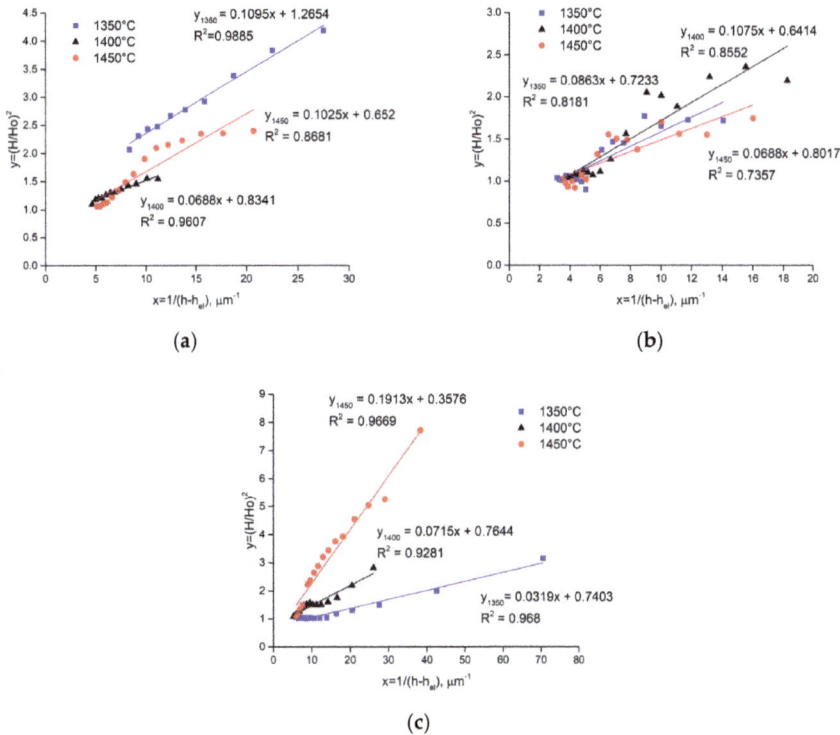

Figure 10. Application of Nix and Gao model to nanohardness experimental data of: Ti_3SiC_2 (a); $TiC+Ti_5Si_3$ (b) and SiC (c) phases of composites SPSed at 1350, 1400 and 1450 °C.

Table 4. Nix and Gao equation fitting parameters of composites SPSed at 1350, 1400 and 1450 °C.

Test Type	Composite Phase	SPS Temperature, °C	Nix and Gao Equation Fitting Parameters		
			H_0, GPa	h^*, μm	R^2
Nanohardness	Ti_3SiC_2	1350	48.5	0.110	0.99
		1400	41.8	0.069	0.96
		1450	51.8	0.103	0.87
	$TiC+Ti_5Si_3$	1350	25.4	0.069	0.81
		1400	29.5	0.108	0.86
		1450	19.8	0.086	0.74
	SiC	1350	56.2	0.032	0.97
		1400	53.0	0.072	0.93
		1450	69.2	0.191	0.97
Microhardness		1350	10.4	0.37	0.91
		1400	11.2	0.40	0.96
		1450	12.5	0.51	0.99

The comparison between the microhardness and nanohardness data fitting to the Nix and Gao model is shown in Figure 11. The results indicate the high values of H_0 and h^* for the microindentation cases, which are the result of the low dislocation density and difficulties to generate and move dislocations in the crystals. We observe regular mechanical behavior of the composite during indentation, and only a small effect of the sintering temperature. Therefore, the ISE of microhardness is not sensitive to the composite structure and is not controlled by dislocation-based mechanisms.

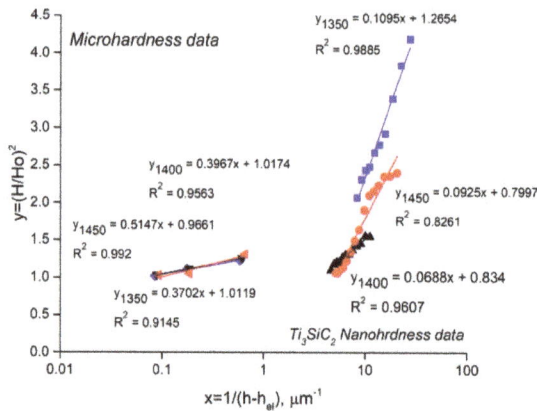

Figure 11. Application of the Nix and Gao model to composites and Ti$_3$SiC$_2$ phase of composites SPSed at 1350, 1400 and 1450 °C.

In contrast, the influence of the sintering temperature on the ISE of nanohardness is considerable for the Ti$_3$SiC$_2$ and SiC-based phases (Figure 10), because the phase composition and structure greatly depend on the diffusion process. An increase in the sintering temperature results in an increase in the Ti content in the SiC-based phase (former SiC particles), which is shown in Table 2. For this reason, the ISE is maximal for the high sintering temperature of 1450 °C (Figure 10c). However, the plastic deformation recourse of this phase remains minimal, and nanoindentation results in considerable distortion of the nanoindent (Figure 12a), as well as a small pile-up effect (Figure 12b). One can observe that crack initiation during loading (Figure 12a) results in the appearance of specific areas on the load-displacement nanoindentation diagrams (Figure 9b), leading to errors in calculating true hardness H_0 and characteristic depth h^* (Equation (4)).

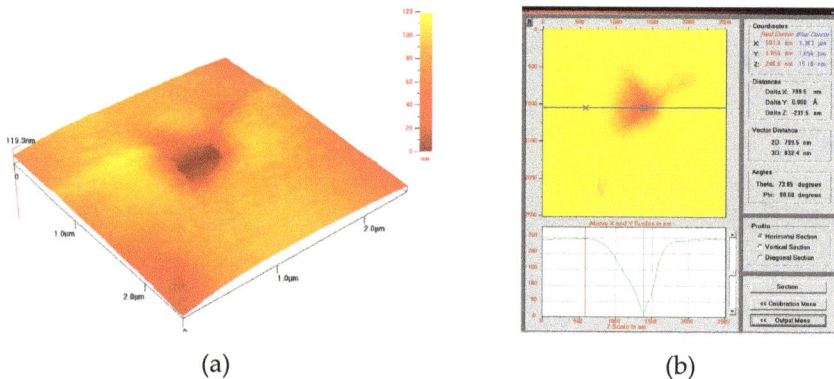

(a)   (b)

Figure 12. Indents (**a**) and material pile-up (**b**) of SiC-based phase of composite SPSed at 1450 °C.

The adequacy of the deformation behavior of the Ti$_3$SiC$_2$ MAX phase to SGP theory can be determined based on SGP equations (Equations (3) and (5)) and the experimentally defined parameters H_0 and h^* by calculating the densities of the GND and SS dislocations. Because $\rho_G = 4\gamma/bD$ [28], it is possible to state that the hardness square (Equation (3)) is inversely related to indent size D. Here γ is the average shear strain. The density of the SS dislocations may be determined based on Equation (5) and the resulting GND and SS dislocation dependence on the indentation depth for the Ti$_3$SiC$_2$ MAX phase will be defined in ongoing work.

4. Conclusions

This work presents the synthesis of Ti_3SiC_2 MAX phase-based composites using a Ti/SiC HEBMed and SPSed powder. The Ti_3SiC_2 MAX phase along with the $TiC+Ti_5Si_3$ phase were generated during SPS in the temperature range of 1300–1450 °C. The effects of SHS were found. It was shown that a higher sintering temperature of up to 1450 °C led to an increase in the Ti_3SiC_2 grain size. The effects of the Ti/SiC powder initial structure controlled by the HEBM process and the phase reactions during SPS on the mechanical properties of the composites were studied. The interdiffusion behavior of the atoms in the phases and at the interface of the phases significantly affected the mechanical properties of the composites. It was found that a thin $TiC+Ti_5Si_3$ layer was formed in situ on the SiC-Ti matrix interface via fast interdiffusion during the SPS process, which acts as a barrier hindering the diffusion of Si atoms from SiC to the Ti matrix. Diffusion of the Ti atoms into the large SiC particles in the applied sintering temperature range was found as well. The ISE was studied to define the deformation behavior of the SPSed Ti-SiC composite phases. It was determined that the ISE on the nanoscale is due to a change in the density of geometrically necessary dislocations in the Ti_3SiC_2 MAX phase, which is in agreement with SGP theory. The other $TiC+Ti_5Si_3$ and SiC phases do not exhibit deformation behavior based on SGP theory and have a low fracture toughness. Improvement of the Ti-SiC structure and properties is possible by optimizing the HEBM and SPS parameters to make a more uniform and controlled reaction sintering process.

Author Contributions: Conceptualization, V.L. and A.C. and P.M.; methodology, D.G. and V.L. and A.C.; investigation, D.G. and P.S.; resources, A.C.; data curation, D.G.; writing—original draft preparation, D.G. and V.L.; writing—review and editing, D.G. and V.L.; visualization, D.G.; supervision, V.L.

Funding: This work was funded by the European Union through Grant No. 604344.

Conflicts of Interest: The authors declare no conflict of interest.

Nomenclature

λ	Constant of Taylor equation (0.5)
b	Burgers vector
D	HEBMed powder agglomerate, μm
F	Indentation load, mN
h	Indentation depth, nm
H	Hardness parameter, GPa
HV_{30}	Vickers hardness measured under the load of 294.2 N
K_{1c}	Fracture toughness, MPa·m$^{1/2}$
l	Total length of cracks, μm
μ	Shear modulus
ρ	Dislocation density
σ	Normal stress
T	Temperature, °C
τ	Shear strength
Θ	Angle between the side surface of the pyramidal indenter and the base plane

References

1. Schaffer, G.B.; McCormick, P.G. Anomalous combustion effects during mechanical alloying. *Metal. Trans. A* **1991**, *22*, 3019–3024. [CrossRef]
2. Shahin, N.; Kazemi, S.; Heidarpour, A. Mechanochemical synthesis mechanism of Ti_3AlC_2 MAX phase from elemental powders of Ti, Al and C. *Adv. Powder Technol.* **2016**, *27*, 1775–1780. [CrossRef]
3. Garay, J.E. Current-Activated, Pressure-Assisted Densification of Materials. *Annu. Rev. Mater. Res.* **2010**, *40*, 445–468. [CrossRef]
4. Liang, B.Y.; Jin, S.Z.; Wang, M.Z. Low-temperature fabrication of high purity Ti_3SiC_2. *J. Alloy. Compd.* **2008**, *460*, 440–443. [CrossRef]

5. Pourebrahim, A.; Baharvandi, H.; Foratirad, H.; Ehsani, N. Effect of aluminum addition on the densification behavior and mechanical properties of synthesized high-purity nano-laminated Ti3SiC2 through spark plasma sintering. *J. Alloy. Compd.* **2018**, *730*, 408–416. [CrossRef]

6. Poletti, C.; Balog, M.; Schubert, T.; Liedtke, V.; Edtmaier, C. Production of titanium matrix composites reinforced with SiC particles. *Compos. Sci. Technol.* **2008**, *68*, 2171–2177. [CrossRef]

7. Figiel, P.; Garbiec, D.; Biedunkiewicz, A.; Wiedunkiewicz, W.; Kochmanski, R.; Wrobel, R. Microstructural, corrosion and abrasive characteristics of titanium matrix composites. *Arch. Metall. Mater.* **2018**, *63*, 2051–2059.

8. Bowen, C.R.; Thomas, T. Macro-porous Ti2AlC MAX-phase ceramics by the foam replication method. *Ceram. Int.* **2015**, *41*, 12178–12185. [CrossRef]

9. Rahman, K.M.; Vorontsov, V.A.; Flitcroft, S.M.; Dye, D. A High Strength Ti–SiC Metal Matrix Composite. *Adv. Eng. Mater.* **2017**, *19*, 1700027.

10. Radovic, M.; Barsoum, M.W. MAX phases: Bridging the gap between metals and ceramics. *Am. Ceram. Soc. Bull.* **2013**, *92*, 20–27.

11. Xu, B.; Chen, Q.; Li, X.; Meng, C.; Zhang, H.; Xu, M.; Li, J.; Wang, Z.; Deng, C. Synthesis of single-phase Ti3SiC2 from coarse elemental powders and the effects of excess Al. *Ceram. Int.* **2019**, *45*, 948–953. [CrossRef]

12. Barsoum, M.W.; Farber, L.; El-Raghy, T. Dislocations, kink bands, and room-temperature plasticity of Ti3SiC2. *Metall. Mater. Trans. A* **1999**, *30*, 1727–1738. [CrossRef]

13. Naguib, M.; Mashtalir, O.; Carle, J.; Presser, V.; Lu, J.; Hultman, L.; Gogotsi, Y.; Barsoum, M.W. Two-Dimensional Transition Metal Carbides. *ACS Nano* **2012**, *6*, 1322–1331. [CrossRef]

14. Xue, M.; Tang, H.; Li, C. Synthesis of Ti3SiC2 Through Pressureless Sintering. *Powder Metall. Met. Ceram.* **2014**, *53*, 392–398. [CrossRef]

15. Yong-Ming, L.; Wei, P.; Shuqin, L.; Jian, C. Synthesis and mechanical properties of in-situ hot-pressed Ti3SiC2 polycrystals. *Ceram. Int.* **2002**, *28*, 227–230. [CrossRef]

16. Shannahan, L.; Barsoum, M.W.; Lamberson, L. Dynamic fracture behavior of a MAX phase Ti3SiC2. *Eng. Frac. Mech.* **2017**, *169*, 54–66. [CrossRef]

17. Kozak, K.; Bućko, M.M.; Chlubny, L.; Lis, J.; Antou, G.; Chotard, T. Influence of composition and grain size on the damage evolution in MAX phases investigated by acoustic emission. *Mat. Sci. Eng. A-Struct* **2019**, *743*, 114–122. [CrossRef]

18. El Saeed, M.A.; Deorsola, F.A.; Rashad, R.M. Optimization of the Ti3SiC2 MAX phase synthesis. *Int. J. Refract. Met. H.* **2012**, *35*, 127–131. [CrossRef]

19. Schubert, W.D.; Neumeister, H.; Kinger, G.; Lux, B. Hardness to toughness relationship of fine-grained WC-Co hardmetals. *Int. J. Refract. Met. H.* **1998**, *16*, 133–142. [CrossRef]

20. Maiti, P.; Bhattacharya, M.; Das, P.S.; Devi, P.S.; Mukhopadhyay, A.K. Indentation size effect and energy balance issues in nanomechanical behavior of ZTA ceramics. *Ceram. Int.* **2018**, *44*, 9753–9772. [CrossRef]

21. Gottselig, B.; Gyarmati, E.; Naoumidis, A.; Nickel, H. Joining of ceramics demonstrated by the example of SiC/Ti. *J. Eur. Ceram. Soc.* **1990**, *6*, 153–160. [CrossRef]

22. Gotman, I.; Gutmanas, E.Y.; Mogilevsky, P. Interaction between SiC and Ti powder. *J. Mater. Res.* **1993**, *8*, 2725–2733. [CrossRef]

23. Zhao, X.; Duan, L.; Wang, Y. Fast interdiffusion and Kirkendall effects of SiC-coated C/SiC composites joined by a Ti-Nb-Ti interlayer via spark plasma sintering. *J. Eur. Ceram. Soc.* **2019**, *39*, 1757–1765. [CrossRef]

24. He, X.; Bai, Y.; Li, Y.; Zhu, C.; Kong, X. In situ synthesis and mechanical properties of bulk Ti3SiC2/TiC composites by SHS/PHIP. *Mat. Sci. Eng. A-Struct* **2010**, *527*, 4554–4559. [CrossRef]

25. Cabrero, J.; Audubert, F.; Pailler, R. Fabrication and characterization of sintered TiC–SiC composites. *J. Eur. Ceram. Soc.* **2011**, *31*, 313–320. [CrossRef]

26. Fleck, N.A.; Muller, G.M.; Ashby, M.F.; Hutchinson, J.W. Strain gradient plasticity: Theory and experiment. *Acta Metall. Mater.* **1994**, *42*, 475–487. [CrossRef]

27. Nix, W.D.; Gao, H. Indentation size effects in crystalline materials: A law for strain gradient plasticity. *J. MECH. PHYS. SOLIDS* **1998**, *46*, 411–425. [CrossRef]

28. Ashby, M.F. The deformation of plastically non-homogeneous materials. *Philos. Mag. A J. Theor. Exp. Appl. Phys.* **1970**, *21*, 399–424. [CrossRef]

29. Tabor, D. *Hardness of Metals*; Clarendon Press: Oxford, UK, 1951.

30. Goldbaum, D.; Ajaja, J.; Chromik, R.R.; Wong, W.; Yue, S.; Irissou, E.; Legoux, J.-G. Mechanical behavior of Ti cold spray coatings determined by a multi-scale indentation method. *Mat. Sci. Eng. A-Struct* **2011**, *530*, 253–265. [CrossRef]
31. Zong, Z.; Lou, J.; Adewoye, O.O.; Elmustafa, A.A.; Hammad, F.; Soboyejo, W.O. Indentation size effects in the nano- and micro-hardness of fcc single crystal metals. *Mat. Sci. Eng. A-Struct* **2006**, *434*, 178–187. [CrossRef]

materials

MDPI

Article

The Tribaloy T-800 Coatings Deposited by Laser Engineered Net Shaping (LENS™)

Tomasz Durejko [1], Magdalena Łazińska [1], Julita Dworecka-Wójcik [1,*], Stanisław Lipiński [1], Robert A. Varin [2] and Tomasz Czujko [1,*]

1   Department of Advanced Materials and Technologies, Military University of Technology,
    2 Gen. Sylwestra Kaliskiego Street, 00-908 Warsaw, Poland; tomasz.durejko@wat.edu.pl (T.D.);
    magdalena.lazinska@wat.edu.pl (M.Ł.); stanislaw.lipinski@wat.edu.pl (S.L.)
2   Department of Mechanical and Mechatronics Engineering, University of Waterloo, Waterloo, ON N2L 3G1,
    Canada; robert.varin@uwaterloo.ca
*   Correspondence: julita.dworecka@wat.edu.pl (J.D.-W.); tomasz.czujko@wat.edu.pl (T.C.);
    Tel.: +48-261-839-445 (J.D.-W.); +48-261-839-445 (T.C.)

Received: 9 February 2019; Accepted: 24 April 2019; Published: 26 April 2019

Abstract: A Tribaloy family of alloys (CoMoCrSi) are characterized by a substantial resistance to wear and corrosion within a wide range of temperatures. These properties are a direct result of their microstructure including the presence of Laves phase in varying proportions. Tribaloy T-800 exhibits the highest content of Laves phase of all other commercial Tribaloy alloys, which provides high hardness and wear resistance. On the other hand, a large content of the Laves phase brings about a high sensitivity to brittle fracture of this alloy. The main objective of this work was a development of the Tribaloy T-800 coatings on the Ni-based superalloy substrate (RENE 77), which employs a Laser Engineered Net Shaping (LENS™) technique. Technological limitations in this process are susceptibility of T-800 to brittle fracture as well as significant thermal stresses due to rapid cooling, which is an inherent attribute of laser techniques. Therefore, in this work, a number of steps that optimized the LENS™ process and improved the metallurgical soundness of coatings are presented. Employing volume and local substrate pre-heating resulted in the formation of high quality coatings devoid of cracks and flaws.

Keywords: tribaloy-type alloy; CoCrMoSi alloy coatings; T-800 alloy; Laves phase; Laser Engineered Net Shaping (LENS™)

---

1. Introduction

Nickel-based (Ni-based) superalloys exhibit a combination of high strength and corrosion/oxidation resistance at elevated temperatures.

These virtues make the Ni-based superalloys widely used for producing components of various systems working at extreme service conditions such as gas turbines in jet engines, pumps, and pipes for chemical and petrochemical industries and rocket engines. A long service at extreme service conditions eventually leads to severe wear of Ni-based superalloy parts, particularly at the surface layer, which limits their useful service life [1]. Therefore, new materials and technologies are being sought, which could improve efficiency and service life of highly stressed components working at high temperatures. In order to regenerate those components and bring them to the original condition, at least partially, protective coatings are applied to extend their useful service life [2,3]. The regenerative coatings also allow us to control certain properties like the increase of corrosion/oxidation and wear resistance. These coatings also prevent the effects of rapid temperature fluctuations. For the components made of Ni-based superalloys that work under severe wear conditions, the improvement of their wear resistance is of paramount importance. A good solution in this case is applying a coating that would be more wear resistance than the substrate being simultaneously highly corrosion/oxidation resistant. Suitable materials are quite commonly used such as cobalt (Co)-based alloys that have trademark names Stellite and Tribaloy (trademarks of the Deloro Stellite Company, Inc., Koblenz, Germany). These alloys are characterized by high strength, hardness, corrosion/oxidation, and wear resistance at a wide range of temperatures [4]. These properties are due to a unique microstructure, which consists of hard particles (either carbides or intermetallic phases) embedded in the Co solid solution matrix. According to the producer specifications, the Stellite alloys are recommended for components requiring corrosion/oxidation, erosion, and wear resistance at temperatures up to 800 °C. In turn, the Tribaloy alloys are recommended for the components requiring wear and corrosion/oxidation resistance at high temperatures, particularly, for applications where a direct lubrication is impossible [5].

Since Ni-based alloy components are frequently in service at temperatures exceeding 800 °C, it is justified to coat them with Tribaloy coatings since it was already implemented by Rolls Royce for turbine blades [6].

The principal alloying element of the Tribaloy alloys is Co in addition to molybdenum (Mo), chromium (Cr), and silicon (Si). Their properties depend on the content of each alloying element.

The most common commercial Co-based Tribaloy alloys are T-400, T-800, and T-900 [7]. Other alloys are T-400C and T-401, whose compositions are modifications of a base alloy T-400 [8–10]. The Tribaloys are usually hypereutectic (except T-401) and have a cast microstructure, which consists of hard, primary intermetallic Laves phases (about 40–60 vol.%) and are uniformly distributed in the eutectic matrix (fine intermetallic dispersoids within the Co solid solution) [11,12]. In Tribaloys, the Laves phases are intermetallic phases of the C-14-type ($MgZn_2$) like Co_3Mo_2Si and/or $CoMoSi$. They are characterized by high hardness approaching 1000-1200 HV [8] and high melting point of about 1560 °C [10,13]. The fraction of these phases in Tribaloy alloys is determined by a large content of Mo and Si as compared to other Co-based alloys like Stellite alloys. A minimal amount of carbon in Tribaloy alloys prevents formation of the M_7C_3-type precipitates that are found in Stellite alloys.

Tribaloy T-800, which contains the largest content of Laves phase at about 60 vol.% in its microstructure, exhibits the highest hardness while a T-400 alloy contains only about 40 vol.% of the Laves phase [12]. In addition, a large content of Laves phase in T-800 provides high adhesive and abrasive wear resistance while it, simultaneously, reduces plasticity and impacts toughness at an ambient temperature. Furthermore, T-800 is also characterized by increased oxidation resistance at a wide range of temperatures attributable to its high chromium content (nearly twice as much as that in T-400).

Initially, the Tribaloy coatings were deposited by thermal processing such as thermal spraying [14,15]. Due to a narrow heating zone, those methods did not provide a good adherence of the deposited layer via metallurgical bonding with the substrate [16]. Therefore, more recently, reports in the literature state the laser techniques employed for the Tribaloy coatings [17]. It is expected that these processes would result in coatings with good metallurgical properties, fine-grained with only minor structural modifications within the deposit-substrate volume, and precisely mirroring the desired shape.

For T-800, the presence of Laves phases in a large quantity guarantees high hardness and wear resistance, although a high content of this brittle phase is simultaneously a drawback, since it favors the brittle crack formation and propagation. That inherent propensity to brittle behavior makes the coating processes more difficult since cracking must be avoided especially in small components [18,19]. The propensity of brittle fracture of Tribaloy T-800 is a problem for laser techniques for which substantial thermal stresses arise during rapid cooling, which exacerbate the propensity for brittle fracture. New Tribaloy alloys have been implemented with the objective of ameliorating this problem. A modification of the base composition results in an improvement of plasticity, resistance to brittle fracture, and, simultaneously, lowers hardness. For example, T-900 having an increased Ni content (a lower content of Co and Mo) has better plasticity and fracture resistance than T-800. Alternatively, alloying of T-800 with iron (Fe) can also improve some properties. If the Fe content exceeds 10 wt.%, the microstructure undergoes a substantial alteration such that dendrites are refined, Laves phases disappear, and the alloy becomes more fracture-resistant while its hardness and wear resistance are lowered [17].

Navas et al. [18] reported that an intelligent choice of processing parameters can also alleviate the problems described above. The technology of Laser Cladding permitted us to obtain good quality T-800 coatings on an AISI 304 substrate. The obtained coating exhibited a high hardness about 850 HV0.3 and measurably improved wear resistance of the substrate [18]. In addition, in order to reduce thermal stresses in the coating, which can be a source of brittle cracking, it could be beneficial to pre-heat the substrate just before depositing the coating [19,20].

Due to its characteristics, T-800 seems to be a very attractive alloy for the components exposed to high service temperatures and wear conditions, like turbine blades. In these applications, besides wear resistance, a substantial oxidation resistance is also required. Yuduo et al. [21] reported that T-800 has high oxidation resistance at elevated temperatures, which can be additionally enhanced by alloying with 4 wt.% rhenium (Re).

In this work, an attempt is made to investigate varying processing parameters for producing coatings of Tribaloy T-800 on the Ni-based superalloy, RENE 77, substrate. In order to obtain metallurgically sound deposits, devoid of microcracks and flaws, exhibiting a gradual distribution in the chemical composition and microhardness at the interface deposit-substrate, the coatings were produced by the LENS^{TM} (Laser Engineered Net Shaping) technique. Due to local substrate heating, this technique minimizes the heat affected zone. This, in turn, reduces "thermal distortions" and the amount of substrate material in the deposit. In addition, LENS^{TM} allows depositing coatings with controlled thickness on any randomly selected area of the substrate.

2. Materials and Methods

2.1. Characterization of the Initial Powder

The initial powder was Tribaloy T-800 provided by LPW Technology Ltd. (Widnes, UK). The powder particles analyzed with an electron microscope Quanta 3G FEM Dual Beam (FEI, Hillsboro, OR, USA) shows a spherical morphology (Figure 1a). A powder cross section reveals a miniscule porosity (Figure 1b) and a dendritic microstructure within the particles (Figure 1c).

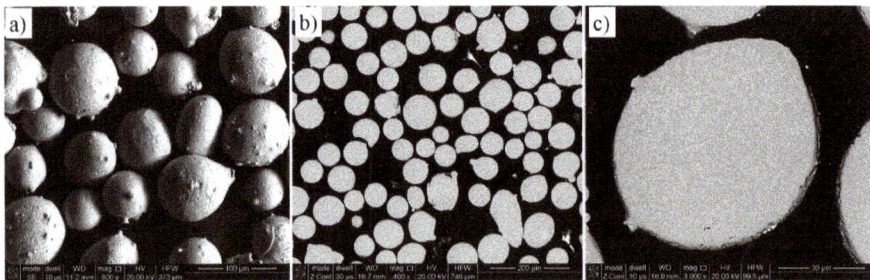

Figure 1. Initial T-800 powder: (**a**) 3D view, (**b**) metallographic cross-section, and (**c**) microstructure of the particle.

Chemical composition measurements by energy dispersive X-ray spectroscopy (EDAX, Mahwah, NJ, USA) confirmed the nominal chemical composition from the producer (Table 1). The particle size distribution analysis showed that about 99% particle sizes was within the range of 44 to 150 μm (Cumulative Volume Fraction in Figure 2). The T-800 powder has high microhardness of 1180 HV0.1.

Table 1. Chemical composition of T-800 powder.

Element Wt.%	Co	Mo	Cr	Si	Fe	Ni
			T-800 Powder			
Spherical particle	Bal.	27	16.9	3.0	0.5	0.8
Nominal (manufacturer)	Bal.	27–30	16.5–18.5	3–3.8	Fe + Ni max 3	

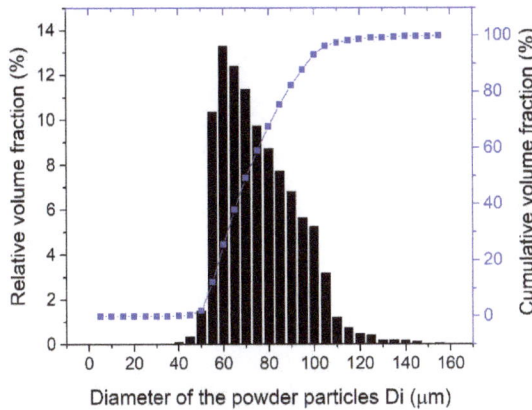

Figure 2. Particle size distribution of the T-800 powder employed in this work.

2.2. Experimental Procedures

Deposits of the T-800 alloy were produced using a Laser Engineered Net Shaping (LENS™) technique. This is one of the large family 3D printing techniques, whose great advantage is a production of components with pre-determined geometries and nearly any shape [22–24].

Currently, LENS is used for manufacturing components made from a wide range of materials, like stainless steels, titanium, nickel, or cobalt-based alloys [25,26]. A control of process parameters such as the powder flow rate, the laser power, and the travel of the working table allows us to obtain a desired microstructure of the components and modification of their surfaces, which increases wear and corrosion resistance [27]. It can also be employed in the regeneration of worn out components [28].

Fabrication of the T-800 coatings was carried out using the LENS 850-R system (Optomec, Albuquerque, NM, USA). It contains a movable table (moving along the X and Y axes), on which a substrate is placed. Inside the work chamber, there is a laser and nozzles supplying the powder under an argon protective atmosphere. Powder is directed into the laser beam focus zone where it is melted and deposited on the substrate. The entire process is computer controlled and results in a controlled thickness of a deposited coating.

The coatings were deposited on the polished and sand blasted RENE 77 alloy substrate with the composition in Table 2. The dimensions of the substrate plates were as follows: 35 mm × 15 mm × 13 mm (length × width × height).

Table 2. Chemical composition of RENE 77 alloy.

Element	Ni	Cr	Ti	Mo	Al	Co
			RENE 77 Alloy			
wt.%	bal.	15.2	3.3	3.8	3.7	16.4

The dimensions of the deposit made by the controlled geometry LENS technique are as follows: width 10 mm, length 10 mm, and height 2.5 mm.

The metallurgical soundness of each deposit was investigated by an optical (Nikon Eclipse MA2000, Amsterdam, The Netherlands) and stero (Nikon SMZ1500, Amsterdam, The Netherlands) microscopes. Both surface and cross-sectional metallographic samples were investigated. Porosity of deposits was assessed by stereological image analysis and expressed as a ratio of the total area of pores to the total area of selected surface, according to the following formula:

$$P = \frac{S_p}{S_{cz}} \times 100\% \tag{1}$$

where: P—porosity (%); Sp—the total pore area (μm^2); Scz—the total area of selected surface (μm^2).

X-ray computed microtomography (μCT) (Nikon/METRIS XT H 225 ST, Leuven, Belgium) was employed on selected deposits to assess the presence of flaws and cracks in the material.

Microstructural observations and chemical analysis (point, linear, and element mapping) were carried out with a scanning electron microsope FEI Quanta 3G FEM Dual Beam (Hillsboro, OR, USA) equipped with an EDS attachment. Phase analysis was carried out in a Rigaku ULTIMA IV diffractometer (Neu-Isenburg, Germany) utilizing CoK$_\alpha$ radiation ranging from 20 to 140° in 2θ with a step size of 0.02° and exposure time of about 3s. An acceleration voltage of 40 kV and 20 mA current were applied.

The Vickers microhardness profile in the substrate-coating was investigated with a microhardness tester Shimadzu HMV (Duisburg, Germany) under a load of 100 G and 10-s dwell time. The indents were made every 100 μm starting from the substrate side.

3. Results and Discussion

3.1. Selection of LENS Parameters

For coating fabrication by the LENS technique, the following process parameters were always fixed: carrier gas flow in a powder supply unit and in a central nozzle was 4 and 25 l/min, respectively. The content of oxygen and water vapor in the working chamber was controlled during the process at the level of 22.7 and 8 ppm, respectively. The thickness of each deposited layer was 0.5 mm. In order to find out the optimal LENS process parameters, a number of trials was carried out where the laser power, powder feeding rate, laser head travel rate, and substrate temperature were varied accordingly. Table 3 shows the working parameters of fabricated coatings. The powder feeding rate was 7.6 g/min for sample 1, 5.2 g/min for sample 6, and 4.6 g/min for other samples. The laser head travel rate ranged from 4 to 16 mm/s. Laser power was within a range of 300 to 500 W even though, for the first two layers in samples 15–18, the laser power was 300 W and the remaining layers were fabricated with the power of 500 W. In addition, the substrate temperature was also regulated. Some coatings were deposited at ambient temperature (samples 1–3) and others after pre-heating the working table to 300 °C. For selected coatings besides pre-heating of the working table, the surface was heated by a laser with a power within a range of 150 to 300 W.

Microscopic observations (not shown) revealed that the coatings fabricated without surface and/or volume substrate pre-heating contained numerous surface and volume cracks and flaws that, most likely, were generated by thermal stresses due to a rapid cooling rate. Those coatings that were fabricated with the substrate volume pre-heating up to 300 °C showed a greatly reduced number of cracks and, in the case of additional point laser heating, the cracks disappeared. Figure 3a,b shows an X-ray computer tomography image for a volume cracked sample and the one which is metallurgically sound, respectively.

Table 3. The set of the LENS process parameters used during deposition of T-800 coatings.

Sample Number	Energy Density * (J/mm^2)	Powder Feeding Rate (g/min)	Substrate Temperature	Remarks
1	25	7.6		
2	33.3	4.6	Ambient temperature	
3	29.2	4.6		
4	29.2	4.6	Pre-heating working table to 100 °C	Numerous surface and volume cracks
5	19.4	4.6	Pre-heating working table to 200 °C	
6	14.6	5.2	Pre-heating working table to 200 °C	
7	50.0	4.6	Pre-heating working table to 200 °C	
8	55.6	4.6	Pre-heating working table to 200 °C	
9	83.3	4.6	Pre-heating working table to 300 °C	
10	83.3	4.6	Pre-heating working table to 300 °C with a laser of 150 W (energy density: 25 J/mm^2)	No surface cracks. Coating dimensions diverge from the assumed geometrical model.
11	83.3	4.6	Pre-heating working table to 300 °C with a laser of 250 W (energy density: 41.7 J/mm^2)	
12	83.3	4.6	Pre-heating working table to 300 °C with a laser of 300 W (energy density: 50 J/mm^2)	
13	83.3	4.6	Pre-heating working table to 300 °C with a laser of 200 W (energy density: 33.3 J/mm^2)	No cracks
14	41.7	4.6	Pre-heating working table to 300 °C with a laser of 200 W (energy density: 16.7 J/mm^2)	Volume cracks
15	Two layers 50, the others 83.3	4.6		Volume cracks and porosity
16 **	Two layers 50, the others 83.3	4.6	Pre-heating working table to 300 °C with a laser of 200 W (final temperature—around 550 °C)	No cracks, porosity
17	Two layers 50, the others 83.3	4.6		Volume cracks
18 **	Two layers 66.7, the others 83.3	4.6		No cracks

* Energy delivery per unit area of material: $E = P/(2r_b V_{beam})$, where P—laser power (W), r_b—beam radius (mm), and V_{beam}—scan speed (mm/s). Applied beam diameter = 1.5 mm. ** The samples were made with the Hatch shrink = 0.3 mm.

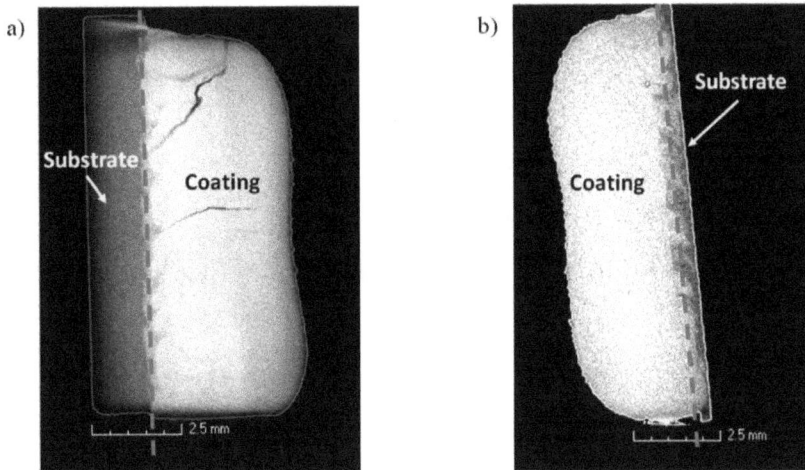

Figure 3. Microfocus X-ray computer tomography image of the LENS-fabricated samples: (**a**) nr 2 and (**b**) nr 18.

Figure 4 shows dependence between substrate temperature and energy density. For samples 1–9, the energy density was calculated using the laser power employed in the depositing process. For samples 10–14, the energy density was calculated using the laser power employed for the substrate heating since it was different than the power for the depositing process. The X marks samples with cracks.

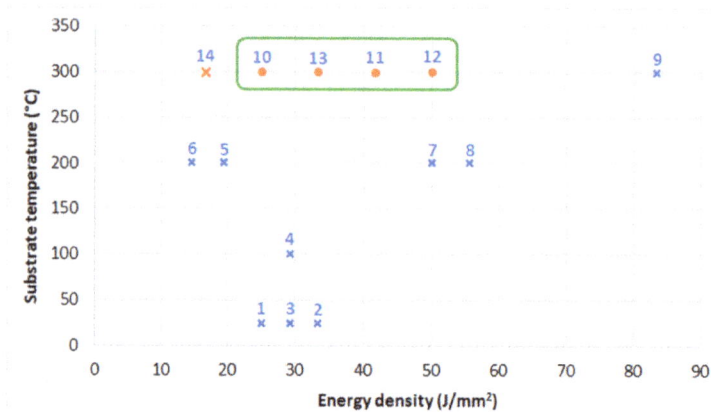

Figure 4. Plot of dependence between substrate temperature and energy density (for samples 1–9, the energy density was calculated using the laser power employed in the depositing process—marked in blue. For samples 10–14, the energy density was calculated using the laser power employed for the substrate heating—marked in orange). The X marks indicate samples with cracks.

Figure 4 shows that it is feasible to obtain metallurgically sound microstructure of Tribaloy T-800 coatings (no cracks) when the substrate is furnace pre-heated to 300 °C with the additional substrate point laser heating if the energy density is within the range of 25 to 50 J/mm². Lower energy density results in coatings' cracking. However, since the geometrical shape of fabricated coatings diverged from the required geometrical model, the coating process was further modified.

The pre-determined geometrical coating dimensions were 10 mm × 10 mm × 2.5 mm with the narrowest dilution zone possible in order to avoid any chemical content changes of the deposit and its properties. Coatings 1–14 diverged greatly from the desired dimensions, which prompted us to change the parameters of fabrication (samples 15–18). The new processes was based on the LENS parameters for sample 13. The laser power was modified such that, when depositing of the first two layers, the power was 300 or 400 W with a further increase to 500 W for the next three to five layers.

Microscopic observations of the cross-sections of samples 13–18 were used to find out the characteristic dimensions of the coatings such as the height, width, dilution zone, and porosity, which are shown in Table 4.

Table 4. Characteristic dimensions and porosity of T-800 coatings deposited with different LENS parameters.

	Coatings T-800			
Sample Number	Height (mm)	Width (mm)	Dilution Zone (mm)	Porosity (%)
13	3.4	11.6	0.88	0.68
14	3.9	10.0	0.23	0.85
15	3.8	10.7	0.07	2.60
16	3.1	10.9	0.12	5.30
17	2.8	11.4	0.25	0.67
18	3.2	10.8	0.10	0.54

The measurements indicate that the characteristic dimensions of Tribaloy T-800 coatings strongly depend on the processing parameters. The coating height nearest the required one (2.5 mm) is exhibited by sample 17, while sample 14 shows depth near the required one (10 mm).

Sample 13 exhibits the largest dilution zone approaching nearly 0.9 mm, while sample 15 has the smallest dilution zone of 0.07 mm, which was fabricated by varying laser power of 300 W for the first two layers and 500 W for the remaining layers. Apparently, using a lower laser power for depositing the first few layers, is beneficial for getting a small dilution zone. Table 4 also includes the results of porosity measurements, which show that samples 15 and 16 has high porosity of 2.6% and 5.3%, respectibvely. The remaining samples have porosity values less than 1%.

Taking into account the results in Tables 3 and 4, the optimal combination of processing parameters, which result in dimensions close to the required ones is met for sample 18.

3.2. Characterization of Obtained Coatings

3.2.1. Microstructure

The microstructure of metallurgically sound sample 18 without cracks and with the lowest porosity (Tables 3 and 4) was investigated by scanning electron microscopy. The SEM images of the areas on the substrate-deposit interface are shown in Figure 5.

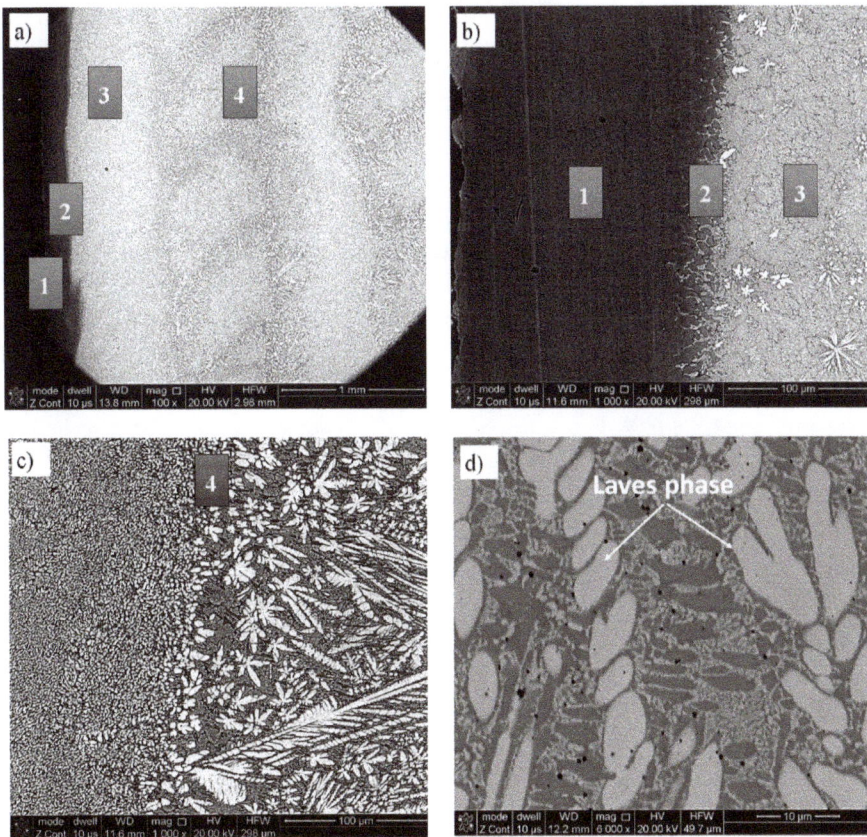

Figure 5. The SEM images of the microstructure of T-800 coating—specimen number 18. Coating/substrate interface (**a,b**) and coatings (**c,d**).

Figure 5a,b show the microstructure of the interface between the deposit and substrate while Figure 5c,d show the microstructure of the deposit bulk.

Figure 5a shows a few areas numbered 1–4. Area #1 is a substrate, area #2 is a deposit melted into the substrate, area #3 is the first deposit layer, and area #4 is the deposit. Higher magnifications of these areas are shown in Figure 5b,c. In addition, in Figure 5c, the microstructure at the cross section of two layers that were deposited transverse to one another, as indicated by the orientation of the bright phase, is observed. Figure 5c,d clearly show a dendritic microstructure of the deposit. Some characteristic phases are observed in Figure 5d at higher magnification.

Elemental distribution maps in Figure 6 show the presence of two phases. Light dendrites are rich in Mo and silicon Si while the interdendritic spaces contain mainly Co and Cr.

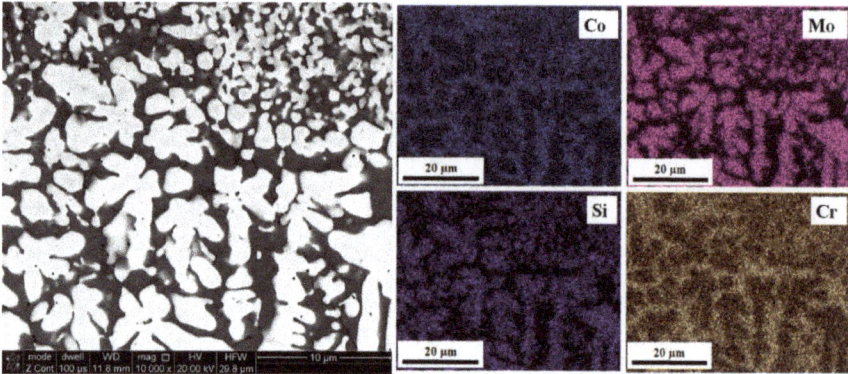

Figure 6. Elemental distribution maps for the T-800 coatings.

An example of the X-ray diffraction pattern in Figure 7 confirms the presence of two phases: Co_3Mo_2Si (#01-082-6068 in DHN PDF 4 database) and a Co-based one (#01-071-4238 in DHN PDF 4 database).

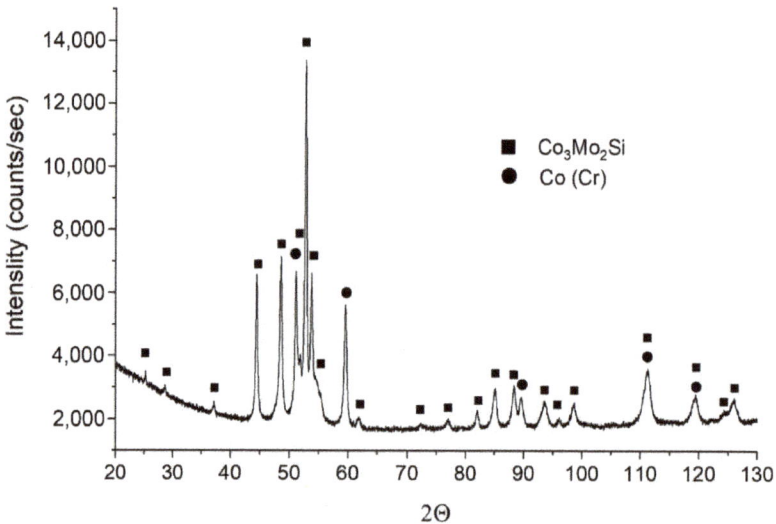

Figure 7. XRD patterns of the T-800 coatings obtained using the LENS technique.

In order to find out a chemical composition profile at the substrate-deposit interface, a linear EDS analysis was carried out. Ideally, the interface should exhibit a continuous compositional change. The EDS analysis started from the substrate toward the deposit (Figure 8a) and the results in Figure 8b indicate a rapid change in the content of Ni, Co, Mo, and Cr. As expected, with increasing distance from the substrate, the content of Ti decreases and traces of Si start appearing. The Ni content initially rapidly drops at the substrate-deposit interface then levels up and, at the distance of 1 mm from the interface, decreases to nearly zero. Within a 1-mm distance from the interface, the dominant elements are Co, Mo, Cr, i, and Si, which correspond well to the original composition of the initial powder.

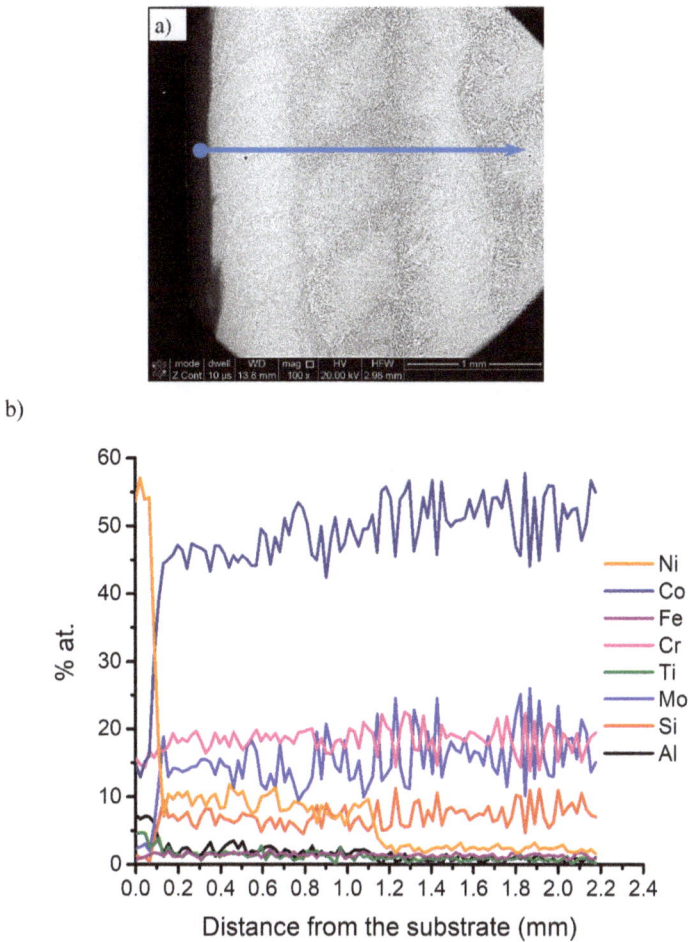

Figure 8. The results of the EDS linear chemical composition analysis of the T-800 coatings specimen 18: (**a**) a macrograph of the deposit with the arrow indicating the direction of linear EDS analysis. (**b**) The chemical content of the elements as a function of the distance from the substrate.

3.2.2. Microhardness

As shown in Figure 9a, the microhardness values show a drop (Figure 9a) at a distance of about 750 μm from the substrate, which, as shown in Figure 9b, is due to a local increase in the Ni content at the same distance. Figure 9b also shows that, at a distance of about 1 mm from the substrate, the Ni content drops to 1–2 at.% while microhardness increases to the 800–900 HV0.1 range and then stabilizes approximately at the same level.

a)

b)

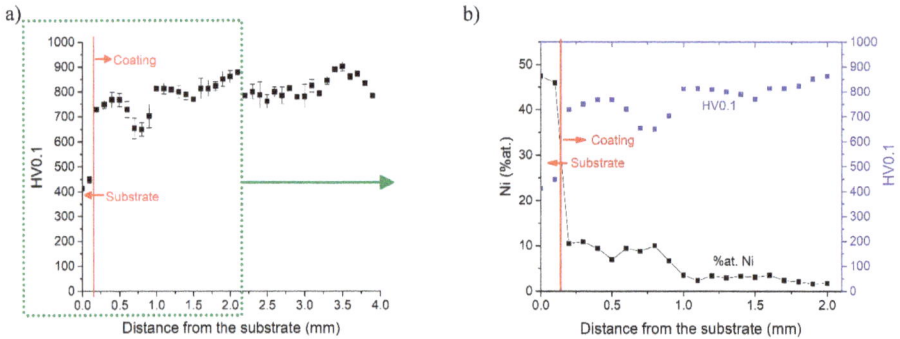

Figure 9. (a) Microhardness distribution of the T-800 coatings deposited on RENE 77 using the LENS process. **(b)** Microhardness distribution and Ni content in sample 18.

The average microhardness of the RENE 77 substrate is about 450 HV0.1 while the microhardness of the deposited Tribaloy T-800 coating is nearly two-fold higher. However, the microhardness of the Tribaloy T-800 coating is nearly 300 HV0.1 lower than that of the initial Tribaloy T-800 powder (1180 HV0.1). The difference in hardness between the T-800 powder and the coating made of this powder is due to the cooling rate used during the manufacturing process. The cooling rate during powder atomization is much higher than the cooling rate used in the LENS process (for example: cooling rate of argon-gas atomized nickel—10^5–10^6 K/s, LENS process—10^3 K/s).

3.3. Discussion

The present work clearly shows that laser techniques and, in particular, the LENS technique, are capable of fabricating the Tribaloy T-800 coatings on the RENE 77 substrate. Testing of various fabrication process parameters has clearly shown that, in order to obtain a metallurgically sound coating, without cracking, a substrate temperature must be judiciously adjusted. In those variants of the process without substarte pre-heating, various surface and volume cracks and flaws were always formed in the deposits, as reported by Diaz [19]. Furthermore, Przybyłowicz i Kusinski [29] recommended heating a Ni-based superalloy substrate (IN718) to 500 °C in order to eliminate cracking in the Tribaloy T-400 coating. In the present work for the T-800—RENE 77 materials, a sole substrate volume pre-heating was not completely effective since cracks still appeared in the sample. A better solution was a combination of substrate volume pre-heating with a point laser heating having energy density within a range of 25 to 50 J/mm². Therefore, by heating the substrate to a temperature of about 550 °C (by heating the working table and laser preheating) and by changing energy density during the fabrication of the coating, we achieve a coating with a good geometrical quality, which exhibits a minimal dilution zone with the substrate and is devoid of cracks.

In this work, we eliminated cracks by pre-heating the substrate. Other solutions are known in the literature to help avoid cracks, such as the ones using intermediate layers [30].

A great advantage of the LENS technique is the possibility of depositing coatings with pre-determined dimensions. Fabrication processes have shown that, through an adjustment of the process parameters, it is feasible to fabricate a metallurgically sound coating, without porosity/cracking and with good adhesion to the substrate with a pre-determined geometry.

A Tribaloy T-800 coating fabricated in the present work is characterized metallurgically by the microstructure typical for the commercial Tribaloy alloys, in which two phases are present. The first phase is an intermetallic Laves phase, which contains predominantly Co, Mo i Si (Co_3Mo_2Si), embedded in the Co solid solution matrix. Navas et al. also reported a similar microstructure [18].

The results of the measurements of linear chemical composition and microhardness changes in the deposited sample show a fluctuation of elemental distribution of alloying elements, in particular

Ni, from the substrate to the deposit. The effect of the Ni distribution content on microhardness is still observed up to about 1 mm distance from the substrate.

The Tribaloy T-800 coatings on the RENE 77 substrate increase microhardness of the working surface by about 400 HV with respect to the microhardness of the RENE 77 substrate.

4. Conclusions

The research carried out in this work shows that it is feasible to fabricate the metallurgically sound Tribaloy T-800 coatings, devoid of cracks and flaws, on the RENE 77 substrate using the LENS technique. It has been found that the technological coating process parameters greatly influence the metallurgical soundness of coatings. A principal process parameter influencing quality is the substrate temperature. Depositing the Tribaloy T-800 coatings using the LENS technique should be carried out on the pre-heated surface, which greatly reduces thermal stresses and prevents cracking and flaw formation within the cross section of the deposit. Microscopic observations confirm a formation of the dual phase microstructure, which is typical for the commercial Tribaloy alloys. These alloys consist of Laves phase dendrites embedded in the Co solid solution matrix. Linear chemical analysis combined with microhardness measurements indicate chemical composition fluctuations, predominantly Ni, at the interface substrate-coating. Microhardness increases nearly two-fold toward the deposit.

Author Contributions: Conceptualization, T.D. and T.C. Investigation, T.D., M.Ł., J.D.-W., and S.L. Supervision, T.C. Visualization, M.Ł. and S.L. Writing—original draft, J.D.-W. Writing—Review & editing, R.A.V. and T.C. All authors have read and approved the content of the manuscript.

Funding: The Ministry of National Defense Republic of Poland Program-Research Grant MUT Project 13-995 supported this work.

Conflicts of Interest: The authors declare no conflict of interest.

References

1. Zielinska, M.; Sieniawski, J. Surface modification and its influence on the microstructure and creep resistance of nickel based superalloy René 77. *Arch. Metall. Mater.* **2013**, *58*, 95–98. [CrossRef]
2. Rottwinkel, B.; Nölke, C.; Kaierle, S.; Wesling, V. Crack Repair of Single Crystal Turbine Blades Using Laser Cladding Technology. *Procedia CIRP* **2014**, *22*, 263–267. [CrossRef]
3. Wang, H.S.; Huang, C.Y.; Ho, K.S.; Deng, S.J. Microstructure evolution of laser repair welded René 77 nickel-based superalloy cast. *Mater. Trans.* **2011**, *52*, 2197–2204. [CrossRef]
4. Nsoesie, S.; Liu, R.; Jiang, K.; Liang, M. High-temperature hardness and wear resistance of Cobalt-based Tribaloy alloys. *Inter. J. Mater. Mech. Eng.* **2013**, *2*, 48–56.
5. Alloy Families. Available online: https://www.deloro.com/alloy-families (accessed on 1 February 2019).
6. Kathuria, Y.P. Some aspects of laser surface cladding in the turbine industry. *Surf. Coat. Technol.* **2000**, *132*, 262–269. [CrossRef]
7. Brochure. Wear Solutions Components. KENNAMETAL STELLITE. Available online: https://www.kennametal.com (accessed on 1 February 2019).
8. Yao, M.X.; Wu, J.B.C.; Yick, S.; Xie, Y.; Liu, R. High temperature wear and corrosion resistance of a Laves phase strengthened Co–Mo–Cr–Si alloy. *Mater. Sci. Eng. A* **2006**, *435–436*, 78–83. [CrossRef]
9. do Nascimento, E.M.; do Amaral, L.M.; D'Oliveira, A.S.C.M. Characterization and wear of oxides formed on CoCrMoSi alloy coatings. *Surf. Coat. Technol.* **2017**, *332*, 408–413. [CrossRef]
10. Jiang, K.; Liu, R.; Chen, K.; Liang, M. Microstructure and tribological properties of solution-treated Tribaloy alloy. *Wear* **2013**, *307*, 22–27. [CrossRef]
11. Liu, R.; Xu, W.; Yao, M.X.; Patnaik, P.C.; Wu, X.J. A newly developed Tribaloy alloy with increased ductility. *Scr. Mater.* **2005**, *53*, 1351–1355. [CrossRef]
12. Halstead, A.; Rawlings, R.D. The fracture behaviour of two Co-Mo-Cr-Si wear resistant alloys ("Tribaloys"). *J. Mate. Sci.* **1985**, *20*, 1248–1256. [CrossRef]
13. Xu, W.; Liu, R.; Patnaik, P.C.; Yao, M.X.; Wu, X.J. Mechanical and tribological properties of newly developed Tribaloy alloys. *Mater. Sci. Eng. A* **2007**, *452–453*, 427–436. [CrossRef]

14. Sahraoui, T.; Feraoun, H.I.; Fenineche, N.; Montavon, G.; Aourag, H.; Coddet, C. HVOF-sprayed Tribaloy©-400: Microstructure and first principle calculations. *Mater. Lett.* **2004**, *58*, 2433–2436. [CrossRef]

15. Bolelli, G.; Cannillo, V.; Lusvarghi, L.; Montorsi, M.; Mantini, F.P.; Barletta, M. Microstructural and tribological comparison of HVOF-sprayed and post-treated M–Mo–Cr–Si (M = Co, Ni) alloy coatings. *Wear* **2007**, *263*, 1397–1416. [CrossRef]

16. Rakhes, M.; Koroleva, E.; Liu, Z. Improvement of corrosion performance of HVOF MMC coatings by laser surface treatment. *Surf. Eng.* **2011**, *27*, 729–733. [CrossRef]

17. Tobar, M.J.; Amado, J.M.; Álvarez, C.; García, A.; Varela, A.; Yáñez, A. Characteristics of Tribaloy T-800 and T-900 coatings on steel substrates by laser cladding. *Surface and Coatings Technology* **2008**, *202*, 2297–2301. [CrossRef]

18. Navas, C.; Cadenas, M.; Cuetos, J.M.; de Damborenea, J. Microstructure and sliding wear behaviour of Tribaloy T-800 coatings deposited by laser cladding. *Wear* **2006**, *260*, 838–846. [CrossRef]

19. Díaz, E.; Amado, J.M.; Montero, J.; Tobar, M.J.; Yáñez, A. Comparative Study of Co-based Alloys in Repairing Low Cr-Mo steel Components by Laser Cladding. *Phys. Procedia* **2012**, *39*, 368–375. [CrossRef]

20. Lin, W.C.; Chen, C. Characteristics of thin surface layers of cobalt-based alloys deposited by laser cladding. *Surf. Coat. Tech.* **2006**, *200*, 4557–4563. [CrossRef]

21. Yuduo, Z.; Zhigang, Y.; Chi, Z.; Hao, L. Effect of Rhenium Addition on Isothermal Oxidation Behavior of Tribaloy T-800 Alloy. *Chin. J. Aeronaut.* **2010**, *23*, 370–376. [CrossRef]

22. Durejko, T.; Ziętala, M.; Polkowski, W.; Czujko, T. Thin wall tubes with $Fe3Al/SS316L$ graded structure obtained by using laser engineered net shaping technology. *Mater. Des.* **2014**, *63*, 766–774. [CrossRef]

23. Durejko, T.; Ziętala, M.; Łazińska, M.; Lipiński, S.; Polkowski, W.; Czujko, T.; Varin, R.A. Structure and properties of the Fe_3Al-type intermetallic alloy fabricated by laser engineered net shaping (LENS). *Mater. Sci. Eng. A* **2016**, *650*, 374–381. [CrossRef]

24. Karczewski, K.; Dąbrowska, M.; Ziętala, M.; Polański, M. Fe-Al thin walls manufactured by Laser Engineered Net Shaping. *J. Alloy. Compd.* **2017**, *696*, 1105–1112. [CrossRef]

25. Fu, J.W.; Yang, Y.S.; Guo, J.J.; Tong, W.H. Effect of cooling rate on solidification microstructures in AISI 304 stainless steel. *Mater. Sci. Tech.* **2013**, *24*, 941–944. [CrossRef]

26. Dittrick, S.; Balla, V.K.; Davies, N.M.; Bose, S.; Bandyopadhyay, A. In vitro wear rate and Co ion release of compositionally and structurally graded CoCrMo-Ti6Al4V structures. *Mater. Sci. Eng. C* **2011**, *31*, 809–814. [CrossRef]

27. Das, M.; Bysakh, S.; Basu, D.; Sampath Kumar, T.S.; Balla, V.K.; Bose, S.; Bandyopadhyay, A. Microstructure, mechanical and wear properties of laser processed SiC particle reinforced coatings on titanium. *Surf. Coat. Tech.* **2011**, *205*, 4366–4373. [CrossRef]

28. Hedges, M.; Calder, N. Rapid Manufacture of Perfomance Materials via LENS, In Cost Effective Manufacture via Net-Shape Processing. *Meeting Proceedings RTO-MP-AVT-139*. Available online: https://www.researchga te.net/ (accessed on 26 May 2006).

29. Przybylowicz, J.; Kusinski, J. Laser cladding and erosive wear of Co–Mo–Cr–Si coatings. *Surf. Coat. Tech.* **2000**, *125*, 13–18. [CrossRef]

30. Sahasrabudhe, H.; Harrison, R.; Carpenter, C.; Bandyopadhyay, A. Stainless steel to titanium bimetallic structure using LENS™. *Addit. Manuf.* **2015**, *5*, 1–8. [CrossRef]

materials

MDPI

Article

In Situ Study on Fracture Behavior of Z-Pinned Carbon Fiber-Reinforced Aluminum Matrix Composite via Scanning Electron Microscope (SEM)

Yunhe Zhang [1,*], Sian Wang [1], Xiwang Zhao [1], Fanming Wang [1] and Gaohui Wu [2]

[1] College of Mechanical and Electrical Engineering, Northeast Forestry University, P.O. Box 310, Hexing Road No.26, Harbin 150040, China; daowsa@nefu.edu.cn (S.W.); zhaoxiwang1216@126.com (X.Z.); 18512490598@163.com (F.W.)

[2] School of Materials Science and Engineering, Harbin Institute of Technology, Harbin 150080, China; wugh@hit.edu.cn

* Correspondence: yunhe.zhang@nefu.edu.cn; Tel.: +86-451-8219-2843 or +86-156-6352-6798; Fax: +86-451-8219-2843

Received: 30 April 2019; Accepted: 31 May 2019; Published: 17 June 2019

Abstract: Inside a scanning electron microscope (SEM) chamber, we performed an in situ interlaminar shear test on a z-pinned carbon fiber-reinforced aluminum matrix composite (Cf/Al) fabricated by the pressure the infiltration method to understand its failure mechanism. Experiments show that introducing a stainless-steel z-pin increases the interlaminar shear strength of Cf/Al composite by 148%. The increase in interlaminar shear strength is attributed to the high strength of the stainless-steel z-pin and the strong bonding between the z-pin and the matrix. When the z-pin/matrix interface failed, the z-pin can still experience large shear deformation, thereby enhancing delamination resistance. The failure mechanism of composite includes interfacial debonding, aluminum plough, z-pin shear deformation, frictional sliding, and fracture. These results in this study will help us understand the interlaminar strengthening mechanism of z-pins in the delamination of metal matrix composites.

Keywords: electron microscopy (in situ SEM); delamination; metal matrix composites (MMCs); z-pinning

1. Introduction

Carbon fiber-reinforced aluminum matrix composites (Cf/Al) have drawn great attention in automotive and aerospace applications due to their high specific strength, high specific modulus, high thermal conductivity, and good fatigue properties [1–3]. However, a major concern with the application of Cf/Al laminated composites is the delamination that results from through-thickness stresses or impact loads, which lead to the reduction of the in-plane mechanical properties and even structural instability of the component [4,5]. Z-pinning is an effective technique that enhances delamination resistance and interlaminar strength of laminates by inserting metallic or fibrous rods (i.e., z-pins) in the through-thickness direction of laminated composites during their fabrication process [6]. A great number of studies have shown that delamination fracture toughness, interlaminar shear strength, and damage tolerance of composite laminates can increase significantly with z-pins [7–9].

Zhang et al. studied the influence of z-pinning on the interlaminar mechanical properties of Cf/Al composites fabricated by the pressure infiltration method [10]. They found that the stainless-steel z-pin creates a strong bond with the composite due to an interface reaction between the z-pin and the aluminum matrix, improving the interlaminar shear strength of the composite by at least 70%. Although their study determined the capacity of the z-pin for improving delamination resistance, it did not give a clear understanding of the failure mechanism in z-pinned Cf/Al composites. Thus, before using z-pinned Cf/Al composites in aerospace structures, it is necessary to understand their mechanical performance and strengthening mechanism under interlaminar shear load.

Researchers have performed several experimental studies to investigate the interlaminar strengthening mechanisms of z-pinned carbon fiber-reinforced polymer matrix composites, such as double cantilever beam (DCB) fracture tests, end notch flexure (ENF) fracture tests, z-pin pull-out tests, and interlaminar shear tests [4,11,12]. They have effectively exploited the in situ scanning electron microscope (SEM) technique to observe the fracture process of composites, since it can help us observe the fracture behavior of other types of composite structures in real time [13–15]. Nevertheless, the SEM technique has not hitherto been used to investigate the interlaminar fracture process of z-pinned composite structures.

This study builds on the recent work by Zhang et al. [10], which investigated the effect of z-pinning on microstructure and interlaminar shear strength of Cf/Al composites. In the current study, unpinned and z-pinned Cf/Al composites were fabricated by pressure infiltration method and double-notched interlaminar shear tests were applied to the unpinned and z-pinned Cf/Al specimens. Using the in situ SEM technique, the different interlaminar fracture behaviors of the z-pinned Cf/Al specimen were observed, such as the initiation and growth of crack, interfacial debonding, and the deformation of the z-pin and the matrix. Finally, the impact of z-pinning on fracture processes of Cf/Al composites were discussed.

2. Experimental

We used M40 carbon fibers (purchased from Toray Industries Inc., Tokyo, Japan) and AISI 321 stainless steel (purchased from Shanghai Baosteel Group Corporation, Shanghai, China) to reinforce 5A06 Al matrix (purchased from Northern Light Alloy Company Ltd., Harbin, China). Table 1 outlines the mechanical properties of 5A06 Al, M40 carbon fiber, and AISI 321. Tables 2 and 3 list the chemical compositions of AISI 321 and 5A06 Al, respectively.

Table 1. Basic properties of carbon fibers, z-pin and 5A06 Al alloy.

Materials	Tensile Strength (MPa)	Elastic Modulus (GPa)	Elongation to Fracture (%)	Density (g/cm³)
M40	4410	377.0	1.2	1.76
AISI 321	1905	198.0	2.0	7.85
5A06 Al	314	66.7	16.0	2.64

Table 2. Chemical composition of 5A06 Al alloy (wt%).

Material	Mg	Mn	Si	Fe	Zn	Cu	Ti	Al
5A06 Al	5.8–6.8	0.5–0.8	0.4	0.4	0.2	0.1	0.02–0.1	Bal.

Table 3. Chemical composition of AISI 321 (wt%).

Material	Cr	Ni	Ti	Mn	Si	C	S	P	Fe
AISI 321	17–19	8–11	0.5–0.8	<2.0	<1.0	<0.12	<0.03	<0.035	Bal.

We used the pressure infiltration method to fabricate the z-pinned and unpinned Cf/Al composites. The carbon fiber bundles were unidirectionally winded by a CNC winding machine to obtain the preform of carbon fibers, and an AISI 321 z-pin with a diameter of 0.6 mm was inserted into the preform by an ultrasonic tool. The preform of carbon fibers with the z-pin was preheated at $500 \pm 10\,°C$. The melted 5A06 Al alloy was infiltrated into the preforms under pressure. During the infiltration process, a pressure of 0.5 MPa was applied and maintained for 2 h, and then the z-pinned Cf/Al composites were solidified in air. An unpinned Cf/Al composite was fabricated in the same route as the reference material. The composites had a carbon fiber volume fraction of about 55%.

The double-notched interlaminar shear specimens were machined by electric discharge using an electric discharge machining method. The distance between the two notches was 6 mm, and the depth of each notch was half of the specimen thickness (i.e., 2.5 mm) to make sure that the plane between the ends of the two notches was subjected to pure interlaminar shear loads. The single z-pin used to reinforce the specimen was equidistant from the two notches. Since the SEM technique can only observe the surface of the specimens, we cannot directly capture the deformation of the z-pin, crack initiation, and propagation of interfacial zone during in situ SEM observations. Hence, to obtain this information, we cut the specimen for in situ observation by an electric discharge along the central axis of the stainless-steel z-pin and half of the z-pin was still bonded with the specimen. It is worth noting that the cutting process does not induce the initial delamination. Figure 1a shows the dimensions of the specimen for in situ observation with the location of half z-pin. Figure 1b shows the three-dimensional model of the specimen for in situ observation with the fixture.

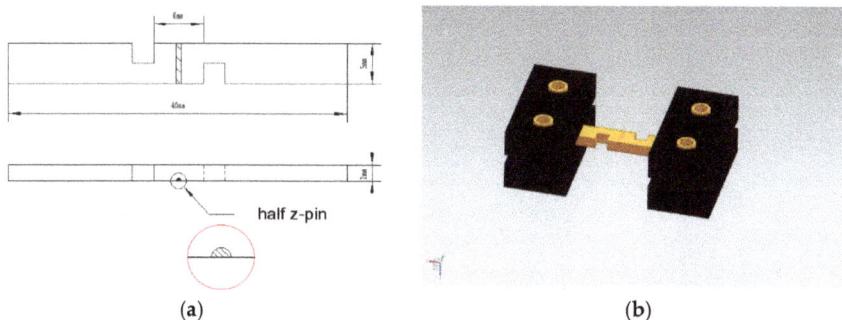

half z-pin

(a)   (b)

Figure 1. (**a**) The dimensions of the specimen for in situ observation. (**b**) The three-dimensional model of the specimen for in situ observation with the fixture.

We conducted an in situ double notch interlaminar shear test in an S-4700 SEM equipped with a tensile stage (SEM, Royal Dutch Philips Electronics Ltd., Amsterdam, The Netherlands), which can be used to apply interlaminar shear loads. The test was performed using a cross-head speed of 0.5 mm/min at room temperature according to the ASTM D3846-2008. During the tests, we gathered the computer-generated load-displacement curves. The loading can be paused at any time to allow in situ observation. We carried out in situ SEM observations mainly on the notch and the interfacial zone between the z-pin and the aluminum matrix where the initial damage can be expected due to stress concentration.

3. Results and Discussion

Figure 2 shows the typical tensile shear curve for Cf/Al composites reinforced with half of the stainless-steel z-pin with a diameter of 0.6 mm. It also shows the representative curve for the unpinned Cf/Al composites for comparison. Table 4 summarizes the interlaminar shear strengths for the z-pinned specimen with a half z-pin for in situ observation, the z-pinned specimen with the entire z-pin, and the control specimen. The resulting numbers indicate that the interlaminar shear strength of the z-pinned Cf/Al composite is higher than the Cf/Al composite without a z-pin reinforcement by 148%. While the measured interlaminar shear strength of Cf/Al composites reinforced with half of the stainless-steel z-pin is slightly lower than that of Cf/Al composites reinforced with the entire stainless-steel z-pin in the previous work, the tensile shear curve for Cf/Al composites reinforced with half of the stainless-steel z-pin is consistent with that of Cf/Al composites reinforced with the entire stainless-steel z-pin with a diameter of 0.6 mm [10]. This shows that the cutting of the specimen and the z-pin does not change the fracture behavior of z-pinned Cf/Al composites. Hence, direct observation of a half z-pin-reinforced specimen via in situ tensile tests is an effective method for observing the fracture process of z-pinned Cf/Al composites.

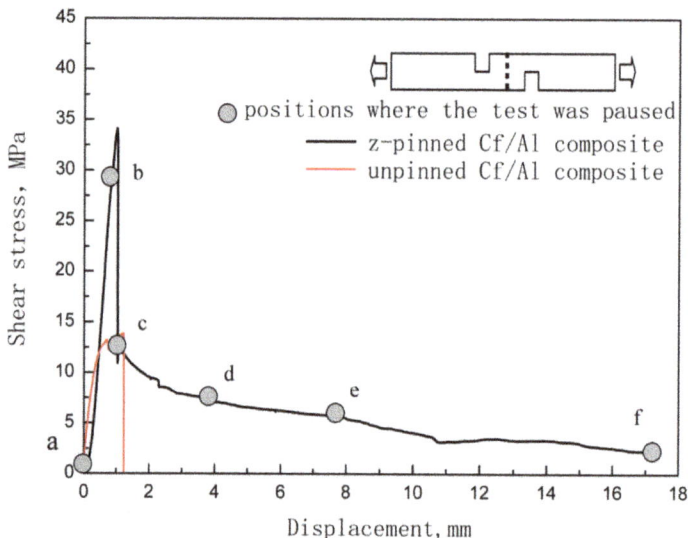

Figure 2. Typical tensile shear curves of carbon fiber-reinforced aluminum matrix composites (Cf/Al) with and without a z-pin.

Table 4. Interlaminar shear strengths of controls and z-pinned specimens.

Specimen	Interlaminar Shear Strength	Standard Deviation
control specimen	13.6 MPa	0.78
z-pinned specimen with half z-pin	33.7 MPa *	1.94
z-pinned specimen with entire z-pin [10]	38.4 MPa	1.96

*: The experimental values were 31.2, 34.1, and 38.4 MPa, respectively. The value of 34.1 MPa is discussed in detail in the present research.

Figure 3 shows the SEM images of the specimen prior to the double-notch tensile test corresponding to point A in the curve. As shown in Figure 3a, the z-pin is almost perpendicular to the loading direction and carbon fiber (it is impossible to obtain a specimen with a perfectly perpendicular z-pin). As Figure 3b,c indicate, the z-pinned Cf/Al specimen shows the extent of the interface reaction layer at z-pin/matrix interface, which is about 10 µm in thickness. The main component of the interface reaction layer was confirmed to be $FeAl_3$ by energy disperse spectroscopy in previous research [10]. This suggests that the z-pins hold a strong bond to the matrix. Moreover, no cracks are found in the matrix and the z-pin/matrix interface layer, as shown in Figure 3d.

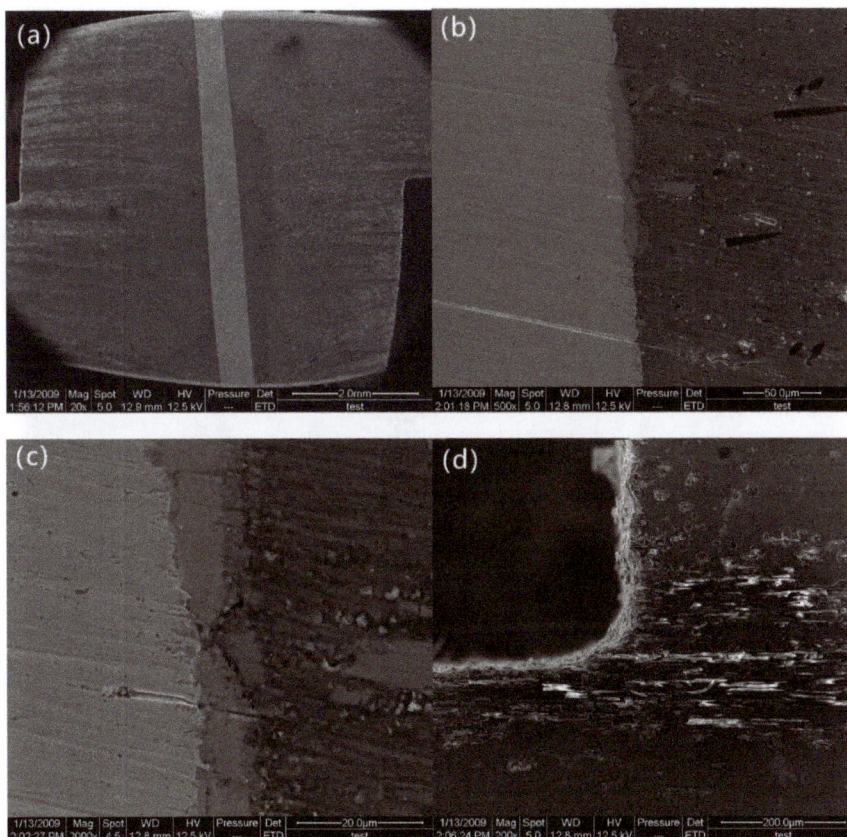

Figure 3. Morphology of the specimen prior to test: (**a**) the z-pin, (**b**) the z-pin/matrix interface in low magnification, (**c**) the z-pin/matrix interface in high magnification, and (**d**) the notch.

Initially, the z-pinned composite exhibits an initial elastic region until it reaches maximum load. As the stress builds up to 28.3 MPa (marked as "b" in Figure 2), a clear deformation of the z-pin is not observed, which suggests that the z-pin only experiences elastic deformation (Figure 4a). Micro-cracks are observed in the z-pin/matrix interface at the mid-plane and they are located in the interfacial layer bonded with the stainless-steel z-pin (Figure 4b,c). However, the interface is not cracked and a good interfacial bonding also remains. As shown in Figure 4d, the specimen is not damaged, which reveals that delamination did not initiate. Therefore, it can be concluded that in spite of the presence of micro-cracks, the interlaminar shear stress is effectively transferred from the matrix to the z-pin through the z-pin/matrix interface. The stainless steel can carry the applied load due to its high strength and stiffness, resulting in the improved interlaminar shear strength. Hence, the interlaminar shear strength of the Cf/Al composite is enhanced owing to the high shear strength and stiffness of the introduced stainless-steel z-pin and the strong interfacial bonding between the z-pin and the matrix.

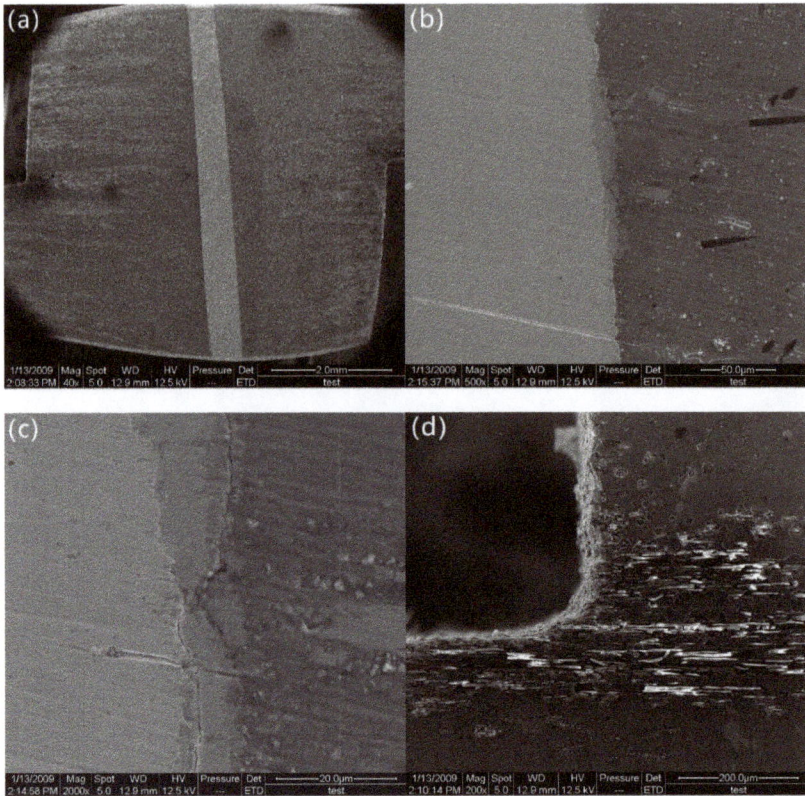

Figure 4. Morphology of specimen in later elastic deformation stage: (**a**) the z-pin, (**b**) micro-cracks in z-pin/matrix interface in low magnification, (**c**) micro-cracks in z-pin/matrix interface in high magnification, and (**d**) the notch.

After reaching the maximum load (34.1 MPa), the stress rapidly decreases until it reaches 11.5 MPa. As the stress decreases to 11.5 MPa (marked as "c" in Figure 2) micro-cracks propagate along the pin/matrix interface, which result in partial debonding at the midplane (Figure 5a,b). As a result, a partial interface cannot be effectively transferred from the matrix to the stainless-steel z-pin. At this point, the carbon fiber/matrix interface becomes the carrier that bears the applied interlaminar shear load. However, the carbon fiber/matrix interface near the notches that are in the stress concentration area cannot sustain the applied load due to low carbon fiber/matrix interfacial strength, which leads to debonding initiation of the carbon fiber/aluminum matrix. For this reason, delamination initiates at the plane between the ends of the two notches (Figure 5c,d) but still does not propagate throughout the entire plane (Figure 5e,f). The shear stress exhibits a load drop in the tensile shear curve as a result of delamination initiation. As shown in Figure 5e, the z-pin experiences permanent S-shaped shear deformation due to shear sliding displacement of the composite laminate, which in turn resists further delamination of the composite [16]. With the gradual rotation of the deformed z-pin axis, the partial interlaminar shear stress applied to the z-pin is decomposed into tensile stress along the z-pin direction. The z-pin loading mode transitions from shear to a combination of shear and tension. This allows the deformed z-pin to withstand relatively high stresses due to its high tensile strength. In addition, the z-pin/matrix interface near the z-pin ends still creates strong bonds, which suggests that the interface still retains a certain capacity to transfer the load. Since the z-pin/matrix interface is not debonded and the z-pin is not damaged, the specimen still has a relatively high carrying capacity (11.5 MPa), which is close to the maximum interlaminar shear stress value of the unpinned Cf/Al composite during the entire test (13.8 MPa). Hence, it can be also summarized that the strong bonding between the z-pin and the matrix plays a significant role in determining the interlaminar shear strength.

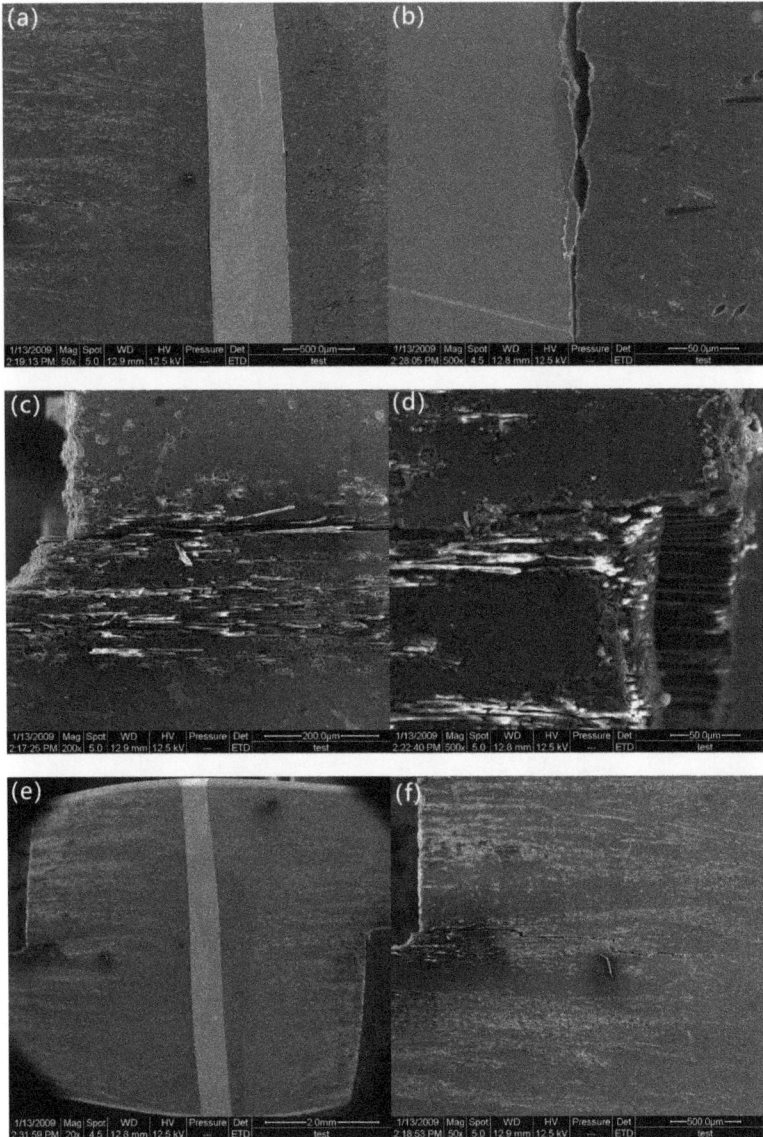

Figure 5. Morphology of the specimen after delamination initiates: (**a**) partial debonding at low magnification, (**b**) partial debonding at high magnification, (**c**) the notch at low magnification, (**d**) the notch at high magnification, (**e**) delamination crack growth at low magnification, and (**f**) delamination crack growth at high magnification.

After the load drops to 11.5 MPa, it begins to decrease slowly. At a stress of 7.6 MPa (marked as "d" in Figure 2), the delamination crack develops from the notches to the z-pin, where delamination propagation is suppressed due to the presence of the z-pin (Figure 6a,b). In addition, the z-pin is severely deformed and pressed laterally into the specimen matrix due to further interlaminar shear displacement of the composite. For this reason, the composite matrix adjacent to the stainless-steel z-pin is severely deformed (Figure 6c,d). At a stress of 6 MPa (marked as "e" in Figure 2), the segment of the z-pin below the delamination surface is pulled slightly outward from the aluminum matrix under interlaminar shear loading. The other segment of the z-pin above the delamination surface apparently slides along the interface due to the pull-out of the z-pin's lower part, as shown in Figure 7a. All these observations indicate that the z-pin/matrix interface has completely failed. Nevertheless, it is hard to know the precise moment at which the failure occurred. This is because even if the interface is completely debonded, the z-pin and the matrix will still be close together due to the compression force applied by the matrix before the z-pin is pulled out. The friction between the sliding z-pin and the surrounding matrix during the pull-out process contributes to the interlaminar strengthening of z-pinned composites [17]. In addition, as seen in Figure 7b, the aluminum matrix close to the delamination surface is significantly ploughed out due to the further lateral deflection of the z-pin, which significantly enhances the interface friction between the z-pin and aluminum matrix and thereby improves the delamination resistance at this stage. The strengthening process that involves lateral compression deformation and the plough of the matrix induced by the z-pin is called the snubbing effect [18,19]. Hence, the interlaminar shear strength of the z-pinned Cf/Al composite during this stage is attributed to the snubbing effect and lateral deflection of the z-pin.

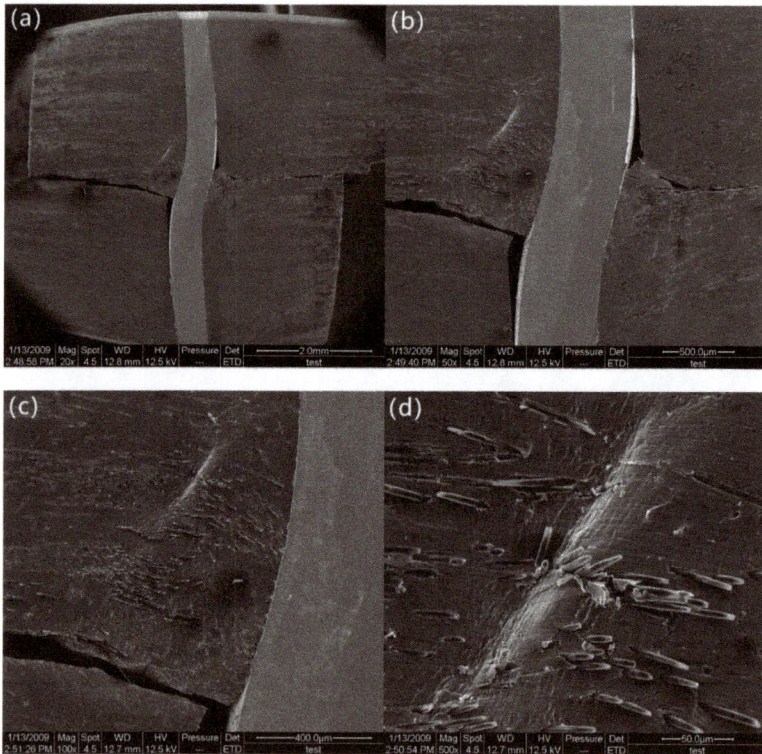

Figure 6. Morphology of the specimen when the matrix is deformed: (**a**) deformed z-pin at low magnification, (**b**) deformed z-pin at high magnification, (**c**) deformed matrix at low magnification, and (**d**) deformed matrix at high magnification.

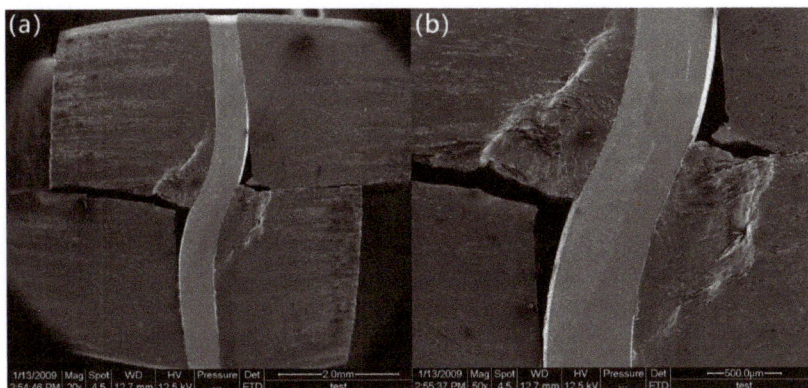

Figure 7. Morphology of the specimen when z-pin/matrix interface is fully debonded: (**a**) the z-pin sliding and pulling out, (**b**) the ploughed matrix.

Finally, at a stress level of 2.3 MPa, the z-pin fails by the outward pull-out, resulting in the final failure of the z-pinned Cf/Al specimen, as shown in Figure 8. It must be noted that if the specimen had not been cut along the central axis of the stainless-steel z-pin and the entire z-pin had still been surrounded by the aluminum matrix, the outward pull-out of the z-pin would have been restrained by the surrounding aluminum matrix. Therefore, the z-pin would have been intended to rupture with minimal pull-out rather than the outward pull-out reported in this study [10].

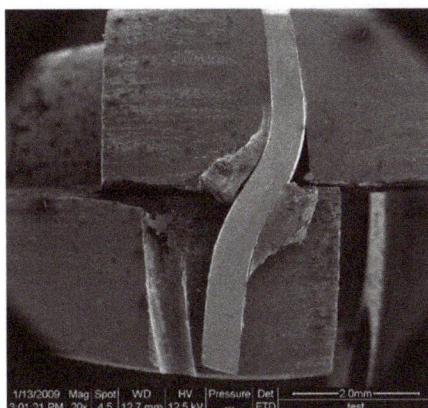

Figure 8. Morphology of the failed specimen for in situ observation.

Previous studies on the mechanical properties of Cf/Al composites have revealed that the interlaminar shear strength of the unpinned Cf/Al composites depends on the interface strength between carbon fiber and the aluminum matrix [20]. Since the carbon fiber/matrix interface strength is poor, low interlaminar shear load causes debonding in the carbon fiber/aluminum interface, resulting in delamination initiation. With the increase in displacement, the delamination gradually develops along the debonded carbon fiber/aluminum interface, leading to the ultimate failure of the Cf/Al specimen. Thus, an unpinned Cf/Al composite shows low interlaminar shear strength.

In the present study, interlaminar strengthening mechanism of z-pinning in the delamination of metal matrix composites is determined by an in situ double-notched interlaminar shear test. Figure 9 shows the schematic of failure process of the z-pinned composites. The failure process of a z-pinned composite subjected to interlaminar shear load is defined in four stages. Initially, micro-cracks are observed in the z-pin/matrix interface due to stress concentration, which has a negligible effect on load transfer through the interface. Thus, the z-pin can carry the main load because of the strong interface bonding and the high strength of the z-pin itself, which improves the interlaminar shear strength of the z-pinned Cf/Al specimen, as shown in Figure 9a. Secondly, the z-pin/matrix interface is partially debonded due to the gradual development of the interface crack. Thus, delamination initiates near the notches and the load drops sharply until its value is close to the interlaminar shear strength of the unpinned Cf/Al composites, as shown in Figure 9b. Then, with the increase in displacement, the z-pin/matrix interface is fully debonded, and the interlaminar cracks extend to the vicinity of the z-pin. Following the interface debonding, the z-pin experiences a small amount of frictional slide along the interface and an irreversible shear deformation. The aluminum matrix is deformed and ploughed out due to lateral deflection of the z-pin, which results in an enhanced frictional region near the delamination fracture surface. The enhanced frictional region partially contributes to the interlaminar shear strength of this stage, as shown in Figure 9c. Finally, the interlaminar shear stress causes the failure of the z-pin by rupture, which eventually leads to the failure of the specimen, as shown in Figure 9d.

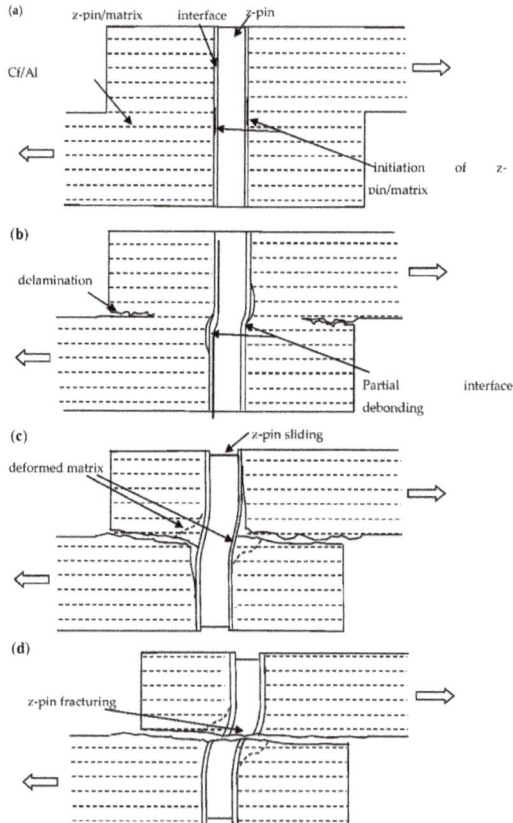

Figure 9. The schematic of the failure process of the z-pinned composites: (**a**) The z-pin/matrix interface crack initiated (**b**) the z-pin/matrix interface was partially debonded and delamination initiated (**c**) z-pin and matrix were deformed (**d**) the z-pin ruptured and the specimen failed.

4. Conclusions

In this study, we carried out an in situ double-notched interlaminar shear test to investigate the delamination suppression mechanisms of the z-pin in carbon fiber-reinforced aluminum alloy composites (Cf/Al) composites. The experimental results determined that z-pinning is highly effective in enhancing the delamination resistance of Cf/Al composites. Introducing the stainless-steel z-pin improved the interlaminar shear strength of Cf/Al composite by 148%. The interlaminar strengthening effect depends on the interface bonding situation. Initially, the z-pinned specimen showed an improved interlaminar strength due to the strong interfacial bonding and the z-pin's high strength. When the interface was debonded, the enhanced frictional region exerted by the deformed z-pin contributed to the interlaminar shear strength of the specimen. The failure process of the z-pinned composite is complex and consisted of carbon fiber/aluminum interfacial debonding, z-pin/aluminum interfacial debonding, aluminum plough, z-pin shear deformation, z-pin frictional sliding, z-pin rupture, and composite delamination.

Author Contributions: Resources, G.W.; writing—original draft preparation, S.W.; writing—review and editing, Y.Z., X.Z. and F.W.; project administration, Y.Z.

Funding: This research was funded by the National Natural Science Foundation of China (Grant No. 51305075), the Science and Technology Innovation Foundation for Harbin Talents (Grant No. 2017RAYXJ021), and the Heilongjiang Province Scientific Research Foundation for Postdoctoral Scholars (Grant No. LBH-Q18004).

Acknowledgments: We would like to acknowledge the financial support to this work by the National Natural Science Foundation of China (Grant No. 51305075), the Science and Technology Innovation Foundation for Harbin Talents (Grant No. 2017RAYXJ021), and the Heilongjiang Province Scientific Research Foundation for Postdoctoral Scholars (Grant No. LBH-Q18004).

Conflicts of Interest: The authors declare no conflict of interest.

References

1. Li, D.G.; Chen, G.Q.; Jiang, L.T.; Xiu, Z.Y.; Zhang, Y.H.; Wu, G.H. Effect of thermal cycling on the mechanical properties of Cf/Al composites. *Mater. Sci. Eng. A* **2013**, *586*, 330–337. [CrossRef]
2. Pei, R.; Chen, G.; Wang, Y.; Zhao, M.; Wu, G. Effect of interfacial microstructure on the thermal-mechanical properties of mesophase pitch-based carbon fiber reinforced aluminum composites. *J. Alloy. Compd.* **2018**, *756*, 8–18. [CrossRef]
3. Zhang, Y.H.; Wu, G.H. Interface and thermal expansion of carbon fiber reinforced aluminum matrix composites. *Trans. Nonferrous Met. Soc. China* **2006**, *20*, 1509–1512. [CrossRef]
4. Li, M.; Matsuyama, R.; Sakai, M. Interlaminar shear strength of C/C-composites: The dependence on test methods. *Carbon* **1999**, *37*, 1749–1757. [CrossRef]
5. Pingkarawat, K.; Wang, C.H.; Varley, R.J.; Mouritz, A.P. Self-healing of delamination cracks in mendable epoxy matrix laminates using poly[ethylene-co-(methacrylic acid)] thermoplastic. *Compos. Part A* **2012**, *43*, 1301–1307. [CrossRef]
6. Mouritz, A.P. Review of z-pinned composite laminates. *Compos. Part A* **2007**, *38*, 2383–2387. [CrossRef]
7. Wang, S.; Wang, Y.H.; Wu, G.H. Interlaminar shear properties of z-pinned carbon fiber reinforced aluminum matrix composites by short-beam shear test. *Materials* **2018**, *11*, 1874. [CrossRef] [PubMed]
8. Pegorin, F.; Pingkarawat, K.; Daynes, S.; Mouritz, A.P. Mode II interlaminar fatigue properties of z-pinned carbon fibre reinforced epoxy composites. *Compos. Part A* **2014**, *67*, 8–15. [CrossRef]
9. Koh, T.M.; Isa, M.D.; Feih, S.; Mouritz, A.P. Experimental assessment of the damage tolerance of z-pinned T-stiffened composite panels. *Compos. Part B* **2013**, *44*, 620–627. [CrossRef]
10. Zhang, Y.H.; Yan, L.L.; Miao, M.H.; Wang, Q.W.; Wu, G.H. Microstructure and mechanical properties of z-pinned carbon fiber reinforced aluminum alloy composites. *Mater. Des.* **2015**, *86*, 872–877. [CrossRef]
11. Knaupp, M.; Baudach, F.; Franck, J.; Scharr, G. Mode I and pull-out tests of composite laminates reinforced with rectangular z-pins. *J. Compos. Mater.* **2014**, *48*, 2925–2932. [CrossRef]
12. Cartié, D.D.R.; Cox, B.N.; Fleck, N.A. Mechanisms of crack bridging by composite and metallic rods. *Compos. Part A* **2004**, *35*, 1325–1336. [CrossRef]

13. Naseem, K.; Yang, Y.; Luo, X.; Huang, B.; Feng, G. SEM in situ study on the mechanical behaviour of Si Cf/Ti composite subjected to axial tensile load. *Mater. Sci. Eng. A* **2011**, *528*, 4507–4515. [CrossRef]
14. Boesl, B.; Lahiri, D.; Behdad, S.; Agarwal, A. Direct observation of carbon nanotube induced strengthening in aluminum composite via in situ tensile tests. *Carbon* **2014**, *69*, 79–85. [CrossRef]
15. Su, J.; Wu, G.H.; Li, Y.; Gou, H.S.; Chen, G.H.; Xiu, Z.Y. Effects of anomalies on fracture processes of graphite fiber reinforced aluminum composite. *Mater. Des.* **2011**, *32*, 1582–1589. [CrossRef]
16. Zhang, B.; Allegri, G.; Yasaee, M.; Hallett, S.R. Micro-mechanical finite element analysis of Z-pins under mixed-mode loading. *Compos. Part A* **2015**, *78*, 424–435. [CrossRef]
17. Yasaee, M.; Lander, J.K.; Allegri, G.; Hallett, S.R. Experimental characterisation of mixed mode traction-displacement relationships for a single carbon composite Z-pin. *Compos. Sci. Technol.* **2014**, *94*, 123–131. [CrossRef]
18. Cox, B.N. Snubbing effects in the pullout of a fibrous rod from a laminate. *Mech. Adv. Mater. Struct.* **2005**, *12*, 85–98. [CrossRef]
19. Cui, H.; Li, Y.; Koussios, S.; Zu, L.; Beukers, A. Bridging micromechanisms of Z-pin in mixed mode delamination. *Compos. Struct.* **2011**, *93*, 2685–2695. [CrossRef]
20. Zhang, Y.; Wu, G. Comparative study on the interface and mechanical properties of T700/Al and M40/Al composites. *Rare Met.* **2010**, *29*, 102–107. [CrossRef]

MDPI

St. Alban-Anlage 66

4052 Basel

Switzerland

Tel. +41 61 683 77 34

Fax +41 61 302 89 18

www.mdpi.com

Materials Editorial Office

E-mail: materials@mdpi.com

www.mdpi.com/journal/materials